study guide and review manual of

BASIC HUMAN ANATOMY AND PHYSIOLOGY

M. H. LINDSAY GIBSON, B.Sc., Ph.D.

Associate Professor, Department of Anatomy,
Faculties of Medicine and Dentistry, University of Manitoba;
and Coordinator of Anatomy and Physiology,
Schools of Nursing and Respiratory Technology,
Health Sciences Centre, Winnipeg, Canada.

1978

W. B. SAUNDERS COMPANY   Philadelphia • London • Toronto

W. B. Saunders Company: West Washington Square
Philadelphia, PA. 19105

1 St. Anne's Road
Eastbourne, East Sussex BN21 3UN, England

1 Goldthorne Avenue
Toronto, Ontario M8Z 5T9, Canada

Study Guide and Review Manual
of Basic Human Anatomy and Physiology ISBN 0-7216-4109-1

 Made in the United States of America. Press of W. B. Saunders Company. Library of Congress Catalog card number 78-55025

Last digit is the print number: 9 8 7 6 5 4 3 2 1

PREFACE

This Study Guide and Review Manual has been written with the purpose of guiding the student through a first course of human anatomy and physiology.

Each chapter of this guide is begun by a set of learning objectives, which were developed to bring into sharp focus the key points to be learnt and to serve as the basis for developing the questions found within each chapter and the review examinations. The student's learning in each chapter is guided by a series of related questions in which the answer and explanations are provided. Finally, at the end of each part the student is provided with a number of questions compiled in an examination format to test his or her cumulative knowledge and to provide practise writing multiple choice examinations.

This guide is keyed to some of the current leading texts of basic human anatomy and physiology so that the student can quickly and easily read the appropriate chapters in the text being used in their course. The student may also want to consult one or more of the other keyed texts for their study. This guide is in no way intended to replace any given text, but rather to supplement a lecture course in anatomy and physiology. It is also intended to encourage the student to study beyond a given set of notes and to develop self-study habits.

I would like to thank Doctor Keith L. Moore, Professor and Chairman of the Department of Anatomy at The University of Toronto, for his initial encouragment to undertake the writing of this study guide and Doctor T.V.N. Persaud, Professor and Head of the Department of Anatomy at The University of Manitoba, for his interest and continued support during the writing of this study guide.

My special thanks go to Robert E. Wright, Helen L. Dietz, and Walter E. Bailey of the W. B. Saunders Company for their great assistance in the preparation of this book.

I am very grateful to my wife, Trish, for proofreading the manuscript and to Brenda Bell DiGaetino for her excellent photographic work and technical assistance. Finally, I would like to give a special thanks to Roslyn Hoad who so cheerfully typed the entire manuscript: without her patience and talents this study guide would never have been completed.

Winnipeg, Canada M. H. Lindsay Gibson

HOW TO USE THIS GUIDE

FOR EACH CHAPTER:

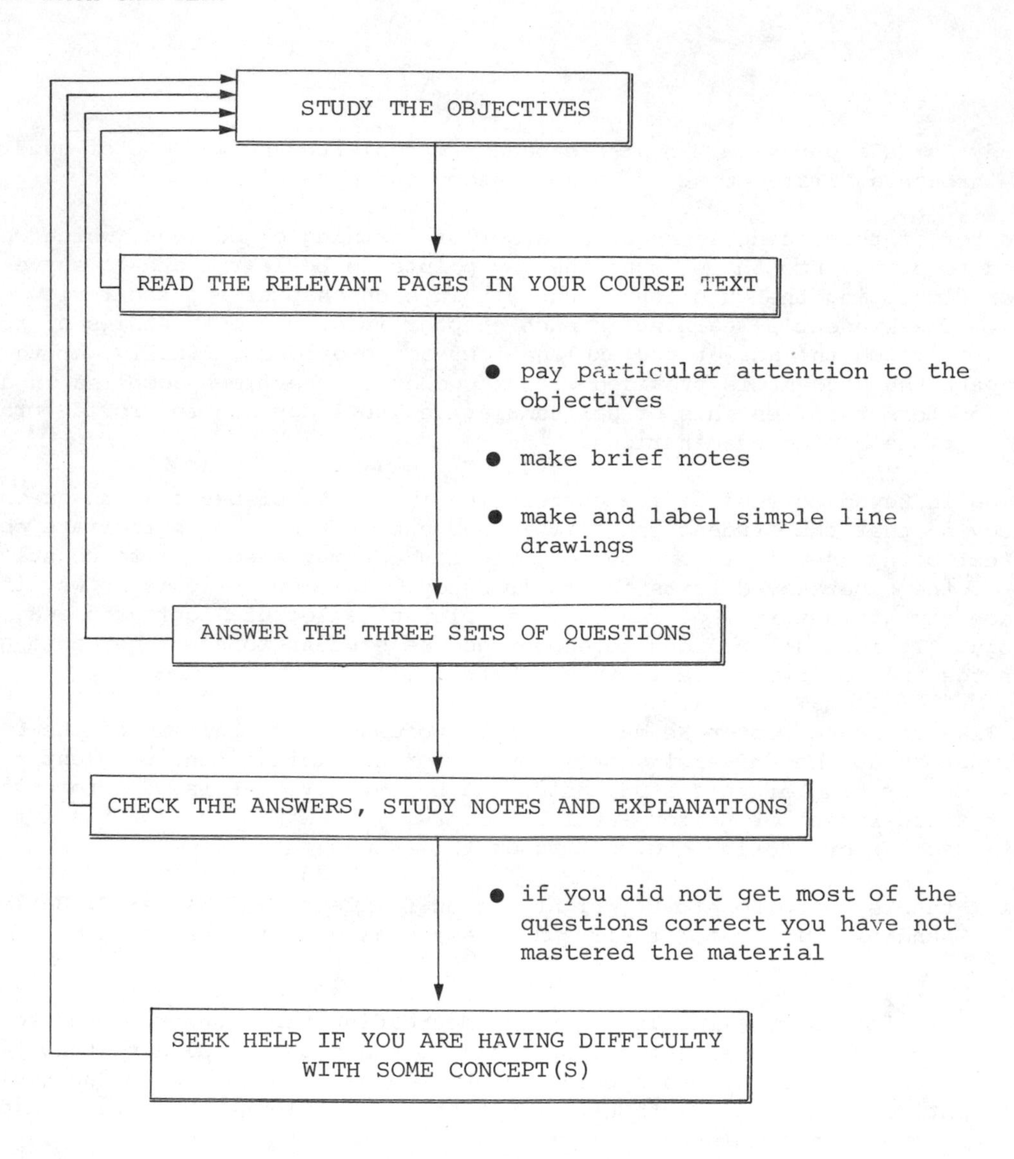

After you have mastered the material in a chapter go on to the next chapter. At the end of each PART you will find a review examination which will test your knowledge of the material covered in a number of chapters. Answer sheets for your use are found in the back of the book on page 341.

CONTENTS

PART 1

PART 2

P A R T 4

REFERENCE BOOKS KEYED

Anthony, C.P., and Kolthoff, N.J.: Textbook of Anatomy and Physiology. 9th ed., Saint Louis, The C.V. Mosby Company, 1975.

Chaffee, E.E., and Greisheimer, E.M.: Basic Physiology and Anatomy. 3rd ed., Philadelphia, J.B. Lippincott Company, 1974.

Crouch, J.E., and McClintic, J.R.: Human Anatomy and Physiology. 2nd ed., New York, John Wiley and Sons, Inc., 1976.

Jacob, S.W., Francone, C.A., and Lossow, W.J.: Structure and Function in Man. 4th ed., Philadelphia, W.B. Saunders Company, 1978.

Landau, B.R.: Essential Human Anatomy and Physiology. 1st ed., Glenview, Illinois, Scott, Foresman and Company, 1976.

Tortora, G.J.: Principles of Human Anatomy. 1st ed., New York, Harper and Row, Publishers, Inc., 1977.

TO TRISH

AND OUR CHILDREN IAN AND JILL

PART ONE

INTRODUCTION TO THE BODY

THE CELL AND ITS PHYSIOLOGY

BODY TISSUES

1. INTRODUCTION TO THE BODY

LEARNING OBJECTIVES

Be Able To:

- Define the sciences of physiology and anatomy; giving the subspecialities of anatomy.
- Describe the body with reference to its anatomical position, relative directions, planes, abdominal regions, and cavities.
- Define and discuss the term homeostasis with reference to the organs involved in its maintenance, its regulating factors and its interaction.
- Discuss the cell, tissues, organs, and systems as structural units of the body.
- List the systems of the human body and give the general functions of each.

RELEVANT READINGS

Anthony & Kolthoff - Chapter 1.
Chaffee & Greisheimer - Chapters 1 and 2.
Crouch & McClintic - Chapter 1.
Jacob, Francone & Lossow - Chapter 1.
Landau - Chapters 1 and 3.
Tortora - Chapter 1.

FIVE-CHOICE COMPLETION QUESTIONS

DIRECTIONS: Each of the following questions or incomplete statements is followed by five suggested answers or completions. SELECT THE SINGLE BEST ANSWER in each case and then circle the appropriate letter at the lower right of each question.

1. THE ___________ SYSTEM IS INVOLVED IN MAINTAINING POSTURE AND PRODUCING HEAT.
   A. Endocrine
   B. Muscular
   C. Respiratory
   D. Skeletal
   E. Urinary

   A B C D E

SELECT THE SINGLE BEST ANSWER

2. THE STUDY OF THE ACTIVITIES OF INDIVIDUAL CELLS IS CALLED:
   A. Cellular physiology
   B. Electron microscopy
   C. Microscopic Anatomy
   D. Neuroanatomy
   E. None of the above

   A B C D E

3. THE HYPOCHONDRIAC REGION IS INDICATED BY ____________.

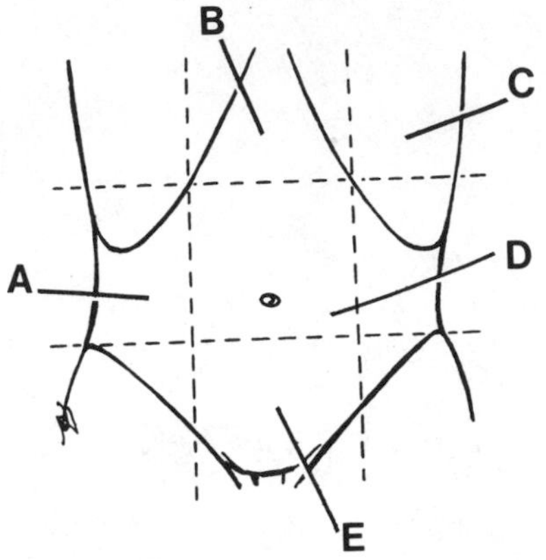

   B C A D E

   A B C D E

4. THE SCAPULAE LIE __________ TO THE CHEST WALL.
   A. Anterior
   B. Inferior
   C. Medial
   D. Lateral
   E. Posterior

   A B C D E

5. THE STUDY OF A DEVELOPING ZYGOTE INTO A MATURE ORGANISM IS CALLED _____________.
   A. Cytology
   B. Histology
   C. Pathology
   D. Embryology
   E. Physiology

   A B C D E

6. IN THE ANATOMICAL POSITION, THE LITTLE FINGER IS __________ TO THE THUMB.
   A. Medial
   B. Distal
   C. Lateral
   D. Inferior
   E. Proximal

   A B C D E

7. HISTOLOGY IS THE SCIENTIFIC STUDY OF:
   A. Cellular function
   B. Cellular structure
   C. Tissue structure
   D. Abnormal tissue
   E. Developmental structure

   A B C D E

8. THE ANKLE IS _______ IN POSITION TO THE KNEE.
   A. Medial
   B. Distal
   C. Lateral
   D. Superior
   E. Anterior

   A B C D E

------------------------ ANSWERS, NOTES AND EXPLANATIONS ------------------------

1. B, The muscular system is not only responsible for the maintenance of body posture and the production of heat, but also provides for the movement of the whole body or its parts. The skeletal system protects and supports the body, allows for leverage, and provides sites for the production of blood cells and the storage of minerals. The endocrine system regulates many of the activitives

of the body through the action of hormones; the urinary system eliminates cellular wastes, maintains the acid-base balance, and helps regulate body fluids. The respiratory system is responsible for obtaining oxygen for the tissues of the body, eliminating carbon dioxide, and helps in regulating the acid-base balance of the body.

2. A, Cellular physiology is the study of the functions or activities of individual cells, along with the physical and chemical factors and processes involved.

3. C, For descriptive purposes the abdomen may be divided into nine regions by two vertical (sagittal) and two horizontal (transverse) planes. The superior horizontal plane passes midway between the jugular notch of the sternum and the symphysis pubis of the pelvis. The inferior horizontal plane passes through the tubercle of the iliac crests. The two vertical planes pass through the midpoint of the right and left inguinal ligaments. The nine regions demarcated by these four planes are illustrated: right and left hypochondriac (A), right and left lumbar (B), right and left iliac (C), epigastric (D), umbilical (E), and hypogastric (F) regions.

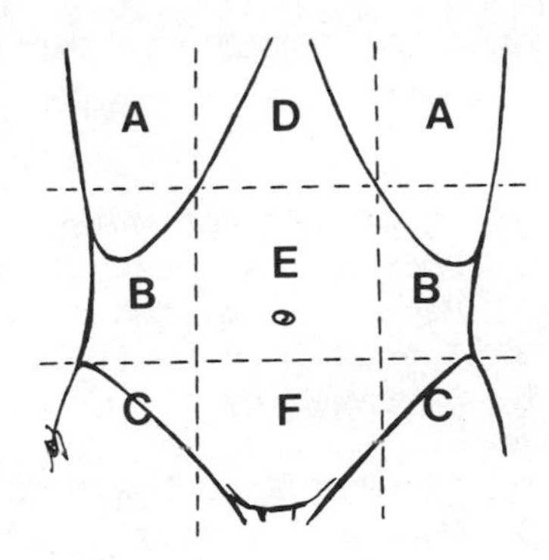

4. E, The scapulae not only lie posterior to the chest wall but may also be correctly described as lying superficial to it and deep to the skin of the back.

5. D, Embryology in its broadest concept is the study of the developing zygote through to its adult form. In the restricted sense, embryology is the study of the prenatal stages of development, especially those occurring during the embryonic period (from the beginning of the second week and continuing through to the end of the seventh week).

6. A, The anatomical position can be defined as the body in the erect position, with the arms at the sides and the head, eyes and palms facing forward. In this position the little finger would be medial to the thumb since the little finger would be closer to the midline of the body than the thumb. Conversely, the thumb would lie lateral to the little finger.

7. C, Histology, or microscopical anatomy, is the scientific study of tissue structure and function, and how the various tissues form the different organs and systems of the body. The studies of cellular function and structure are the sciences of physiology and cytology, respectively. The study of abnormal or diseased tissue is called pathology, whereas the study of developmental structure is called embryology.

8. B, The term distal means the farthest from any point of reference, whereas the term proximal means the opposite. Therefore the ankle is distal to the knee, but the ankle may also be described as being inferior to the knee. On the other hand, the knee may be described as lying proximal or superior to the ankle.

MULTI-COMPLETION QUESTIONS

DIRECTIONS: In each of the following questions or incomplete statements, ONE OR MORE of the completions given is correct. At the lower right of each question, circle A if 1, 2, and 3 are correct; B if 1 and 3 are correct; C if 2 and 4 are correct; D if only 4 is correct; and E if all are correct.

1. THE HOMEOSTATIC MECHANISMS OF THE BODY ARE SUBJECT TO CONTROL BY THE __________ SYSTEM(S).
   1. Circulatory
   2. Endocrine
   3. Muscular
   4. Nervous

   A B C D E

2. ORGAN(S) LOCATED IN THE ABDOMINAL CAVITY INCLUDE THE:
   1. Liver
   2. Kidney
   3. Spleen
   4. Pancreas

   A B C D E

3. WHICH OF THE FOLLOWING STATEMENTS IS(ARE) CORRECT FOR HOMEOSTASIS?
   1. Provides a constant cellular environment
   2. Regulated by the endocrine system
   3. Involved in controlling body temperatures
   4. Operates by a regulating feedback mechanism

   A B C D E

4. WHICH OF THE FOLLOWING STATEMENTS ABOUT TISSUES IS(ARE) CORRECT?
   1. Contains a varying amount of matrix
   2. Performs a particular function
   3. Form larger units called organs
   4. Smallest structural unit of the body

   A B C D E

5. WHICH OF THE FOLLOWING IS(ARE) CONSIDERED A VENTRAL CAVITY OF THE BODY?
   1. Pelvic
   2. Spinal
   3. Pericardial
   4. Cranial

   A B C D E

6. THE ORGAN(S) WHOSE PRIMARY FUNCTION IS THE MAINTENANCE OF HOMEOSTASIS IS(ARE) THE:
   1. Prostate
   2. Liver
   3. Spleen
   4. Kidneys

   A B C D E

7. WHICH OF THE FOLLOWING IS(ARE) A DISCIPLINE OF HUMAN ANATOMY?
   1. Gross
   2. Histology
   3. Embryology
   4. Pathology

   A B C D E

8. THE CIRCULATORY SYSTEM IS:
   1. Primarily responsible for body movements
   2. Directly involved with the digestion of food
   3. Responsible for the transmission of pain
   4. Involved in the regulation of body temperature

   A B C D E

------------------------ ANSWERS, NOTES AND EXPLANATIONS ------------------------

1. C, 2 and 4 are correct. The different systems of the body performing homeostatic mechanisms such as the digestive, circulatory and respiratory systems, are themselves controlled by the nervous and endocrine systems. In both these control systems specific cells react to chemical or hormonal changes in the blood and in turn elicit a cellular response to correct these changes. The speed at which nerve impulses correct a change in the internal environment of the body is very rapid; whereas the speed at which hormones correct such changes is usually much slower.

2. E, All are correct. The abdominal cavity is the largest of the ventral cavities of the body. It is separated superiorly from the thoracic cavity by the diaphragm and is continuous inferiorly with the pelvic cavity. The abdominal cavity contains the liver, gallbladder, spleen, pancreas, organs of the digestive tract and the kidneys with their ureters.

3. E, All are correct. Homeostasis is a condition which provides a relatively constant environment for the cells of the body so that they can function normally. Factors vital to the normal functioning of a cell are: exact concentrations of gases, nutrients, and ions; and optimal temperatures and osmotic pressures. Any stress producing a change in the cellular environment of the body is quickly corrected for by a feedback mechanism. This fine balance of homeostasis is regulated by the endocrine and nervous systems.

4. A, 1, 2, and 3 are correct. The term tissue can be defined as an aggregation of cells separated by a varying amount of matrix (intracellular substance) and performing a particular function. Combinations of the four basic tissues (i.e., epithelial, connective, muscular and nervous) form organs and in turn various organs performing a common function form a system. The smallest structural unit of the body is the cell.

5. B, 1 and 3 are correct. The dorsal cavities of the body consist of the cranial (A in the diagram) and spinal (B) cavities and contain the brain and spinal cord, respectively. The ventral cavities of the body include the thoracic (E), abdominal (D) and pelvic (C) cavities. The thoracic cavity can be further subdivided into a central pericardial cavity, containing the heart, and two lateral pleural cavities, each containing a lung.

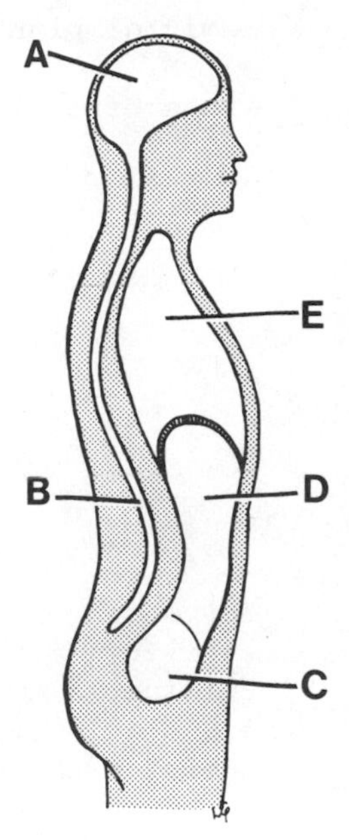

6. C, 2 and 4 are correct. Several organs of the body can best be understood as homeostatic organs, since their primary function is directed towards the maintenance of homeostasis; these organs include the heart, lungs, kidneys, liver, gastrointestinal tract, and the skin. However, all body structures from the cellular to the systemic level, contribute in some way to the maintenance of the internal environment of the body within normal physiological limits.

7. A, 1, 2, and 3 are correct. The discipline of human anatomy deals with the structure of man and has four divisions: neuroanatomy, gross anatomy, histology, and embryology. Neuroanatomy deals with the structure of the nervous system; whereas gross and histology deal with normal macroscopic and microscopic structures of man, respectively. Embryology deals with the normal anatomical development of man, from the zygote through to the adult and on to senility.

8. D, Only 4 is correct. The circulatory system is primarily involved in transporting oxygen, carbon dioxide, hormones, blood cells, and other substances throughout the body. It is also responsible for maintaining the pH and electrolyte composition of the body and helps to regulate body temperature.

FIVE-CHOICE ASSOCIATION QUESTIONS

DIRECTIONS: Each group of questions below consists of a numbered list of descriptive words or phrases accompanied by a diagram with certain parts indicated by letters, or by a list of lettered headings. For each numbered word or phrase, SELECT THE LETTERED PART OR HEADING that matches it correctly. Then insert the letter in the space to the right of the appropriate number. Sometimes more than one numbered word may be correctly matched to the same lettered part or heading.

1. _____ Oblique plane
2. _____ Median plane
3. _____ Coronal plane
4. _____ Sagittal plane
5. _____ Transverse plane

D A B E C

6. _____ Responsible for the absorption of food.
7. _____ Eliminates CO_2 from the body.
8. _____ Includes the special sense of taste.
9. _____ Provides support for the body.
10. _____ Involved in perpetuation of man.

A. Nervous system
B. Skeletal system
C. Respiratory system
D. Reproductive system
E. None of the above

------------------------ ANSWERS, NOTES AND EXPLANATIONS ------------------------

1. C, An oblique plane is one taking a slanted or inclined direction and lies somewhere between a vertical and a horizontal plane.

2. A, The median plane is a vertical plane running through the center of the body, dividing it into right and left halves. This plane may also be called a midsagittal plane.

3. D, Any vertical plane that intersects the median plane at right angles is a frontal or coronal plane. It divides the body into anterior and posterior portions.

4. B, A sagittal plane takes a vertical direction and divides the body into right and left portions. If the two portions were equal in size then the plane would be more precisely called a median or midsagittal plane.

5. E, Any plane that is at right angles to both the median and coronal planes is considered to be transverse. The term transverse implies across the longitudinal axis of the body and may not always be synonymous with horizontal, as seen in a transverse section of the foot.

6. E, The system involved with the absorption of food is the digestive system. To be more specific the small intestine of the gut is involved in the absorption of monosaccharides, glycerol, and amino acids, which are all products of digestion.

7. C, The respiratory system, consisting of the nasal and oral cavities, pharynx, larynx, trachea, bronchi, and lungs, carries oxygen into the body, eliminates carbon dioxide, and helps regulate the acid-base balance of the body.

8. A, The body is highly dependent on its ability to perceive the state of its internal and external environments. This perception of awareness of its environment is one of the most important functions of the nervous system. There are both general sensations such as the perception of cold, heat, pain, touch, and pressure and special sensations such as the perception of smell, taste, sight, and hearing.

9. B, The skeletal system, comprising the bones of the body, provides not only for support of the body but also provides protection, allows for leverage, and stores minerals. In the adult it also provides sites for the production of new blood cells.

10. D, The reproductive system has the sole responsibility of maintaining the human species by producing each new successive generation. In doing to, each generation passes on to the next its genetic material by way of the ovum and the sperm.

2. THE CELL AND ITS PHYSIOLOGY

LEARNING OBJECTIVES

Be Able To:

- Define the term protoplasm and describe the components of a typical cell; giving their functions.
- With the aid of a diagram or a short paragraph describe the passive mechanisms of diffusion, facilitated diffusion, filtration, and osmosis.
- Define the meaning of hypotonic, hypertonic, and isotonic solutions.
- Describe the active mechanisms of active transport, phagocytosis, and pinocytosis.
- Describe the cell cycle; identify the phases of mitosis and list the biochemical events occurring in interphase.
- Compare and contrast the components and functions of deoxyribonucleic and ribonucleic acids.
- Define the term gene and discuss its importance.

RELEVANT READINGS

Anthony & Kolthoff - Chapter 2.
Chaffee & Greisheimer - Chapter 2.
Crouch & McClintic - Chapters 2, 3, 4, and 5.
Jacob, Francone & Lossow - Chapter 2.
Landau - Chapters 1 and 2.
Tortora - Chapter 2.

FIVE-CHOICE COMPLETION QUESTIONS

DIRECTIONS: Each of the following questions or incomplete statements is followed by five suggested answers or completions. SELECT THE SINGLE BEST ANSWER in each case and then circle the appropriate letter at the lower right of each question.

1. DNA SYNTHESIS TAKES PLACE DURING __________ OF THE CELL CYCLE.
   A. Anaphase
   B. Interphase
   C. Metaphase
   D. Prophase
   E. Telophase

   A B C D E

2. THE TAKING OF A PARTICLE INTO A CELL BY A MEMBRANOUS VESICLE IS CALLED:
   A. Pinocytosis
   B. Phagocytosis
   C. Diffusion
   D. Osmosis
   E. None of the above

   A B C D E

3. RED BLOOD CELLS DROPPED INTO A(AN) _________ SOLUTION WOULD LOSE WATER AND SHRINK.
   A. Atonic
   B. Isotonic
   C. Hypotonic
   D. Hypertonic
   E. None of the above

   A B C D E

4. THE MOVEMENT OF A SUBSTANCE INTO A CELL AGAINST A CONCENTRATION GRADIENT IS CALLED __________.
   A. Osmosis
   B. Diffusion
   C. Filtration
   D. Facilitated diffusion
   E. None of the above

   A B C D E

5. A MEMBRANE-BOUND ORGANELLE CONTAINING ACID HYDROLASES IS CALLED A _____________.
   A. Mitochondrion
   B. Ribosome
   C. Golgi complex
   D. Lysosome
   E. Nucleolus

   A B C D E

6. IN TELOPHASE OF MITOSIS THE:
   A. Chromosomes begin to coil
   B. Spindle appears
   C. Nucleolus appears
   D. Nuclear envelope disappears
   E. Plasmalemma disappears

   A B C D E

7. THE ORGANELLE RESPONSIBLE FOR NEW PROTEIN SYNTHESIS IN THE CELL IS THE _____________.
   A. Ribosome
   B. Mitochondrion
   C. Centrosome
   D. Golgi apparatus
   E. Lysosome

   A B C D E

8. THE NITROGEN BASE NOT FOUND IN THE RNA MOLECULE IS _________.
   A. Adenine
   B. Cytosine
   C. Guanine
   D. Thymine
   E. Uracil

   A B C D E

------------------------ ANSWERS, NOTES AND EXPLANATIONS ------------------------

1. B, New synthesis or replication of DNA takes place during interphase of the cell cycle. Interphase is the period of time elapsing between successive mitoses. During this phase all activities of the cell are capable of taking place except cell division. A cell in interphase has chromatin (darker staining) masses in its nucleus which are coiled segments of DNA. The lighter areas represent uncoiled portions of the DNA molecule. During uncoiling, DNA

separates at the points where the nitrogen bases are bonded. Each of the exposed bases then picks up a complementary base along with its associated sugar and phosphate group from the cytoplasm of the cell. This uncoiling and complementary base pairing continues until each of the original DNA strands is matched and joined with two newly formed DNA strands. The net effect is that the nucleus now has a double complement of DNA and is ready to enter mitosis and divide.

2. B, The taking of a particle into a cell by a membranous vesicle is an active process called phagocytosis. This process is important because particles or molecules that would normally be restricted from crossing the cell membrane can enter the cell in this way. This process also allows white blood cells to remove bacteria and other foreign substances from the blood and destroy them by lysosomal activity.

3. D, Red blood cells dropped into a hypertonic solution would lose water and shrink (crenate). This process of osmosis occurs because the hypertonic solution has a higher concentration of solute and a lower concentration of water than the red blood cell. Therefore, water leaves the cell across the cell membrane and causes the cell to crenate. However, if one dropped a red blood cell in a hypotonic solution (a lower concentration of solute and a higher concentration of water) water would then enter the cell causing it to swell. If the cell continued to swell it would eventually burst or hemolyse (losing its hemoglobin). An isotonic solution would have no effect on a cell dropped into it, because the water and solute concentrations would be the same in both the cell and the solution.

Cell in
hypertonic solution → Crenation

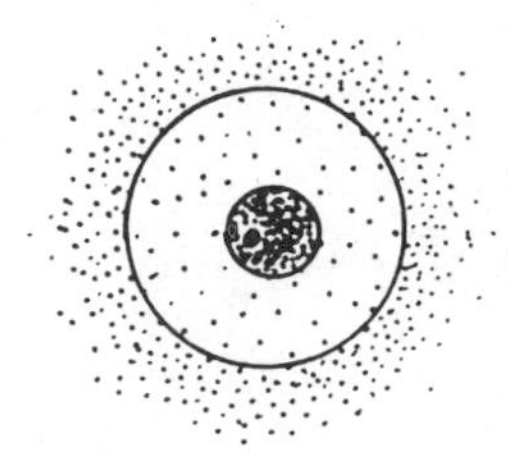

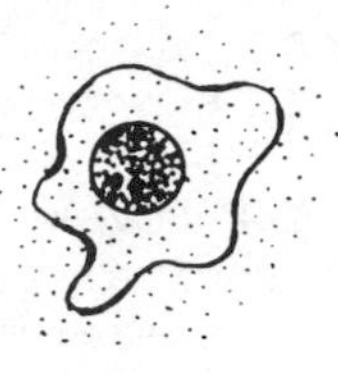

4. E, All of the processes listed (osmosis, filtration, diffusion, and facilitated diffusion) are passive processes involved in moving a substance across the plasma membrane. In passive processes a substance moves across the plasma membrane without any help from the cell and occurs from an area of high concentration to an area of a lower concentration. This movement from a high concentration area to a low concentration area is said to move down the concentration gradient. A substance can also be pushed through a cell membrane by pressure from an area where the pressure is greater to an area where it is less. The movement of a substance into a cell against a concentration gradient requires the utilization of energy by the cell and is therefore considered an active process.

5. D, Lysosomes are membrane-bound organelles containing hydrolytic enzymes capable of breaking down proteins, DNA, RNA, and some carbohydrates under slightly acid pH conditions. These organelles are found in large numbers in white blood cells and are actively involved in intracellular digestion of foreign matter taken into cells by phagocytosis. Lysosomes are also thought to digest or lyse damaged organelles of a normal cell or the cell itself when it has been severely damaged.

6. C, Telophase, the last stage of mitosis, can be considered a reconstruction stage of the cell following cell division. The nuclear membrane appears and encloses the two groups of chromosomes which are becoming thread-like and less distinct. During this stage the nucleolus appears in each new daughter cell

and the spindle fibers disappear. The cytoplasm of the cell cleaves or undergoes cytokinesis and forms two new daughter cells with new cell membranes.

7. A, Ribosomes, spherical particles of ribonucleoprotein, may be either free within the cytoplasm or attached in a linear fashion to membranes making up the endoplasmic reticulum. Ribosomes attached to such membranes are believed to be concerned with the synthesis of proteins to be secreted by the cell. Free ribosomes are believed to be involved in protein synthesis necessary to sustain cell proliferation and other intracellular functions.

8. D, The ribonucleic acid (RNA) molecule contains four nitrogen bases: adenine, cytosine, guanine and uracil. A major difference between RNA and DNA is that: the nitrogen base thymine is used in the synthesis of deoxyribonucleic acid (DNA) instead of uracil. The second major difference between RNA and DNA is that the sugar ribose is found in RNA; whereas deoxyribose is used in the synthesis of DNA.

MULTI-COMPLETION QUESTIONS

DIRECTIONS: In each of the following questions or incomplete statements, ONE OR MORE of the completions given is correct. At the lower right of each question, circle A if 1, 2, and 3 are correct; B if 1 and 3 are correct; C if 2 and 4 are correct; D if only 4 is correct; and E if all are correct.

1. THE NUCLEUS OF A CELL:
   1. Controls metabolic activities
   2. Contains a few mitochondria
   3. Holds both RNA and DNA
   4. Is always present

   A B C D E

2. WHICH OF THE FOLLOWING WOULD INCREASE THE RATE OF DIFFUSION?
   1. Cool the solution
   2. Add more solute
   3. Increase the viscosity
   4. Decrease the molecular size

   A B C D E

3. THE CELL MEMBRANE:
   1. Consists of lipids
   2. Contains numerous pores
   3. Measures about 75 angstroms
   4. Functions as a non-selective permeable structure

   A B C D E

4. WHICH OF THE FOLLOWING IS(ARE) NOT TRUE FOR A GENE?
   1. Consists of a double strand of DNA molecule
   2. Directs the manufacture of a particular kind of RNA
   3. Determines the sequencing of amino acids in a protein
   4. Found in specific segments of the RNA molecule

   A B C D E

5. WHICH OF THE FOLLOWING IS(ARE) TRUE FOR PROTOPLASM?
   1. Composed mostly of water
   2. Contains unique compounds
   3. Is considered living
   4. Reaches a high degree of organization

   A B C D E

6. WHICH OF THE FOLLOWING OCCURS DURING INTERPHASE?
   1. Chromosomes line up at the equatorial plate
   2. The total amount of DNA is reduced
   3. Metabolic functions decline
   4. Synthesis of RNA and proteins takes place

   A B C D E

7. PROPHASE IS CHARACTERIZED BY:
   1. A disappearing nucleolus
   2. Disappearance of chromosomes
   3. Formation of the mitotic spindle
   4. An intact nuclear membrane

A B C D E

8. WHICH OF THE FOLLOWING WOULD BE VISIBLE IN A METAPHASE CELL?
   1. Nucleolus
   2. Centrioles
   3. Nuclear membrane
   4. Chromosomes at the equatorial plate

A B C D E

9. WHICH OF THE FOLLOWING IS(ARE) CORRECT FOR THE GOLGI APPARATUS?
   1. Concentrates secretory products
   2. Well developed in the non-secretory cells
   3. Involved in polysaccharide synthesis
   4. Plays an important role in mitosis

A B C D E

ANSWERS, NOTES AND EXPLANATIONS

1. B, 1 and 3 are correct. The nucleus is an essential component of nearly all cells. The nucleus contains both DNA, in the form of chromosomes, and RNA, in the nucleolus. The nucleus controls the metabolic activities of the cell; in the case of mature red blood cells, which do not have a nucleus, such activities are limited and cell division is impossible. Mitochondria are located within the cytoplasm of the cell but not in the nucleus.

2. C, 2 and 4 are correct. Diffusion, a passive process, takes place when there is a net movement of molecules or ions from an area of high concentration to an area of low concentration. The smaller the molecules or ions involved the faster the rate of diffusion. Similarly, a high concentration of molecules (solute) in an area of microscopic dimensions will also produce an increased rate of diffusion. A heated solution will diffuse more quickly because of its high kinetic energy and a low viscous solution will tend to diffuse more rapidly than a high viscous solution.

3. B, 1 and 3 are correct. The cell membrane (plasmalemma) is a trilaminar structure measuring about 75 angstroms. Its three layers have been interpreted as being two protein layers separated by a lipid layer. It was thought that the cell membrane contained numerous pores but the use of the electron microscope has proven this fact to be wrong. The cell membrane carries out many important activities, one of which is that it acts as a selective permeable structure.

4. D, 4 only is correct. The gene, the key to heredity, consists of about 1,000 paired nucleotides appearing in a specific sequence in the DNA molecule. Each gene directs the synthesis of a particular kind of RNA using the same method of replication as DNA. Messenger RNA, produced by the gene template, leaves the nucleus and at the ribosomes in the cytoplasm itself acts as a template to which transfer RNA molecules attach to form a specific protein. Therefore each gene is responsible indirectly for the making of a particular protein. These proteins are responsible by very intricate processes for the specialized cells of the body and determine the general appearance of our body and the activities it can perform.

5. E, All are correct. The living protoplasm of a cell contains two compartments, the nucleus, containing the nucleoplasm (karyoplasm), and protoplasm surrounding the nucleus, called the cytoplasm. Water is the most abundant compound in

protoplasm but it also contains unique compounds such as proteins, lipids, carbohydrates, and nucleic acids. Protoplasm is never a disorganized group of chemicals and compounds but rather an organized mass which forms definite units, the cells of the body.

6. D, 4 only is correct. The life cycle of a cell (see the diagram) consists of mitosis (with its four phases) and a period of time elasping between successive mitoses called interphase. The first period (G_1) of interphase is measured between the completion of mitosis and the onset of DNA synthesis or the S-phase. During the S-phase the cell replicates its genetic material or DNA. Following this phase there is a short period called G_2 which ends as the cell enters prophase of mitosis. During mitosis all metabolic functions of the cell cease except those involved with the separation of the genetic material. Protein synthesis occurs throughout interphase but has its highest rate during S-phase. The rate of RNA synthesis is constant throughout interphase but ceases during mitosis. Some scientists believe that both RNA and DNA synthesis cannot go on at the same time, that is during the S-phase. After mitosis a daughter cell may either enter the cell cycle again (1) or it may go on to maturate (2) for a specific function. Such a specialized cell will eventually go on to die and be replaced.

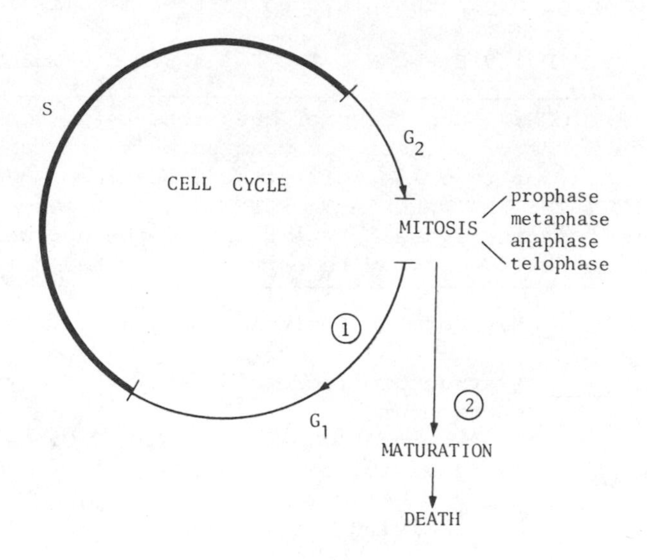

7. B, 1 and 3 are correct. Prophase is the first stage of cell division or mitosis. In it the nucleolus and the nuclear membrane disappear, whereas the chromosomes appear and form two chromatids attached by a centromere. The centrioles begin to migrate to opposite ends of the cell and form the mitotic spindle.

8. C, 2 and 4 are correct. During metaphase, the second stage of mitosis, the chromatid pairs of chromosomes line up at the equatorial plate. During this phase both centromeres and chromatid pairs separate and become attached by spindle fibers to the centrioles. The nucleus and nucleolus are not visible during this phase of cell division.

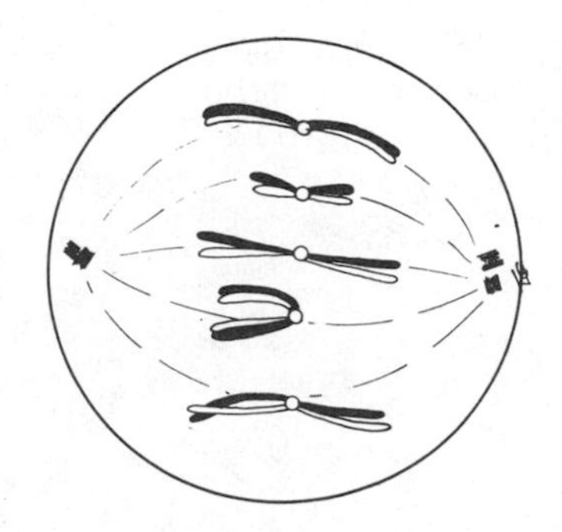

9. A, 1, 2, and 3 are correct. The Golgi apparatus of glandular cells is thought to be responsible for the concentration and packaging of secretory products. The Golgi apparatus is also found in non-secretory cells but its function in these cells has not been elucidated. This cell organelle is also thought to be responsible for the synthesis of some very complex polysaccharides. The Golgi apparatus plays no part in mitosis.

FIVE-CHOICE ASSOCIATION QUESTIONS

DIRECTIONS: Each group of questions below consists of a numbered list of descriptive words or phrases accompanied by a diagram with certain parts indicated by letters, or by a list of lettered headings. For each numbered word or phrase, SELECT THE LETTERED PART OR HEADING that matches it correctly. Then insert the letter in the space to the right of the appropriate number. Sometimes more than one numbered word may be correctly matched to the same lettered part or heading.

1. _____ May form an inclusion body.
2. _____ A structure increasing the surface area of a cell.
3. _____ The division of cytoplasm.
4. _____ Small living structures.
5. _____ A living semifluid material.

A. Cytoplasm

B. Cytokinesis

C. Melanin

D. Microvilli

E. Organelles

6. _____ Involved in cell division.
7. _____ An inclusion body.
8. _____ Contains hydrolytic enzymes.
9. _____ Functions in protein synthesis.
10. _____ Made up exclusively of RNA.
11. _____ Ribosomes may attach to it.
12. _____ Plays an important role in destroying bacteria brought into a cell.

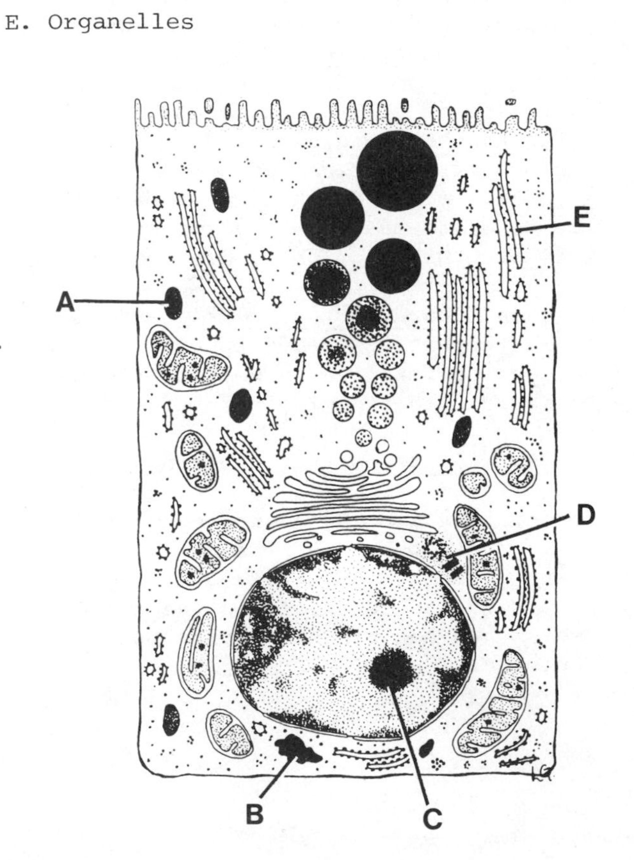

ANSWERS, NOTES AND EXPLANATIONS

1. C, Melanin along with other organic products of the cell such as glycogen, lipids, and hemoglobin crystals (found only in red blood cells) may be stored within the cell in the form of inclusion bodies. Often the presence of these bodies is transitory and may appear or disappear at various times during the life of a cell.

2. D, Microvilli are small finger-like projections of the cell membrane. They function by increasing the surface area of a cell for absorption of materials and are found on most of the epithelial cells lining the small intestine. The diagram at the right shows how villi increase the total surface area. The cell at the top has a one-dimensional surface area as indicated by A-A'. If the same cell has villi, the surface area is increased to B-B'. In actual fact it would be even greater because of the third dimension of the cell not shown.

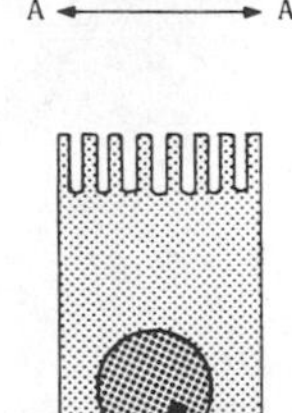

B ← → B'

3. B, Cytokinesis is the name given to the division of cytoplasm and is seen in the latter stage of anaphase and telophase of mitosis.

4. E, Organelles are small living structures of cells involved in the different functions of the cell. They tend to segmentalize the various functions of the cell, giving it order.

5. A, The cytoplasm is the living diffuse matrix-like material of the cell. The organelles are located within the cytoplasm and bound by the unit membrane; forming the structural unit of the body of the cell.

6. D, The centrioles are involved in division of the cell into two daughter cells. Paired centrioles, at right angles to each other, form the centrosomes or diplosome, serving as centers for the organization of microtubules which form the spindle and asters of the dividing cell.

7. B, This inclusion body contains lipids which serves as a local store of energy. The stored lipids are also a source of short carbon chains that can be utilized by the cell to form membranes and other structural components of the cell.

8. A, Lysosomes are membrane-bound structures containing hydrolytic enzymes which are capable of breaking down DNA, RNA, and carbohydrates. Large numbers of lysosomes are found in cells actively engaged in phagocytosis.

9. E, The endoplasmic reticulum functions in many chemical reactions of the cell. The granular type, having ribosomes attached to it, is thought to be actively involved in protein synthesis, whereas the functions of the agranular type are thought to vary depending on the cell type.

10. C, Each interphase nucleus contains one or more nucleoli, consisting of both DNA and RNA. The nucleolus is thought to be involved in the synthesis of ribosomal RNA. The nucleolus probably produces the ribosomal precursors, which then are carried to the cytoplasm where they are assembled into functional ribosomes.

11. E, Endoplasmic reticulum which has ribosomes attached to it is called granular endoplasmic reticulum and is known to be involved with protein synthesis. Some endoplasmic reticulum however, does not have ribosomes attached to it and is called agranular. Agranular endoplasmic reticulum is thought to be involved with the release and recapture of calcium ions in the contraction and relaxation of striated muscle; the biosynthesis of steroid hormones in some endocrine glands; and the metabolism of cholesterol and lipids in the liver.

12. A, Lysosomes play an important role in destroying bacteria or other foreign particles brought into the cell by phagocytosis. Lysosomes also remove worn out or damaged organelles from the cytoplasm of normal cells. White blood cells and other phagocytic cells contain large numbers of lysosomes.

3. BODY TISSUES

LEARNING OBJECTIVES

Be Able To:

- Define the term tissue and describe the four basic types as to their origin, appearance and functions.
- Give the basic features of epithelia and describe the various types of simple, stratified, and glandular epithelia as to their functions.
- Identify simple line drawings of the various epithelia and indicate where they are located in the body.
- Describe the basic components making up any connective tissue.
- Identify simple line drawings of the different types of connective tissues, their functions and give sites where they are located in the body.
- Compare and contrast the general features, functions, and locations of the three types of muscle tissue.
- Describe the function of the two general types of cells found in nervous tissue.
- Describe the differences between allografts, autografts, isografts, and xenografts.

RELEVANT READINGS

Anthony & Kolthoff - Chapter 3.
Chaffee & Greisheimer - Chapter 2.
Crouch & McClintic - Chapter 6.
Jacob, Francone & Lossow - Chapter 3.
Landau - Chapter 3.
Tortora - Chapters 3, 5, 9, and 12.

FIVE-CHOICE COMPLETION QUESTIONS

DIRECTIONS: Each of the following questions or incomplete statements is followed by five suggested answers or completions. SELECT THE SINGLE BEST ANSWER in each case and then circle the appropriate letter at the lower right of each question.

1. A CELL FOUND IN AREOLAR TISSUE RESPONSIBLE FOR THE SYNTHESIS OF HISTAMINE IS THE:
   A. Histiocyte
   B. Mast cell
   C. Fibroblast
   D. Leukocyte
   E. Reticular cell

   A B C D E

2. HYALINE CARTILAGE IS FOUND IN THE:
   A. Auditory tube
   B. Auricle
   C. Intervertebral disks
   D. Fetal skeleton
   E. None of the above

   A B C D E

3. WHICH OF THE FOLLOWING IS NOT A TYPICAL FUNCTION OF A STRATIFIED EPITHELIAL LAYER?
   A. Absorption
   B. Secretion
   C. Protection
   D. Waterproofing
   E. Distension

   A B C D E

4. WHICH OF THE FOLLOWING IS NOT A CHARACTERISTIC OF EPITHELIAL TISSUE?
   A. Anchored by a basement lamina
   B. Has a high regenerative capacity
   C. May have more than one cell layer
   D. Found lining tubular structures
   E. Primarily involved in contraction

   A B C D E

5. NEURONS ARE SPECIALIZED CELLS INVOLVED IN:
   A. Secretion
   B. Conduction
   C. Contraction
   D. Protection
   E. Absorption

   A B C D E

6. THE TISSUE INDICATED AT THE RIGHT IS A COLLECTION OF ______ FIBERS.
   A. Nerve
   B. Cardiac muscle
   C. Collagen
   D. Skeletal muscle
   E. Elastic

   A B C D E

7. A TISSUE TRANSPLANT BETWEEN TWO PATIENTS OF DIFFERENT GENETIC BACKGROUNDS IS CALLED A(AN) __________.
   A. Isograft
   B. Allograft
   C. Autograft
   D. Xenograft
   E. None of the above

   A B C D E

SELECT THE SINGLE BEST ANSWER

8. WHICH OF THE FOLLOWING IS NOT TRUE FOR A CONNECTIVE TISSUE?
   A. Consists mainly of cells
   B. The most widely distributed tissue of the body
   C. Contains protein fibers
   D. Serves to attach individual cells
   E. Cells are responsible for the nature of the intercellular material

A B C D E

9. WHICH OF THE FOLLOWING TISSUES IS NOT CLASSIFIED AS A CONNECTIVE TISSUE?
   A. Areolar
   B. Blood
   C. Bone
   D. Dentin
   E. Muscle

A B C D E

------------------------ ANSWERS, NOTES AND EXPLANATIONS ------------------------

1. B, The mast cell is located in loose connective (areolar) tissue and is responsible for the release of heparin and histamine near blood vessels. Heparin is an anticoagulent and plays an important role in preventing blood from clotting within the blood vessels. Histamine is a chemical dilator (enlarger) of small blood vessels. Other cells that may be found in loose connective tissues are: histiocytes (macrophages), melanocytes (pigment cells), fibrocytes, fat, plasma, and various white blood cells.

2. D, Hyaline cartilage makes up most of the embryonic skeleton. Later in life much of it becomes ossified to form the bony skeleton of the adult. Hyaline cartilage is the most abundant type of cartilage found in the body. It is located at the joint surfaces of bones, the anterior ends of the ribs, and forms part of the nose, larynx, trachea, and bronchi. Elastic cartilage helps to form the auricle of the ear and the auditory tube; whereas fibrocartilage is found in the intervertebral disks.

3. A, Stratified epithelial tissues cover areas in which little, if any, absorption is required. In general, stratified epithelial tissues are found in areas of maximal stress (i.e., palms of the hand and the bladder). It also provides for secretion, protection, waterproofing and distension in various linings or coverings of the body.

4. E, Epithelial tissues cover body surfaces, line body cavities, and form glands. It consists of one or more layers of cells, of which the basal layer is attached to a basement lamina. Cells of the basal layer undergo mitosis and produce new cells to replace those superficial ones that wear out and are eventually shed. There is little intercellular material between epithelial cells and no blood vessels. The blood vessels, found in the underlying connective tissues, provide oxygen and remove cell wastes by simple diffusion. In classifying epithelial tissue a single layer is referred to as simple; whereas two or more layers are designated as stratified.

5. B, Neurons are specialized cells of the body involved in the conduction of electrical impulses. These cells along with neuroglial cells are members of the most highly organized tissue of the body, and make up the nervous system. This system, primarily the neurons, is involved in initiating, controlling, and coordinating the diverse activities of the body so it can adapt to its environment.

6. D, Skeletal muscle is a voluntary type of striated muscle attached to the skeleton, providing for its movement. A skeletal muscle cell or fiber is

multinucleated and is surrounded by a membrane called the sarcolemma. The skeletal muscle cell has regular cross-striations of alternating dark and light bands. These cross-striations, appearing on the myofilaments, are alternating bands of myosin (dark) and actin (light) proteins. The myofilaments run the full length of the individual muscle cell.

7. B, A graft of tissue between two individuals of the same species but of different genotypes is called an allograft (homotransplant). An isograft (isotransplant) is one between genetically identical individuals (i.e., identical twins); whereas a tissue transplant between animals of different species is called a xenograft (heterograft). An autograft (autotransplant) is a grafting of tissue from one part of the body to another site within or on the same individual.

8. A, Connective tissues are the most widely distributed tissue of the body. This primary tissue is characterized by a large amount of different types of intercellular material (matrix), contains relatively few cells, and differing types of protein fibers. This tissue serves to connect structures, parts of structures, or individual cells. The cells found in the different connective tissue types are responsible for the varying forms of the matrix.

9. E, Muscle is not a connective tissue, but a group of highly specialized cells capable of contraction. The other tissues listed (areolar, blood, bone and dentin) are classified as connective tissues. They all contain the three basic elements of any connective tissue, i.e., a preponderance of ground substance, some cells, and protein fibers.

MULTI-COMPLETION QUESTIONS

DIRECTIONS: In each of the following questions or incomplete statements, ONE OR MORE of the completions given is correct. At the lower right of each question, circle A if 1, 2 and 3 are correct; B if 1 and 3 are correct; C if 2 and 4 are correct; D if only 4 is correct; and E if all are correct.

1. EPITHELIAL TISSUE HAS A:
   1. Great deal of intercellular material
   2. Free surface
   3. Connective function
   4. Basement (membrane) lamina

   A B C D E

2. MESOTHELIAL TISSUE:
   1. Forms cellular membranes
   2. Provides support
   3. Is simple squamous
   4. Secretes mucus

   A B C D E

3. WHICH OF THE FOLLOWING IS(ARE) CORRECT FOR AN EXOCRINE GLAND?
   1. Secretes enzymes
   2. Has a duct system
   3. Are epithelial derivatives
   4. Empty into blood vessels

   A B C D E

4. WHICH OF THE FOLLOWING STATEMENTS IS(ARE) TRUE FOR THE TERM TISSUE?
   1. Consists of cells
   2. Makes up organs
   3. Contains some interstitial material
   4. Has many functioning types

   A B C D E

5. REGULARLY ARRANGED DENSE CONNECTIVE TISSUE IS TYPICALLY FOUND IN:
   1. Joint capsules
   2. Fasciae
   3. Sheaths
   4. Tendons

   A B C D E

A	B	C	D	E
1,2,3	1,3	2,4	only 4	all correct

6. WHICH OF THE FOLLOWING IS(ARE) CORRECT FOR MATURE BONE TISSUE?
   1. Contains functioning chondrocytes
   2. Is highly vascular
   3. Has great quantities of elastin
   4. Contains lamellae

A B C D E

7. CARTILAGE HAS:
   1. An extensive vascular network
   2. A firm protein matrix
   3. An elaborate basement membrane
   4. Cells called chondroblasts

A B C D E

8. CARDIAC MUSCLE IS:
   1. Located in the heart
   2. Striated
   3. Innervated by the autonomic nervous system
   4. Forms a branching network

A B C D E

9. WHICH OF THE FOLLOWING IS(ARE) DERIVED FROM ECTODERM?
   1. Epidermis
   2. Liver
   3. Central nervous system
   4. Muscle tissue

A B C D E

------------------------ ANSWERS, NOTES AND EXPLANATIONS ------------------------

1. C, 2 and 4 are correct. Epithelial tissue is a cellular tissue with little, if any, intercellular material. It has a free surface to the external surface of the body, the lumen of a tubular structure, or the cavities of the body. Epithelial tissues are typically anchored to the underlying connective tissues by a noncellular layer called the basement membrane. Epithelial tissues may function in protection, transportation, absorption, secretion, or distension.

2. B, 1 and 3 are correct. Mesothelial tissue consists of a single layer of squamous cells in the form of a membrane. This cellular membrane, in conjunction with a thin layer of loose connective tissue, forms the serous membranes of the peritoneal, pericardial and pleural cavities. These membranes provide a small amount of serous exudate (fluid) which minimizes friction and provides for the free movement of organs within the cavities of the body.

3. A, 1, 2 and 3 are correct. Exocrine glands are epithelial derivatives and maintain a duct which allows secretory products to be released at the surface of the body or into a hollow organ. The secretory products of exocrine glands include enzymes, oil, or sweat. Endocrine glands, on the other hand, secrete only hormones. Endocrine glands are also epithelial derivatives but have no ducts. They empty their hormones directly into the blood stream.

4. E, All are correct. The term tissue implies an aggregation of cells working together to perform a particular function. These cells are bound together by varying amounts of fibrous and amorphous intercellular (interstitial) substances. There are four basic types of tissue: epithelial, connective, muscular, and nervous. These different types of tissue combine in varying amounts to form the numerous organs of the body.

5. D, Only 4 is correct. In regularly arranged dense connective tissue the collagen fibers lie parallel to each other with cells intervening between the parallel

bundles. This arrangement of fibers provides for great tensile strength and is found in tendons and ligaments, where such strength is needed. In irregularly arranged dense connective tissue the fibers run in all directions and is considerately weaker than the regularly arranged dense connective tissue. Irregular dense connective tissue can be found in joint capsules, fasciae, sheaths, and septae.

6. C, 2 and 4 are correct. Bone, like all connective tissues, contains the three basic elements: cells, matrix, and fibers. Bone differs from cartilage, which it is similar to, in that its matrix is impregnated by inorganic salts. The functioning cells of bone are called osteocytes. Bone is surrounded by a connective tissue membrane, similar to the one found in cartilage, and is called the periosteum. Bone is vascular in that Haversian canals (A in the diagram) allow blood vessels to penetrate the bone tissue. These canals are surrounded by concentric rings of bone called lamellae (B) and the osteocytes are located in the lacunae (C).

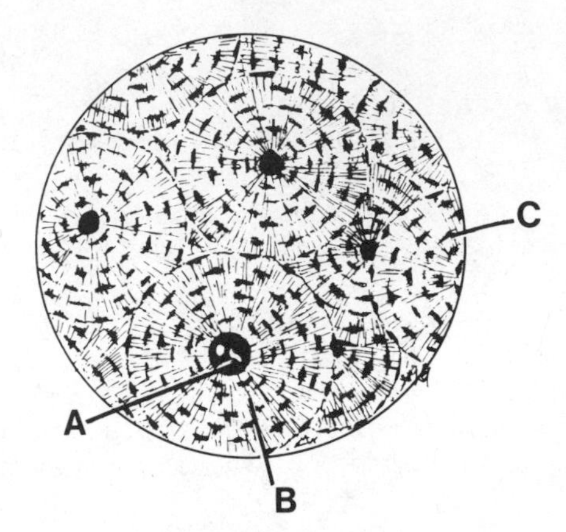

7. C, 2 and 4 are correct. Cartilage, a connective tissue, is a resilient tissue with cells, called chondroblasts, that produce a firm intercellular substance or matrix. Cartilage has no basement membrane but is covered by a dense connective tissue sheath called the perichondrium. The nutritive requirements of cartilage must diffuse from the blood vessels located in the perichondrium, as there are no blood vessels in the cartilage matrix itself. There are three types of cartilage (i.e., hyaline, fibrocartilage, and elastic) and each varies as to the ratio of matrix and the amount or type of fibers found.

8. E, All are correct. Cardiac muscle is located only in the heart, thus its name. Its cells are joined together at sites called intercalated disks and form a branching network or syncytium. Cardiac muscle is like skeletal muscle since they both have cross striations. Cardiac muscle is also similar to smooth muscle in that both are not under the control of the will but instead are innervated by the autonomic nervous system (ANS), or involuntary. Skeletal muscle is under control of the will, or voluntary.

9. B, 1 and 3 are correct. Ectoderm, one of three primary germ layers, gives rise to the epidermis of the skin and all its derivatives. Ectoderm also gives rise to the anterior pituitary, inner ear, and lens of the eye. The central nervous system, retina, pineal body, posterior pituitary, cranial sensory ganglia and nerves, and medulla of the adrenal gland are also ectodermal or more correctly, neuroectodermal derivatives. The liver is an endodermal (entodermal) derivative; whereas muscle tissues are mesodermal in origin. All connective tissues and blood vessels are also mesodermal derivatives.

FIVE-CHOICE ASSOCIATION QUESTIONS

DIRECTIONS: Each group of questions below consists of a numbered list of descriptive words or phrases accompanied by a diagram with certain parts indicated by letters, or by a list of lettered headings. For each numbered word or phrase, SELECT THE LETTERED PART OR HEADING that matches it correctly. Then insert the letter in the space to the right of the appropriate number. Sometimes more than one numbered word or phrase may be correctly matched to the same lettered part or heading.

ASSOCIATION QUESTIONS

1. _____ Found lining the bladder.

2. _____ An epithelial tissue primarily providing protection.

3. _____ A lining which appears to be stratified but is not.

4. _____ An epithelium capable of accommodating distension.

5. _____ Lines the lumen of the small intestine.

6. _____ Cuboidal epithelium.

7. _____ May or may not become keratinized.

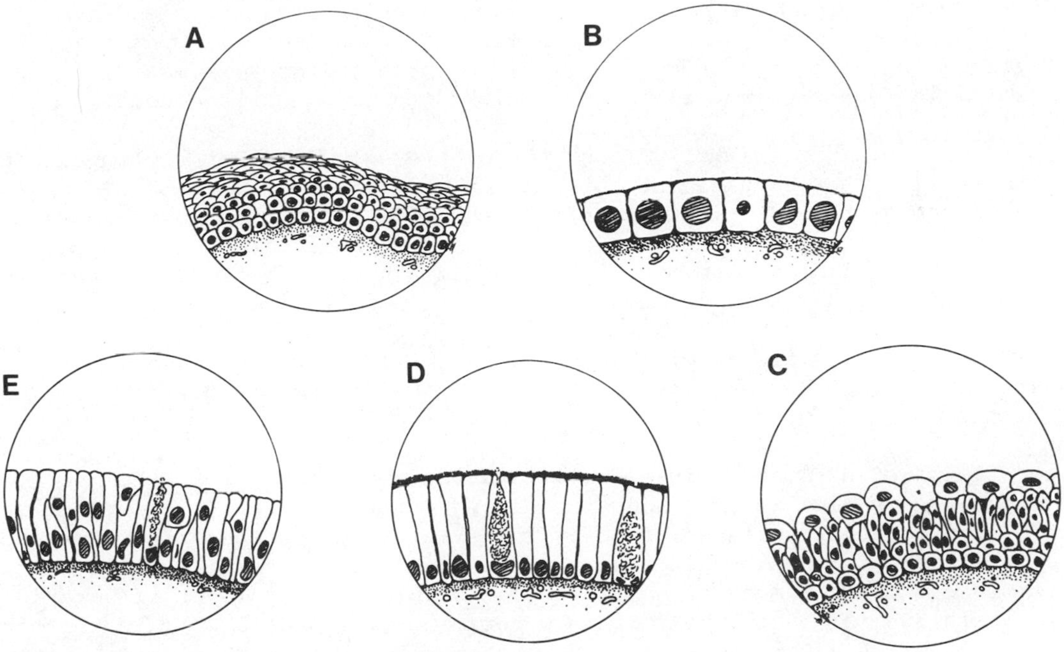

A B

E D C

8. _____ A cell having fat droplets in its cytoplasm.

9. _____ Located primarily in the fetus.

10. _____ Responsible for supporting cells involved in conducting electrical impulses.

11. _____ Function as constrictors of blood vessels.

12. _____ Known to produce antibodies.

A. Smooth muscle cell

B. Neuroglial cell

C. Plasma cells

D. Mesenchymal cells

E. None of the above

------------------------ ANSWERS, NOTES AND EXPLANATIONS ------------------------

1. C, Transitional epithelium is a stratified layer of epithelial cells. The deepest layer of cells is cuboidal or columnar, the middle cell layers are irregular (polyhedral) in shape, whereas the superficial layer consists of large cells with convex surfaces. This type of epithelium is found lining the excretory passages of the urinary system, including the bladder.

2. A, Stratified squamous epithelium is found lining the mouth, esophagus, vagina, part of the conjunctiva and retina of the eye, and part of the female urethra. It also forms the outer or epidermal layer of the skin. Its deepest layer of cells is either cuboidal or low columnar. These cells as they reach the surface, migrating through six or seven layers, become more flattened. The superficial layers have some cells which are devoid of nuclei.

3. E, This epithelium because of the varied shapes of its cells and the appearance of their nuclei at different levels, is often mistakenly thought to be a stratified tissue. However, each cell maintains contact with the basement membrane and so it is more correctly called a simple epithelium. Because of this, this epithelium is called pseudostratified or 'false' stratified epithelium. It is found lining the parotid duct and a portion of the male urethra.

4. C, Transitional epithelium, found lining much of the urinary system, is capable of accommodating the distention of hollow organs such as the bladder. In its distended form this epithelial tissue, lining the bladder, appears as two layers. In the contracted state (i.e., when the bladder is empty), this epithelium is four or five cell layers thick with large surface cells.

5. D, Simple columnar epithelium consists of a single layer of rectangular cells with their nuclei situated at one level near the base. These cells are found lining the gastrointestinal tract from the stomach to the anus and are involved in the absorption of intestinal contents and the secretion of mucus and enzymes. A ciliated form of this epithelial type is found lining the uterus, uterine tubes, and some of the small bronchi of the lungs.

6. B, Cuboidal cells are as wide as they are tall. They are found lining the walls of many glandular ducts and also form the secretory portions of many glands.

7. A, Stratified squamous epithelium lining wet surfaces such as the mouth, esophagus, and vagina is of a nonkeratinizing type. However, this epithelial type found on dry surfaces, such as the skin, is an example of a keratinizing epithelium. In the keratinizing process the more superficial cells of the membrane undergo a metamorphosis into a tough nonliving layer of keratin (protein) which is attached to the underlying living cells. This keratin layer provides a certain amount of waterproofing both for water lost from the body and the minute amount of absorption (imbibing) of water into the body. This layer, because of its nature, helps to protect the underlying epithelial cells from being injured by the ordinary wear and tear to which skin is exposed. This keratin layer is also relatively impervious to bacterial penetration so acts as a first line of defense against infection. Over the soles of the feet and the palms of the hands the keratin layer of the stratified squamous epithelium (epidermis) of the skin becomes very thick because of the pressure and the added wear and tear.

8. E, Fat cells usually located in loose connective tissue contain varying quantities of stored fat. When large numbers of these cells are organized into lobules they are collectively termed adipose tissue. The functions of fat cells or adipose tissue are: the storage of fat as a potential fuel source (per gram has the highest calorie content of all foods); acts as an insulator for the body against the cold; becomes a filler of tissue spaces; and acts as a protective cushion for many organs, i.e., the kidney.

9. D, Mesenchymal cells are cellular components of loose connective tissue found almost exclusively in the fetus. These undifferentiated cells are the embryonic source of all connective tissues and therefore can be called the stem cell of the connective tissues. Occasionally a few undifferentiated mesenchymal cells remain in the loose connective tissues of the adult and if stimulated, may develop and form an isolated area of tissue different from that surrounding it. This occurrence is extremely rare and the number of these cells in the adult is probably very small.

10. B, Neuroglial cells are the support cells of the central nervous system (CNS). These cells, of which there are four types, are responsible not only for supporting the neurons of the CNS, but are also thought to be involved in transporting chemicals to the neurons which are essential for their metabolic functions.

11. A, Smooth muscle cells are found in the walls of many tubed structures, such as blood vessels, and are responsible for their constriction. All muscle cells have the special ability to contract. Smooth and cardiac muscle are considered to contract involuntarily or not under control of the conscious, whereas skeletal muscle is considered to contract voluntarily.

12. C, Plasma cells are found throughout the body, but the majority of them are found in the loose connective tissues of the digestive tract. These cells, in response to the invasion of the body by an infectious agent, produce a protein gamma globulin or antibody. This antibody combines with the disease organism and through a number of reactions leads to the inactivation of the organism. The plasma cells therefore produce the antibody which is the first step involved in the immune response of the body.

REVIEW EXAMINATION OF PART 1

INTRODUCTORY NOTE: This review examination consists of 50 multiple-choice questions based on the learning objectives listed for the following chapters of this Study Guide and Review Manual: 1. INTRODUCTION TO THE BODY; 2. THE CELL AND ITS PHYSIOLOGY; and 3. BODY TISSUES. Before beginning, tear out an answer sheet from the back of the book and read the directions on how to use it. The key to the correct responses is on page 339.

FIVE-CHOICE COMPLETION QUESTIONS

DIRECTIONS: Each of the following questions or incomplete statements is followed by five suggested answers or completions. SELECT THE SINGLE BEST ANSWER in each case and then blacken the appropriate space on the answer sheet.

1. THE MACROSCOPIC STUDY OF THE HUMAN BODY IS CALLED ________.
   A. Pathology
   B. Gross Anatomy
   C. Neuroanatomy
   D. Physiology
   E. Embryology

2. WHICH OF THE FOLLOWING IS NOT FOUND IN ANY CONNECTIVE TISSUE?
   A. Smooth muscle
   B. Reticular fibers
   C. Macrophages
   D. Elastic fibers
   E. Ground substance

3. THE PROCESS OF ______ REFERS SPECIFICALLY TO THE MOVEMENT OF WATER MOLECULES THROUGH A SEMIPERMEABLE MEMBRANE FROM A HIGHER AREA TO A LOWER AREA OF WATER CONCENTRATION.
   A. Active transport
   B. Facilitated diffusion
   C. Filtration
   D. Osmosis
   E. Phagocytosis

4. THE SYSTEM RESPONSIBLE FOR BRINGING OXYGEN INTO THE BODY IS CALLED THE __________ SYSTEM.
   A. Digestive
   B. Urinary
   C. Reproductive
   D. Respiratory
   E. Endocrine

5. STRATIFIED SQUAMOUS EPITHELIUM LINES THE __________.
   A. Bronchi
   B. Mouth
   C. Stomach
   D. Trachea
   E. Ureter

SELECT THE SINGLE BEST ANSWER

6. THE STERNUM LIES _______ TO THE HEART.
   A. Lateral
   B. Medial
   C. Posterior
   D. Superior
   E. None of the above

7. THE ______ PLANE IS A VERTICAL ONE DIVIDING THE BODY INTO EQUAL RIGHT AND LEFT PORTIONS.
   A. Coronal
   B. Median
   C. Oblique
   D. Sagittal
   E. Transverse

8. CELLULAR MEMBRANES LINING THE VESSELS OF THE BLOOD-VASCULAR AND LYMPHATIC SYSTEMS ARE CALLED:
   A. Serous
   B. Endothelial
   C. Mesothelial
   D. Mesenchymal
   E. None of the above

9. THE BASE NOT FOUND IN THE DNA MOLECULE IS __________.
   A. Adenine
   B. Cytosine
   C. Guanine
   D. Thymine
   E. Uracil

10. THE LUNGS ARE LOCATED IN THE CAVITY INDICATED BY ___________.

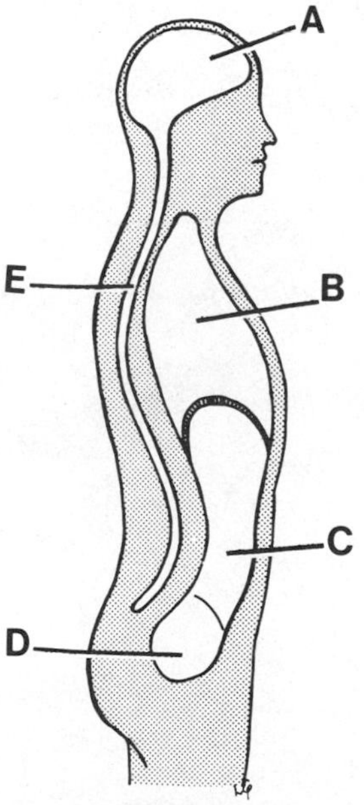

11. A STRUCTURE INVOLVED IN THE PRODUCTION OF ATP AND CONTAINING ENZYMES INVOLVED IN PROTEIN SYNTHESIS AND LIPID METABOLISM IS THE _______.
    A. Endoplasmic reticulum
    B. Mitochondrion
    C. Ribosomes
    D. Golgi apparatus
    E. Lysosomes

12. IN ________ OF MITOSIS THE NUCLEAR MEMBRANE REFORMS.
    A. Anaphase
    B. Interphase
    C. Metaphase
    D. Prophase
    E. Telophase

SELECT THE SINGLE BEST ANSWER

13. MESOTHELIAL TISSUE IS CLASSIFIED AS:
    A. Simple cuboidal
    B. Simple columnar
    C. Pseudostratified
    D. Stratified squamous
    E. None of the above

14. THIS ORGANELLE IS A(AN)____________.
    A. Endoplasmic reticulum
    B. Mitochondrion
    C. Lysosome
    D. Nucleus
    E. Vacuole

15. THE PASSIVE PROCESS BY WHICH A LIPID INSOLUBLE SUBSTANCE PASSES THROUGH THE CELL MEMBRANE IS CALLED _________.
    A. Osmosis
    B. Filtration
    C. Phagocytosis
    D. Facilitated diffusion
    E. None of the above

16. WHICH OF THE FOLLOWING IS NOT A CELL TYPE FOUND IN AREOLAR CONNECTIVE TISSUE?
    A. Fat
    B. Mast
    C. Fibroblast
    D. Plasma
    E. Chondrocyte

17. WHICH OF THE FOLLOWING IS NOT TRUE FOR A TISSUE?
    A. Smallest functional unit of the body
    B. Makes up organs
    C. Has a specific function
    D. Consists of cells
    E. Contains some interstitial material

18. THIS CELL IS IN ______.
    A. Anaphase
    B. Interphase
    C. Metaphase
    D. Prophase
    E. Telophase

19. A TISSUE GRAFT BETWEEN TWO GENETICALLY IDENTICAL INDIVIDUALS IS CALLED A(AN) ______.
    A. Isograft
    B. Allograft
    C. Autograph
    D. Xenograft
    E. None of the above

MULTI-COMPLETION QUESTIONS

DIRECTIONS: In each of the following questions or incomplete statements, ONE OR MORE of the completions given is correct. On the answer sheet, blacken the space under A if 1, 2 and 3 are correct; B if 1 and 3 are correct; C if 2 and 4 are correct; D if only 4 is correct; and E if all are correct.

20. THE SYSTEM(S) INVOLVED IN COORDINATING AND REGULATING HOMEOSTATIC MECHANISMS OF THE BODY IS(ARE) THE:
    1. Circulatory
    2. Digestive
    3. Respiratory
    4. Nervous

21. IN ANAPHASE THE:
    1. Spindle rays break up
    2. Nucleolus reappears
    3. Cytokinesis is completed
    4. Chromatids move to the opposite poles

22. A CORONAL PLANE:
    1. Divides the body into right and left portions
    2. Intersects the median plane at right angles
    3. Crosses the longitudinal axis of the body
    4. Divides the body into anterior and posterior portions

23. AN ENDOCRINE GLAND:
    1. Has a duct
    2. Produces substances called hormones
    3. Secretes onto the skin or into an organ
    4. Is an epithelial derivative

24. THE SKELETAL SYSTEM:
    1. Supports the body
    2. Provides sites for blood formation
    3. Contains great stores of minerals
    4. Protects vital organs of the body

25. SMOOTH MUSCLE:
    1. Constricts blood vessels
    2. Forms a syncytium
    3. Is controlled by the ANS
    4. Has fine striations

26. WHICH OF THE FOLLOWING STATEMENTS IS(ARE) CORRECT FOR A HYPOTONIC SOLUTION WHEN SEPARATED FROM ANOTHER SOLUTION BY A SELECTIVELY PERMEABLE MEMBRANE?
    1. Pressure and volume increases
    2. An osmotic pressure develops
    3. A net osmosis develops
    4. Has a lower potential osmotic pressure

27. THE PROCESS OF PHAGOCYTOSIS INCLUDES THE:
    1. Ingestion of particulate matter
    2. Passive ingestion by a cell
    3. Formation of a digestive vacuole
    4. Involvement of the nuclear membrane

A	B	C	D	E
1,2,3	1,3	2,4	only 4	all correct

28. THE ELBOW JOINT LIES _______ TO THE SHOULDER.
    1. Posterior
    2. Inferior
    3. Proximal
    4. Distal

29. WHICH OF THE FOLLOWING STATEMENTS IS(ARE) TRUE FOR THE CYTOPLASMIC MATRIX?
    1. Location of metabolic and synthetic activities
    2. Contains inclusions
    3. Surrounds the nucleus
    4. Its amount is helpful in identifying some cell types

30. THE GOLGI APPARATUS HAS BEEN SHOWN TO FUNCTION IN THE:
    1. Production of ATP
    2. Synthesis of RNA
    3. Autodigestion of the cell
    4. Packaging of secretory products

31. TISSUES OF THE BODY:
    1. Unite to form organs
    2. Are made up of cells
    3. Can be classified into four distinct types
    4. Have defined functions

32. DEOXYRIBONUCLEIC ACID IS:
    1. Located in the nucleus
    2. Duplicated during mitosis
    3. The carrier of genetic material
    4. Concentrated in lysosomes

33. WHICH OF THE FOLLOWING STRUCTURES IS(ARE) DERIVED FROM ENDODERM?
    1. Retina
    2. Liver
    3. Spleen
    4. Pancreas

34. THE LYMPHATIC SYSTEM:
    1. Helps combate foreign substances that enter the body
    2. Communicates with the venous system
    3. Functions in filtering blood
    4. Is involved in transporting oxygen

35. WHICH OF THE FOLLOWING IS(ARE) COMPATIBLE WITH A LOW RATE OF DIFFUSION?
    1. A heated solution
    2. A higher concentration
    3. A low viscosity
    4. Larger molecules

36. WHICH OF THE FOLLOWING IS(ARE) FOUND IN LOOSE CONNECTIVE TISSUE?
    1. Fibroblast
    2. Plasma cell
    3. Melanocytes
    4. Histiocyte

37. SIMPLE COLUMNAR EPITHELIUM IS LOCATED LINING THE:
    1. Bladder
    2. Stomach
    3. Esophagus
    4. Bronchioles

38. INCLUDED AMONG MESODERMAL DERIVATIVES IS(ARE):
    1. Most of the skeleton
    2. The urogenital system
    3. The serous membranes of the body cavities
    4. The dermis of the skin

FIVE-CHOICE ASSOCIATION QUESTIONS

DIRECTIONS: Each group of questions below consists of a numbered list of descriptive words or phrases accompanied by a diagram with certain parts indicated by letters, or by a list of lettered headings. For each numbered word or phrase, SELECT THE LETTERED PART OR HEADING that matches it correctly. Then blacken the appropriate space on the answer sheet. Sometimes more than one numbered word or phrase may be correctly matched to the same lettered part or heading.

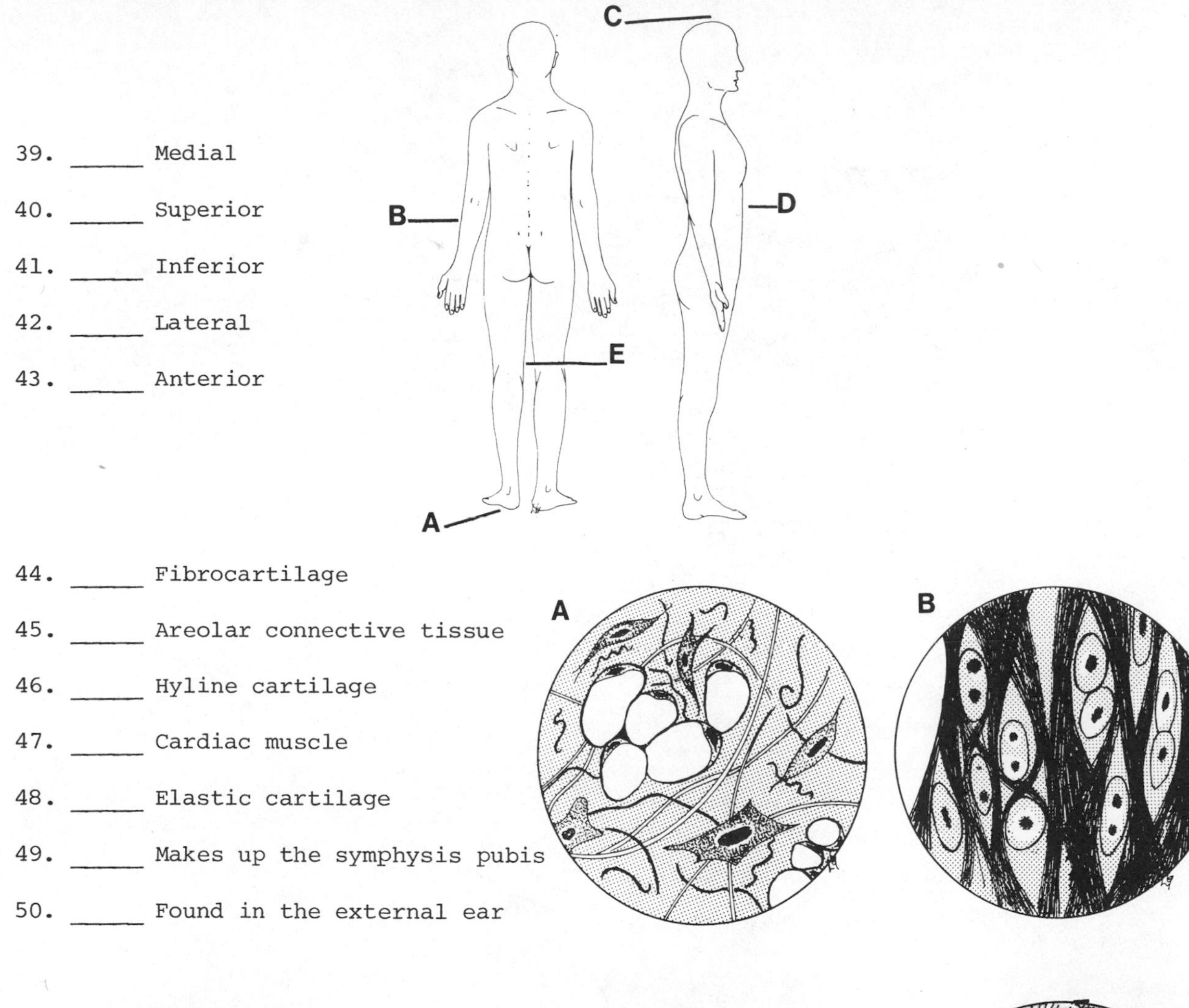

39. _____ Medial

40. _____ Superior

41. _____ Inferior

42. _____ Lateral

43. _____ Anterior

44. _____ Fibrocartilage

45. _____ Areolar connective tissue

46. _____ Hyline cartilage

47. _____ Cardiac muscle

48. _____ Elastic cartilage

49. _____ Makes up the symphysis pubis

50. _____ Found in the external ear

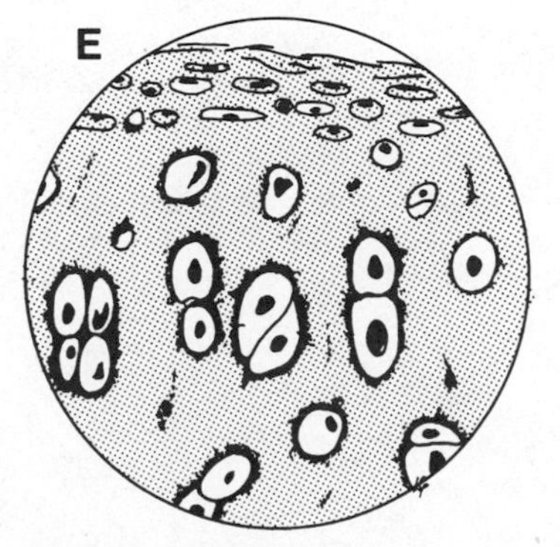

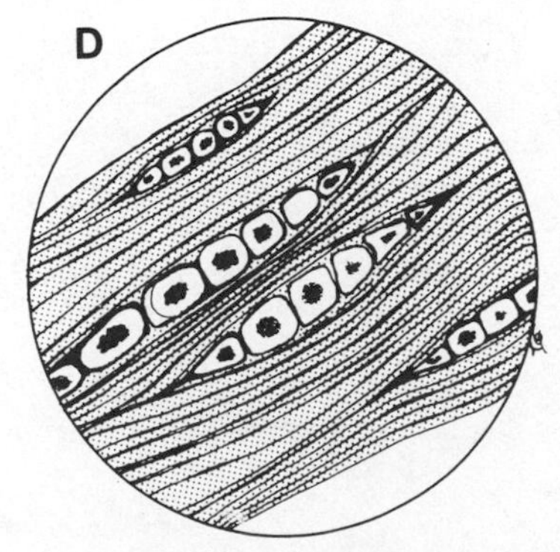

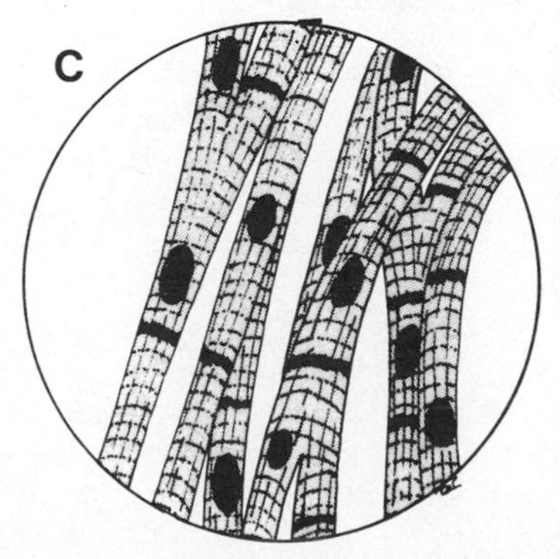

PART TWO

THE SKIN AND OTHER MEMBRANES

THE SKELETAL SYSTEM

JOINTS AND THEIR MOVEMENTS

MUSCLE TISSUE AND ITS PHYSIOLOGY

4. THE SKIN AND OTHER MEMBRANES

LEARNING OBJECTIVES

Be Able To:

- Define the term membrane and describe the properties and functions of the following membranes: mucous, serous, and synovial.
- Compare and contrast the epidermis and dermis of the skin as to their origin, layers, relative thicknesses, and functions.
- Discuss the varied functions of the skin.
- Describe the anatomy and functions of epidermal derivatives such as hair and nails.
- Describe the general structure, location, and functions of sebaceous, sweat, and modified sweat glands.
- Give the factors involved in wound healing and the features differentiating first, second, and third degree burns.
- Define the following pathologies of the skin: acne, decubitus ulcer, furuncle, hives, melanoma, nevus, polyp, psoriasis, and wart.

RELEVANT READINGS

Anthony & Kolthoff - Chapters 4 and 15.
Chaffee & Greisheimer - Chapter 2.
Crouch & McClintic - Chapter 6.
Jacob, Francone & Lossow - Chapter 4.
Landau - Chapters 3 and 26.
Tortora - Chapter 4.

FIVE-CHOICE COMPLETION QUESTIONS

DIRECTIONS: Each of the following questions or incomplete statements is followed by five suggested answers or completions. SELECT THE SINGLE BEST ANSWER in each case and then circle the appropriate letter at the lower right of each question.

1. WHICH OF THE FOLLOWING IS NOT CONSIDERED A FUNCTION OF THE SKIN?
   A. Produces vitamin D
   B. Allows selective absorption
   C. Conducts electrical impulses
   D. Provides protection
   E. Mediates sensation

   A B C D E

2. A CONDITION OF THE SKIN THAT IS ATTRIBUTABLE TO AN ALLERGIC RESPONSE IS CALLED _______.
   A. Acne
   B. Hives
   C. Jaundice
   D. Psoriasis
   E. None of the above

   A B C D E

3. THE MEISSNER'S CORPUSCLES OF THE DERMIS ARE RESPONSIBLE FOR THE SENSATION OF:
   A. Cold
   B. Pain
   C. Warmth
   D. Light touch
   E. Deep pressure

   A B C D E

4. THE LUNULE IS THE:
   A. Actively growing region of the nail
   B. Gland which produces sebum
   C. Extra layer found in thick skin
   D. Outer layer of a hair shaft
   E. Area where the sebaceous gland empties its product

   A B C D E

5. WHICH OF THE FOLLOWING IS NOT TRUE FOR THE EPIDERMIS?
   A. Has four or five distinctive layers
   B. Contains numerous blood vessels
   C. May contain carotene pigment
   D. Is thinner than the dermis
   E. Allows percutaneous absorption

   A B C D E

6. THE FUNCTION OF MELANIN IN THE SKIN IS TO:
   A. Serve as a waterproofing
   B. Enhance sensation of the skin
   C. Aid in the production of keratin
   D. Protect the body against radiation
   E. None of the above

   A B C D E

7. THE PARIETAL PLEURA IS FOUND LINING THE:
   A. Larger bronchi
   B. Fibrous capsule of a joint
   C. Walls of the thoracic cage
   D. Abdominal cavity
   E. Wall of the pelvic cavity

   A B C D E

ANSWERS, NOTES AND EXPLANATIONS

1. C. The skin, in particular the epidermis, provides for a number of different functions. It provides for protection in that it has a surface film, of an acidic pH, derived from sweat and sebaceous glands. This film helps retard bacterial and fungal fungae growth. The keratinized cells of the surface layers of the skin provide a physical protection for the body. The epidermis, once considered impervious to everything, is known to allow some hormones, vitamins,

organic bases, and gases to enter the blood stream by a process called percutaneous absorption. The skin also mediates the sensations of pain, touch, cold, and warmth by specialized nerve endings. Skin plays an important role in thermoregulation, or controlling the amount of heat lost from the body. The role of the skin centers around its ability to act as a radiant surface and to transfer heat from the cutaneous blood vessels to the surface. Another mechanism of thermoregulation directly involving the skin is sweating. The sweat glands secrete a product, which is almost pure water, on to the surface of the skin. This water evaporates and cools the skin surface and its underlying tissues. The hypothalamus of the brain is the center regulating the amount of heat lost from the body. Another function of the skin is that it produces a sterol from dehydrocholesterol called vitamin D. The production of vitamin D is enhanced by the exposure to ultraviolet light. This vitamin, once absorbed by the blood stream, becomes involved in the metabolism of calcium and phosphate in the body.

2. B. Hives (urticaria) is a vascular response of the skin to the sensitivity of an individual to certain foods such as shellfish and strawberries or to certain drugs such as penicillin. In hives the skin exhibits pale or reddened raised patches which are itchy. Acne is a local infection and inflammation of the sebaceous glands. Psoriasis is a chronic inflammatory disease of unknown origin. It is characterized by patches of dry whitish scales which are neither infectious nor contagious. Jaundice is a syndrome characterized by hyperbilirubinemia and deposition of bile pigment in the skin and mucous membranes. A jaundiced patient appears yellow.

3. D. The Meissner's corpuscle of the dermis is responsible for the sensation of light touch in the skin. The skin also has other types of nerve endings: Pacinian corpuscles for the detection of deep pressure; Ruffini corpuscles for heat; Krause corpuscles for cold; and simple nerve endings for pain. In addition to these nerve endings involved with sensation there are other nerves found in the skin which are responsible for the innervation of blood vessels, sweat glands, and arrector muscles.

4. A. The lunule is the white, half-moon shaped area found at the base of the nail (unguis), responsible for its growth. The nail also has a root, a body and free edges which lie superficial to the nail bed. This bed contains capillaries; giving the nail its characteristic pink color. The capillaries do not show through the lunule or basal area of the nail so it appears white. The nails are modified corneal and lucidal areas of the skin. The strength of the nail is increased because of its gentle curves in the longitudinal and transverse planes. The nails along with the skin and its other derivatives (glands and hair) make up the integumentary system.

5. B. The skin covers the surface of the body and consists of a thinner superficial portion called the epidermis and a thicker deeper portion, the dermis. The epidermis has four layers in thin skin and an additional layer, the stratum lucidum in thick skin. The epidermis contains no blood vessels but receives its nourishment by diffusion from capillaries lying within the dermis. The color of skin is primarily because of the pigment granules in the various layers of the skin. For example, melanin is found in the three basal layers of the epidermis of Caucasians and in all five layers in blacks. The pigment carotene is found in the stratum corneum of the epidermis and adipose layers of the dermis in Oriental people. The epidermis is impervious to most substances such as electrolytes and water, but some gases and lipid-soluble substances (i.e., vitamins A, D, and K; estrogens; testosterone; aspirin; oxygen; and carbon dioxide) readily pass through by percutaneous absorption.

6. D. The function of melanin in the skin (epidermis) is to provide protection from the radiation effect of the sun's rays. The more one is exposed to the sun the more melanin is produced and the darker it gets. The amount of melanin in the skin of an individual is genetically determined. However, other factors such as the volume of blood and the amount of unoxygenated hemoglobin in the blood contribute significantly to moderate the effect of melanin on skin color. The skin color ranges from pale white to black. In addition, an increase in adrenocorticotropic (ACTH) and melanocyte-stimulating (MSH) hormones of the anterior pituitary gland will increase both the amount and the color of melanin produced.

7. C. There are four important types of membranes in the body: mucous, serous, synovial, and cutaneous. The mucous membranes are found lining the respiratory passages, the gastrointestinal tract, and the genitourinary tract; whereas synovial membranes are located lining the synovial joints of the body. The cutaneous membrane or skin covers the external surface of the body. The serous membranes are found lining the cavities of the body, i.e., thoracic, abdominal, and pelvic. The serous membrane lining the cavity is called the parietal layer and often it is reflected upon the organs of the cavity where it is then called the visceral layer. The parietal pleura, for example, would line the thoracic cage and become the visceral pleura on the surface of the lungs. The serous membrane lining the abdominal and pelvic cavities is called the peritoneum, and that lining the fibrous sac containing the heart is called the serous layer of the pericardium. All of these serous membranes secrete a fluid which protects the surface, reduces the friction of moving parts, or both.

MULTI-COMPLETION QUESTIONS

DIRECTIONS: In each of the following questions or incomplete statements, ONE OR MORE of the completions given is correct. At the lower right of each question, circle A if 1, 2, and 3 are correct; B if 1 and 3 are correct; C if 2 and 4 are correct; D if only 4 is correct; and E if all are correct.

1. SWEAT GLANDS OF SKIN SECRETE:
   1. Water
   2. Mineral salts
   3. Nitrogenous wastes
   4. Carbon dioxide

   A B C D E

2. WHICH OF THE FOLLOWING MEMBRANES PRODUCE A SEROUS FLUID?
   1. Cutaneous
   2. Synovial
   3. Mucous
   4. Pleural

   A B C D E

3. WHICH OF THE FOLLOWING STATEMENTS IS(ARE) TRUE FOR THE ROOT OF A HAIR?
   1. Lies below the surface of the skin
   2. Is restricted to the epidermal region
   3. Receives new cells from the matrix
   4. Forms the outer root sheath

   A B C D E

4. WHICH OF THE FOLLOWING IS(ARE) TRUE FOR A SECOND DEGREE BURN?
   1. Regeneration of the epidermis is impossible
   2. Dermal layer is damaged
   3. Skin is usually insensitive
   4. Fluid-filled blisters are common

   A B C D E

5. THE DERMIS OF THE SKIN:
   1. Is a mesodermal derivative
   2. Contains many blood vessels
   3. Produces cleavage lines in the epidermis
   4. Usually is thinner than the epidermis

   A B C D E

A	B	C	D	E
1,2,3	1,3	2,4	only 4	all correct

6. WHICH OF THE FOLLOWING IS(ARE) CORRECT FOR PROCESSES INVOLVED IN THE HEALING OF A SKIN WOUND?
   1. Blood vessels become dilated
   2. Granulation tissue is formed
   3. Fibroblasts undergo mitosis
   4. Fibrin protein is present

   A B C D E

7. EPIDERMAL DERIVATIVES INCLUDE:
   1. Hair shaft
   2. Arrector pili muscle
   3. Sweat glands
   4. Blood vessels

   A B C D E

8. THICK SKIN IS LOCATED ON THE:
   1. Back
   2. Thigh
   3. Anterolateral abdominal wall
   3. Soles of the feet

   A B C D E

------------------------ ANSWERS, NOTES AND EXPLANATIONS ------------------------

1. E, All are correct. Perspiration (sweat) secreted by the sweat glands of the skin is clear with a faint characteristic odor. It contains water, sodium chloride, mineral salts, small amounts of nitrogenous wastes, carbon dioxide and other compounds in trace amounts. Hormonal activity, external temperature and humidity, and fluid intake all influence the composition of sweat. Some sweat glands secrete organic compounds which on bacterial decomposition produce an offensive odor.

2. D, Only 4 is correct. Membranes are thin connective tissue sheets which either line the cavities or cover the surface of the body. Those membranes lining cavities secrete a small amount of serous fluid which serves to lubricate and allow organs to move one upon the other. A serous membrane is found lining the thoracic cavity (pleura), the abdominal and pelvic cavities (peritoneum) and the sac containing the heart (pericardium). Not only does the serous membrane line these cavities but it also is reflected over the organs found within these cavities. The serous membrane covering the organs is called the visceral layer, whereas the serous membrane layer lining the walls of the cavity is called the parietal layer. The synovial membrane of a synovial joint secretes a fluid, similar to a dialysate of blood, which provides for the lubrication of the articulating surfaces. Mucous membranes line the gastrointestinal tract, the respiratory passages, and the genitourinary tract. They consist, as do the serous membranes, of an epithelial layer backed by a connective tissue layer. Mucous membranes secrete a mucus material for protection and they also may be involved in absorption of substances into the body.

3. B, 1 and 3 are correct. Hair arises in a tubular invagination of the epidermis, the hair follicles, and extends down into the dermis. The active follicle has a terminal expansion or bulb, which contains a concavity in its base called the papilla. The papilla contains loose connective tissue and many blood vessels which nourish the follicle. The papilla is covered by epidermal matrix cells of the hair and the root sheath. The cells from the matrix develop into different types of cells, forming the layers of both the hair shaft and inner root sheath. The hair shaft consists of: 1) a central core or medulla; 2) a concentric layer of cortex; and 3) an outer concentric layer or cuticle. All three of these layers are keratinized. The free end of the shaft protrudes beyond the surface of the skin. The hair does not grow continuously but rather in spurts, with periods of rest in between. The root sheath is formed by two

components: 1) the inner root sheath which like the hair itself is keratinized and is derived from the matrix, and 2) the outer root sheath which is a continuation of the epidermis down into the follicle.

4. C, 2 and 4 are correct. A burn may be described in two ways: first by the area of the body burnt and second by the depth or degree of the destruction. The area involved in a burn can be determined by using the "rule of nines", in which different areas are considered to represent a percentage of the total body area (i.e., the upper limb = 9%, the head and neck = 9%). The degree of a burn is dependent on such things as the skin layers involved, the sensitivity remaining, the regenerative capacity, and the character of the remaining tissues. A first degree burn involves only the surface layers of the epidermis, appears red, and is tender to the touch (e.g., a sunburn). A second degree burn is where most of the epidermis is destroyed, there is some involvement of the dermis, there is redness, and blisters are usually present. A second degree burn is also very painful because of the irritation to the bared nerves. Third degree burns involve all skin layers and there are no epidermal remnants left. Also there is general insensitivity and the muscle, bone and other deeper structures may be charred. The healing of third degree burns usually requires some type of skin grafting. Once healed, fibrous scar tissues may prevent nearby joints from moving through their normal range of motion. Immediate concerns of severe burns are the loss of the normal functions of the skin, such as the protection it provides against infection, and the loss of body fluids, electrolytes, and blood proteins.

5. A, 1, 2, and 3 are correct. The dermis of the skin is a mesodermal derivative, whereas the epidermis along with its derivatives are ectodermal in origin. The dermis is a tough, flexible, elastic layer of irregular dense connective tissue and is thicker than the epidermis. The elasticity of the dermis produces the characteristic cleavage lines found over the total body. The dermis consists of an outer layer of loose connective tissue called the papillary layer which projects up into the epidermis. This layer contains blood vessels, forming anastomotic channels that are able to control the amount of blood flowing through the dermis. This factor is an important element in temperature regulation of the body. The deeper reticular layer of the dermis consists of regular dense loose connective tissue elements running parallel to the surface of the skin. This deep layer of the dermis is responsible for attaching the skin to the subcutaneous tissue.

6. E, All are correct. Skin tissue, like other epithelia, replace destroyed cells under normal physiological conditions. The first visible response to trauma of the skin is bleeding into the wound surface. In a short time fibrin found in blood produces a clot which seals off the wound. The area becomes quickly engorged by blood carrying oxygen and other nutrients because of a general vasodilation of the blood vessels. Phagocytes of the circulating blood invade the area and remove broken cells and debris. Newly formed fibroblasts, endothelial tissue, and new blood vessels from the subcutaneous tissue invade the wound and form granulation tissue. The basal layer of the epidermis produces new cells which eventually cover the area and proliferate to form the new skin. Other epidermal derivatives such as hair, nails, and glands if destroyed are not replaced.

7. B, 1 and 3 are correct. The skin (integument) is derived from both ectodermal and mesodermal germ layers. The epidermis and its derivatives such as nails, glands, and hairs come from the ectoderm. The underlying dermis and its components come from the mesodermal layer. Included among the mesodermal derivatives are blood and lymphatic vessels, smooth muscle fibers of the arrector pili muscle, and loose connective tissue elements of the dermis.

8. D, Only 4 is correct. Thick skin is located primarily on the soles of the feet and the palms of the hands. The skin is thicker in these areas because of the increased stress applied to the palms and soles. All other areas of the body are covered by thin skin. Histologically, the differences between thick and thin skin are that in thin skin the individual layers are not as deep and often the stratum lucidum is missing.

FIVE-CHOICE ASSOCIATION QUESTIONS

DIRECTIONS: Each group of questions below consists of a numbered list of descriptive words or phrases accompanied by a diagram with certain parts indicated by letters, or by a list of lettered headings. For each numbered word or phrase, SELECT THE LETTERED PART OR HEADING that matches it correctly. Then insert the letter in the space to the right of the appropriate number. Sometimes more than one numbered word may be correctly matched to the same lettered part or heading.

1. _____ The stratum spinosum.

2. _____ The reticular layer.

3. _____ The layer where mitotic figures are most commonly found.

4. _____ The keratinized layer of the epidermis.

5. _____ Cell layer containing keratohyalin granules.

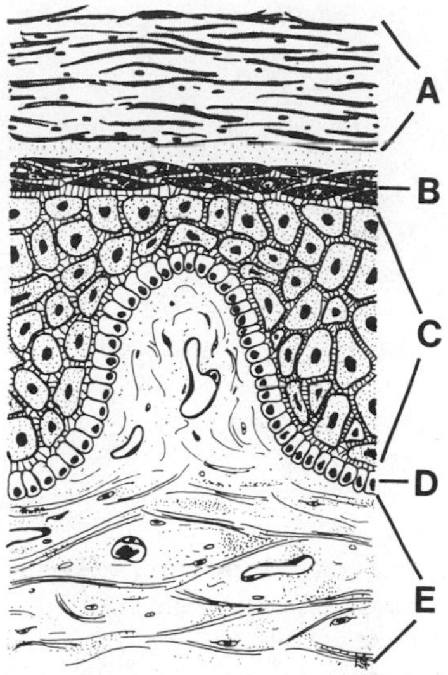

6. _____ An ulcer formed because of continual pressure on the skin.

7. _____ A nonmaligant pigmented area of the skin.

8. _____ An abcess resulting from the infection of a hair follicle.

9. _____ A nonmalignant skin tumor caused by a virus.

10. _____ A malignant tumor of melanocytes.

A. Wart

B. Decubitus

C. Nevus

D. Furuncle

E. None of the above

------------------------ ANSWERS, NOTES AND EXPLANATIONS ------------------------

1. C. The stratum spinosum is the layer of the epidermis lying immediately above the stratum basale (germinativum) (D in the diagram). This layer gets its name from the fact that its cells are prickle-like in appearance. The stratum spinosum consists of eight to 10 rows of cells bearing short spines. These spines are firmly attached to the spines of adjacent cells by desmosomes. Some mitotic figures may be located in this layer.

2. E. The reticular layer along with the papillary layer, which projects up into the epidermis, form the dermis of the skin. These two layers cannot be clearly separated but the reticular layer consists of dense connective tissue (collagen bundles) running parallel to the surface; whereas the papillary layer consists of loose connective tissue elements. At various levels of the dermis there are hair follicles, sweat, and sebaceous glands, which are epidermal derivatives extending down into the dermis. The dermis also contains blood vessels, nerves, and nerve endings. Immediately below the dermis a subcutaneous layer of loose connective tissue or hypodermis is found.

3. D. The stratum germinativum (basale), located adjacent to the dermis, consists of a single row of cuboidal or columnar cells. From this layer new cells are produced by mitosis to replace those cells which are eventually shed from the surface or stratum corneum (A in the diagram). Mitotic figures are also found in the stratum spinosum. This process of change (cytomorphosis) from the basal to the surface layers takes from 15 to 20 days, depending upon the region of the body and a number of other factors.

4. A. The stratum corneum, of thick skin, consists of many layers of lifeless cells which have lost their nuclei. In these cells the protein keratin replaces the cytoplasm. The most peripheral cells of this layer are constantly being desquamated. This hard keratinized layer serves to protect the underlying tissues. The layer immediately beneath the stratum corneum is called the stratum lucidum (not labelled in the diagram). This layer is thought to have cells which are in the process of becoming keratinized. The stratum lucidum appears as a clear refractile zone containing a substance called eleidin.

5. B. The stratum granulosum found immediately below the stratum lucidum consists of three to five layers of cells. These cells contain basophilic granules of keratohyalin, which are believed to be involved with the formation of keratin.

6. B. An ulcer is a loss of cells from a cutaneous or mucous surface, causing gradual disintegration and necrosis. A decubitus ulcer (pressure or bed sore) is caused by a prolonged pressure on an area of skin of a patient who is confined to bed for an extended period of time. By turning the patient frequently during this time the pressure is reduced on the skin and this helps to prevent such ulcerations.

7. C. A nevus (mole) is normally a round pigmented area of skin which may be present at birth or may develop later in life. They range in color from yellow-brown to black. A nevus is usually not malignant (tending to grow and eventually causing death) but may, by irritation, be stimulated to become malignant. A nonpigmented mole is called an amelanotic nevus. Nevi may also be derived from vascular or nervous tissues.

8. D. A furuncle (boil) is a painful nodule formed in the corneum and subcutaneous tissues of the skin by bacteria. The bacteria enter the skin through hair follicles and produce a central core which is surrounded by an area of necrotic tissue.

9. A. A wart is a nonmalignant epidermal tumor (growth) produced by a virus. Since warts are of viral origin they are contagious and relatively common.

10. E. A malignant-melanin pigmented tumor is called a melanoma. Such tumors may develop either from existing nevi or spontaneously from melanocytes found within the skin. Melanomas have a great tendency to form metastases which spread throughout the body, usually causing death.

5. THE SKELETAL SYSTEM

LEARNING OBJECTIVES

INTRODUCTION TO BONE

Be Able To:

- Give the functions of the skeletal system.
- Describe the cell types found in bone, giving their functions.
- Classify the bones of the body based on their shape and give examples of each type.
- Compare and contrast intramembranous and endochondral (cartilaginous) bone formation.
- Discuss the factors involved in the homeostasis, repair, and growth of bone.
- Define the features of comminuted, transverse, pathologic, greenstick, compression, simple, and compound fractures.
- Describe the components of a long bone and the histology of spongy and compact bone.
- Define the following descriptive terms pertaining to bone: condyle, crest, foramen, fossa, head, meatus, sinus, spinous process, trochanter, tubercle and tuberosity.
- Define the terms axial and appendicular skeleton, assigning the bones of the body to their appropriate group.

THE SKULL

Be Able To:

- List and identify on diagrams the single and paired bones forming the cranial and facial portions of the human skull.
- On a diagram of a fetal or adult skull identify the following sutures: coronal, sagittal, lambdoidal, and squamosal.

- Describe the function of a fontanel in the fetal skull and give the location of the following fontanels: anterior, posterior, posterolateral, and anterolateral.

- Give the location and function of the auditory ossicles and the hyoid bone.

- List the bones forming the lateral wall and nasal septum of the nose.

- List and identify on a diagram the bones forming the orbits of the skull, identifying the superior and inferior orbital fissures and the optic foramen.

- On a diagram of the inferior view of the skull or the floor of the cranial cavity identify the following: optic canal, superior orbital fissure, foramen rotundum, foramen ovale, internal acoustic meatus, jugular foramen, hypoglossal canal, foramen magnum, occipital condyles, pterygoid plates, styloid process, posterior nares, external acoustic meatus, palatine process of the maxilla, and the palatine bone.

- Describe the locations and functions of the paranasal and mastoid sinuses.

- Given a diagram of the mandible identify the following: condylar process, coronoid process, ramus, angle, body, mental protuberance, and the mental and mandibular foramina.

THE THORACIC CAGE AND VERTEBRAL COLUMN

Be Able To:

- Define the bones forming the thoracic inlet and outlet.

- Describe the named components of a typical rib and give the differences between true, false, and floating ribs.

- Draw the components of the sternum, indicating where the costal cartilages and the clavicles articulate.

- Describe the function, the total number of vertebrae, the number of vertebrae in each of the five segments, and the location of the primary and secondary curves of the vertebral column.

- Define the exaggerated abnormal curves of scoliosis, lordosis, and kyphosis.

- Discuss the structure and function of the intervertebral disc; giving the location and function of the intervertebral foramen and the following ligaments: anterior and posterior longitudinal, ligamentum flavum, ligamentum nuchae, supraspinous and interspinous.

- Describe the components making up a 'typical' vertebra.

- Compare and contrast the atlas and axis of the vertebral column.

- Compare and contrast the 'typical' vertebrae of the cervical, thoracic and lumbar regions of the vertebral column with regard to their overall size, shape, planes of their articulating surfaces, and the directions and shapes of their spinous processes.

THE UPPER AND LOWER LIMBS

Be Able To:

- Describe the bony components forming the pectoral (shoulder) and pelvic girdles.

- On a diagram of the scapula identify the following: acromial process, coracoid process, spine, glenoid fossa, supraspinous and infraspinous fossae, scapular notch, superior and inferior angles, and the medial and lateral borders.

- Identify on a diagram the following bony landmarks of the humerus: head, intertubercular groove, lesser and greater tubercles, deltoid tuberosity, medial and lateral epicondyles, capitulum, trochlea, and the coronoid and olecranon fossae.

- Locate on a diagram of the ulna the following: olecranon process, semilunar and radial notches, coronoid process, ulnar tuberosity, styloid process and the articular surface of the ulnar head.

- On a diagram of the radius locate the following: head, neck, radial tuberosity, ulnar notch, and styloid process.

- On a diagram of the hand identify the phalanges, metacarpal and carpal bones.

- On a diagram of the hip bone (os coxae) identify: ilium, ischium, pubis, iliac crest, anterior superior and anterior inferior iliac spines, acetabulum, acetabular notch, obturator foramen, lesser and greater sciatic notches, ischial tuberosity, symphysis pubis, and posterior superior and posterior inferior iliac spines.

- Describe the structural differences between the female and male pelves.

- Locate on a diagram of the femur the following: head, neck, greater and lesser trochanters, intertrochanteric line and crest, lateral and medial epicondyles, adductor tubercle, and lateral and medial condyles.

- On a diagram of the tibia and fibula, identify the following on the tibia: medial and lateral condyles, intercondylar eminence, tuberosity, and medial malleolus; the head and lateral malleolus on the fibula.

- On a diagram of the foot identify the phalanges, metatarsal, and tarsal bones.

- List the bones making up the medial and lateral longitudinal arches of the foot.

RELEVANT READINGS

Anthony & Kolthoff - Chapter 5.
Chaffee & Greisheimer - Chapter 3.
Crouch & McClintic - Chapter 7.
Jacob, Francone & Lossow - Chapter 5.
Landau - Chapter 4.
Tortora - Chapters 5, 6, and 7.

FIVE-CHOICE COMPLETION QUESTIONS

DIRECTIONS: Each of the following questions or incomplete statements is followed by five suggested answers or completions. SELECT THE SINGLE BEST ANSWER in each case and then circle the appropriate letter at the lower right of each question.

1. A SUBSTANCE NOT INVOLVED IN THE HOMEOSTASIS OF BONE IS:
   A. Parathyroid hormone
   B. Vitamin D
   C. Calcitonin
   D. Gonadal hormones
   E. Renin

   A B C D E

2. A ________ IS AN OPENING IN BONE THROUGH WHICH BLOOD VESSELS AND NERVES MAY PASS.
   A. Foramen
   B. Fossa
   C. Meatus
   D. Sinus
   E. Tubercle

   A B C D E

3. WHICH OF THE FOLLOWING IS A COMPONENT OF THE AXIAL SKELETON?
   A. Clavicle
   B. Femur
   C. Scapula
   D. Sacrum
   E. Tibia

   A B C D E

4. A MEMBRANE CONTAINING OSTEOBLASTS AND FOUND LINING THE MEDULLARY CAVITY IS THE:
   A. Endosteum
   B. Lamellae
   C. Trabeculae
   D. Periosteum
   E. None of the above

   A B C D E

5. THE CENTRAL SPONGY BONE PORTION OF THE FLAT BONES OF THE SKULL IS CALLED THE ________.
   A. Table
   B. Diplöe
   C. Epiphysis
   D. Medulla
   E. None of the above

   A B C D E

6. WHICH OF THE FOLLOWING IS NOT A BASIC FUNCTION OF BONE?
   A. Provides attachments for muscle
   B. Serves as storage sites for some minerals
   C. Produces new blood cells
   D. Involved in the production of heat
   E. Protects the soft structures of the body

   A B C D E

7. AN EXAMPLE OF AN IRREGULAR BONE IS THE:
   A. Radius
   B. Clavicle
   C. Mandible
   D. Occipital
   E. Ulna

   A B C D E

8. A FRACTURE IN WHICH THE BONE IS SPLINTERED OR CRUSHED INTO SMALL PIECES IS CALLED ________.
   A. Comminuted
   B. Compound
   C. Greenstick
   D. Pathologic
   E. Simple

   A B C D E

ANSWERS, NOTES AND EXPLANATIONS

1. E. A number of sustances are involved in the homeostasis of bone but renin is not one of them. There is a constant interchange of calcium between the bones of the body and the blood, resulting in a fairly constant calcium ion concentration in the blood. Calcitonin, a hormone of the thyroid, lowers the amount of calcium ions in the body by inhibiting the breakdown of bone and by accelerating the uptake of calcium by the bones. Parathyroid hormone works to oppose the effects of calcitonin, that is, it stimulates the breakdown of bone and thus enhances the number of calcium ions released into the blood. The growth hormone (somatotropin) is involved in stimulating the growth of bone. Thyroxine, another hormone from the thyroid gland, is a moderate stimulator of bone growth and along with the gonadal hormones is responsible for the maturation of bone. Vitamin D is a nutritional requirement of the body so that calcium and phosphorus can be absorbed by the intestine and can be utilized in the growth of bone. A lack of vitamin D in the diet may lead to a condition known as rickets. A deficiency of vitamin C in the diet leads to a poor production of collagen and bone matrix, whereas a deficiency of vitamin A produces a retardation of the growth rate.

2. A. Various structures, such as nerves and blood vessels, pass through an opening in bone called a foramen. This term is not confined to bone but may also be given to an opening in a membranous structure.

3. D. The bones of the human skeleton can be categorized into belonging to the axial or appendicular skeleton. The axial skeleton consists of all midline bones such as the skull (A in the diagram), including the mandible and hyoid bone; the vertebral column (B); the ribs (C); and the sternum. All other bones of the skeleton belong to the appendicular skeleton, or the bones of the upper and lower limbs.

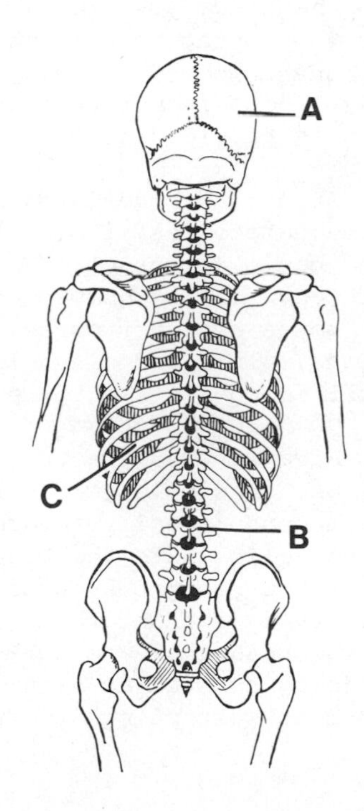

4. A. The endosteum (A in the diagram) is the membrane lining the medullary (marrow) cavity (B) of a long bone and the small cavities of spongy bone. This membrane contains cells called osteoblasts that are capable of producing new bone. The endosteum is similar in structure to the periosteum (C) which is located on the outer surface of the bone. Lamellae are concentric layers of compact bone, whereas trabeculae are thin spicules of bone found in spongy bone.

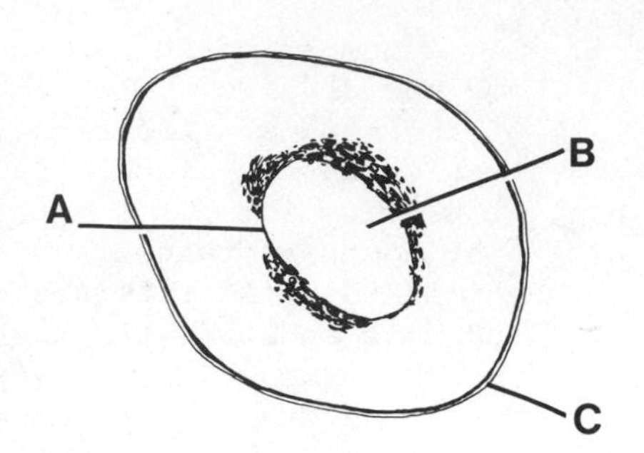

5. B. The flat bones of the skull are made up of a central portion consisting of spongy (cancellous) bone, whereas the outer and inner layers are made up of compact (cortical) bone. The central portion is called the diplöe and the outer and inner compact layers are called tables.

6. D. The skeletal system performs several important functions for the body. It provides for the support and protection of the soft tissues and forms the basis for the erect posture of the body. Bones also serve as levers and provide sites for the storage of minerals such as calcium and phosphorus, and the red marrow of bones produce a steady supply of new blood cells as they are needed by the body. Bone is not involved in the production of heat but this important function is provided by the skeletal muscles of the body.

7. C. Irregular bones are those of the skeleton which do not fit into the other three categories, i.e., flat, long, and short. The irregular bones have the same basic construction as all bones: a central core of spongy bone and an outer thin covering of compact bone. Besides the mandible, other examples of irregular bones are: vertebrae, ossicles of the ear, hyoid, sacrum, and ethmoid. The radius, ulna, and clavicle are long bones, whereas the occipital bone of the skull is considered to be a flat bone.

8. A. A comminuted fracture is one in which the bone is splintered into small pieces. A simple (closed) fracture produces no open wound in the skin, whereas a compound (open) fracture produces an open wound in the skin. A greenstick fracture is where the bone is broken on one side and is bent on the other. It is like a young branch of a tree that bends but does not break completely through. This type of fracture is common in young children. A pathologic fracture occurs as the result of a mild injury to a bone which has preexisting involvement with pathologies such as a tumor or an infection. This pathology makes the bone weak and more susceptible to fracture.

MULTI-COMPLETION QUESTIONS

DIRECTIONS: In each of the following questions or incomplete statements, ONE OR MORE of the completions given is correct. At the lower right of each question, circle A if 1, 2, and 3 are correct; B if 1 and 3 are correct; C if 2 and 4 are correct; D if only 4 is correct; and E if all are correct.

1. AMONG THE FLAT BONES ARE INCLUDED THE:
   1. Occipital bone
   2. Patella
   3. Ribs
   4. Carpal bones

   A B C D E

A	B	C	D	E
1,2,3	1,3	2,4	only 4	all correct

2. WHICH OF THE FOLLOWING STATEMENTS IS(ARE) TRUE FOR THE OSSIFICATION OF BONES?
   1. The secondary centers appear in the diaphysis
   2. Involves the breakdown of cartilage
   3. Begins to occur in bones shortly after birth
   4. Ends in the late teens or early twenties

   A B C D E

3. AN OSTEOCLAST IS A CELL WHICH:
   1. Has lost its ability to form new bones
   2. Lays down fine collagen protein
   3. Is trapped within a lacuna
   4. Is responsible for bone resorption

   A B C D E

4. WHICH OF THE FOLLOWING BONY PROJECTIONS IS(ARE) NOT INVOLVED AS A SITE FOR MUSCLE ATTACHMENT?
   1. Condyle
   2. Crest
   3. Head
   4. Tuberosity

   A B C D E

5. VOLKMANN'S CANALS:
   1. Are found in spongy bone
   2. Connect Haversian canals
   3. Contain proliferating osteoblasts
   4. Carry nerves projecting from the periosteum

   A B C D E

6. THE GROWTH OF THE LONG BONES IS:
   1. Controlled by a hormone from the anterior hypophysis
   2. Found in the epiphyseal plate
   3. Usually terminated in about the 20th year in the male
   4. Due to a stimulation of chondrocytes

   A B C D E

7. INTRAMEMBRANOUS BONE FORMATION:
   1. Takes place in the flat bones of the skull
   2. Produces trabeculae that are surrounded by red bone marrow
   3. Begins in a cluster of osteoblasts
   4. Uses fibrocartilage as a model

   A B C D E

8. WHICH OF THE FOLLOWING BONES ARE CONSIDERED TO BE EXAMPLES OF SHORT BONES?
   1. Capitate
   2. Clavicle
   3. Cuboid
   4. Coccyx

   A B C D E

------------------------ ANSWERS, NOTES AND EXPLANATIONS ------------------------

1. B, 1 and 3 are correct. Frontal, occipital, parietal, and temporal bones of the skull, ribs, and sternum are all examples of flat bones. Flat bones consist of a central core of spongy bone covered by an inner and outer thin layer of compact bone. The spongy bone of the flat bones of the skull is called the diploe, whereas the inner and outer layers of compact bone are referred to as the tables. The patella is considered to be a sesamoid bone and the carpal bones are all short bones.

2. C, 2 and 4 are correct. The ossification of long bones begins long before birth and first appears as a primary ossification center in the diaphysis. The process involves the proliferation of cartilage cells which undergo hypertrophy,

breakdown, and are eventually replaced by osteoblasts and finally the deposition of calcium phosphate. Single secondary ossification centers appear in each epiphysis at the ends of a long bone. The primary and secondary ossification centers, by their active destruction of cartilage and the formation of bone, spread out until only two thin strips of cartilage remain, one at each end of the bone (i.e., epiphyseal plate). The formation of bone continues until all the cartilage is replaced by bone and then the epiphyses are said to be closed. The closing of the epiphyses takes place at about 18 years in a female and about 20 years in a male.

3. D, Only 4 is correct. The osteoclast is a giant cell containing anywhere from 15 to 20 nuclei. This cell is closely associated with the areas of new bone formation and the remodelling of bone. Osteoclasts are thought to secrete proteolytic enzymes which dissolve bone material. Osteoblasts, on the other hand, form new bone and are found on the advancing surfaces of developing and growing bone, whereas osteocytes are osteoblasts that have become entrapped by their own bony matrix. The osteocytes lie in a space within the matrix called a lacuna. Osteocytes are more likely to be found in mature bone.

4. B, 1 and 3 are correct. A condyle is a projection of a bone which is usually involved in a joint and therefore would be covered by articular cartilage. Condyles are located at the distal end of the femur and the proximal end of the tibia. A head of a bone is also involved in a joint but it is usually located beyond a narrow neck-like portion, e.g., the head of the femur or the humerus. A crest and a tuberosity are bony projections which provide sites for muscle attachments. Structurally a crest is a long narrow ridge-like projection, whereas a tuberosity is a large round projection. An example of a crest is seen as the intertrochanteric crest of the femur. The tibial tuberosity of the tibia is a good example of a tuberosity. Other types of bony projections involved as sites for muscle attachment are: spines, trochanters, and tubercles.

5. C, 2 and 4 are correct. The Volkmann's canals are found in compact bone connecting adjacent Haversian canals, coursing to the free surface, or to the medullary cavity. They usually run at right or oblique angles to the Haversian canals. These canals transmit blood vessels throughout the compact bone and similarly accommodate nerves that project inward from the periosteum.

6. E, All are correct. The growth of a long bone is in two directions, one increasing its length and the other increasing its diameter. The increase in the total length occurs because of a proliferation of the chondrocytes at the epiphyseal plates, whereas the increase in the diameter results from a deposition of new membrane bone beneath the periosteum. The growth of bone is under the control of a hormone secreted from the anterior hypophysis called somatotropin. This hormone stimulates the cartilage cells of the epiphyseal plate to divide. Growth continues until all the cartilage cells of the epiphyseal plate are transformed into bone. When all growth ceases the epiphyses are said to be closed. This usually occurs at about 18 years in females and at about 20 years in males.

7. A, 1, 2, and 3 are correct. Intramembranous bone formation takes place in a well vascularized layer of connective tissue, not fibrocartilage. This type of new bone formation, where a connective tissue membrane is replaced by bone, is found in the flat bones of the skull--the frontal, parietal, occipital and temporal bones and part of the mandible. This ossification process begins when the osteoblasts lay down a matrix which then in turn is impregnated by calcium phosphate. Trabeculae are formed in this way, and some of the osteoblasts become trapped forming osteocytes. The outer layer of this spongy bone will become compact bone, forming the outer and inner tables of the skull bones. The central area of spongy bone (diplöe) between the two tables of compact bone consists of thickened trabeculae with red bone marrow (hemopoietic) tissue in between.

8. B, <u>1 and 3 are correct</u>. The capitate and cuboid bones are considered to be short bones and are found in the hand and foot, respectively. The clavicle is classified as a long bone and the coccyx is made up of small (four or five) vertebrae which are considered to be irregular bones.

FIVE-CHOICE ASSOCIATION QUESTIONS

DIRECTIONS: Each group of questions below consists of a numbered list of descriptive words or phrases accompanied by a diagram with certain parts indicated by letters, or by a list of lettered headings. For each numbered word or phrase, SELECT THE LETTERED PART OR HEADING that matches it correctly. Then insert the letter in the space to the right of the appropriate number. Sometimes more than one numbered word may be correctly matched to the same lettered part or heading.

1. _____ Haversian canal
2. _____ Lamella
3. _____ Canaliculus
4. _____ Osteocyte
5. _____ Lacuna

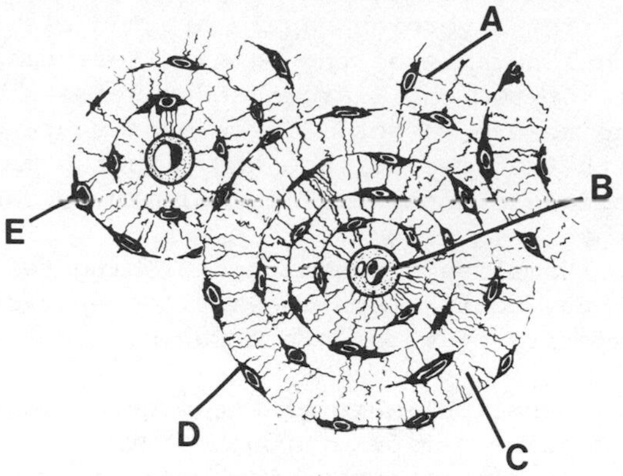

6. _____ Metaphysis
7. _____ Covered in part by articular cartilage
8. _____ Contains the medullary cavity
9. _____ Consists almost exclusively of compact bone
10. _____ Is made up of dividing cartilage cells

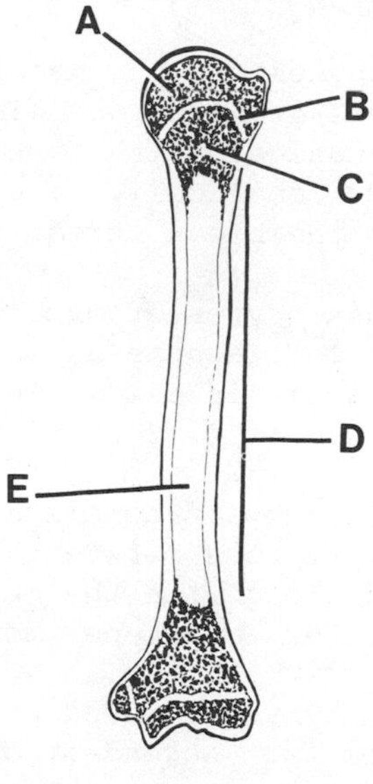

ASSOCIATION QUESTIONS

11. _____ Inhibits the breakdown of bone

12. _____ Involved in stimulating the growth of bone

13. _____ Stimulates osteoclasts to destroy bone

14. _____ Hyperactivity of this hormone during childhood will produce giantism

A. Mineralocorticoids

B. Parathyroid hormone

C. Thyrocalcitonin

D. Somatotropin

E. None of the above

------------------------ ANSWERS, NOTES AND EXPLANATIONS ------------------------

1. B. The Haversian canals are tunnels found in compact bone, running parallel to the long axis of the bone. These canals contain loose connective tissue and one or two capillaries and venules which supply the bony matrix with nutrients and remove wastes. These canals are connected with one another and communicate with the free surface and with the marrow (medullary) cavity via transverse or oblique channels called Volkmann's canals.

2. C. Lamellae are a distinguishing feature of compact bone and are not found in spongy bone. Lamellae are concentric rings of mineralized bone matrix, the majority of which are arranged about the Haversian canals.

3. A. Canaliculi are minute canals radiating out from the lacunae, forming an extensive network throughout the concentric lamellae of compact bone. These canals are essential for the nutrition of bone cells and provide avenues for the exchange of metabolites between cells and the nearest perivascular space.

4. E. Osteocytes are the major cells of fully formed bone. They are derived from osteoblasts and are found in the lacunae of the calcified bone matrix. Osteocytes communicate with each other by cellular processes passing through the connecting canaliculi. The osteocytes are involved in releasing calcium from bone into the blood stream under the control of the parathyroid hormone.

5. D. The osteocytes of mature compact bone are found in small cavities or lacunae. The lacunae are uniformly spaced throughout the bone matrix and each contains a single cell. The lacunae communicate with each other by small interconnecting canals or canaliculi.

6. C. The metaphysis consists of columns of spongy (cancellous) bone, forming a transitional zone between the epiphyseal plate and the diaphysis of a long bone. This area along with the epiphyseal plate constitutes the growth zone and it is here that the long bone increases its length.

7. A. The epiphysis, found at either end of a long bone, consists of spongy bone and arises from secondary ossification centers. These expanded ends of a long bone extend as far as the epiphyseal or growth plate. The spongy bone has a thin covering of compact bone, which is itself covered by hyaline cartilage. This cartilage provides for the articulating surface of the bone.

8. B. The medullary (marrow) cavity (labelled E in the diagram) is located in the diaphysis (shaft) of a long bone. This cavity is bound at either end by the spongy bone of the metaphysis and surrounded by the thick compact bone of the shaft. In the adult, the medullary cavity of long bones contains yellow marrow, consisting primarily of adipose tissue. Red marrow, responsible for the

production of some blood cells, is located in spongy bone of the vertebrae, diplöe of cranial bones, sternum, ribs, and the proximal epiphyses of the humerus and femur.

9. D. The diaphysis (shaft) of a long bone is a hollow cylinder of compact bone arising from the primary ossification center. The inner wall of the marrow cavity may consist of a thin layer of spongy bone especially towards the distal and proximal metaphyseal regions. The marrow cavity in the adult contains yellow marrow. This tissue is primarily fat but is thought to retain some hemopoetic potential (the ability to produce some types of blood cells in an emergency).

10. B. An increase of bone length is because of the proliferation of cartilage cells located in the epiphyseal plate. As growth comes to an end in the adult skeleton the final stages of ossification take place. The cartilage cells eventually degenerate and are replaced by osteoblasts that form bony tissue. At this point the epiphyses of the bone are closed and further longitudinal growth is impossible.

11. C. Thyrocalcitonin, a hormone produced by the thyroid gland, is involved in the homeostasis of the blood calcium level. This hormone lowers the amount of calcium in the blood by inhibiting bone breakdown and by accelerating the absorption of calcium by the bones. The blood calcium level, by a negative feedback mechanism, controls the secretion of thyrocalcitonin by the thyroid gland. Secretion of this hormone is thought not to be under control of the hypophysis.

12. D. Somatotrophic (growth) hormone, from the anterior pituitary, is involved in maintaining the adult size and increasing the growth rate of soft and hard tissues, including bone. The growth hormone stimulates cells to grow and multiply by an anabolic process whereby the rate by which amino acids enter cells is increased; producing a buildup of proteins. The growth hormone is also involved in fat metabolism.

13. B. The parathyroid hormone, secreted by the small parathyroid glands, stimulates the osteoclasts of bone. Thus stimulated, these cells break down bone and release calcium into the blood stream. This parathyroid gland is stimulated to release its hormone by low calcium levels in the blood, and like thyrocalcitonin is not under control of the hypophysis. These two hormones have the opposite effects on the blood calcium level.

14. D. An abnormal increase in the amount of somatotrophic (growth) hormone produced during childhood will result in giantism, which is an abnormal increase in the length of the long bones. Hypersecretion of this anterior pituitary hormone during adulthood will produce acromegaly.

FIVE-CHOICE COMPLETION QUESTIONS

DIRECTIONS: Each of the following questions or incomplete statements is followed by five suggested answers or completions. SELECT THE SINGLE BEST ANSWER in each case and then circle the appropriate letter at the lower right of each question.

1. THE OPTIC FORAMEN IS INDICATED BY ________.

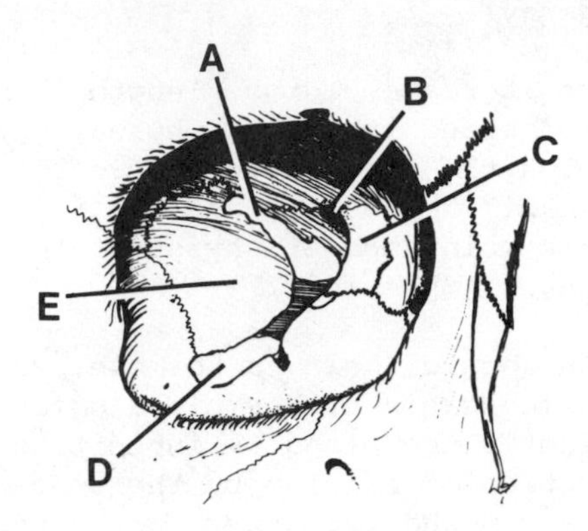

A B C D E

2. THE ______ IS A BONE MAKING UP THE FACE OF THE SKULL.
   A. Occipital
   B. Parietal
   C. Palatine
   D. Sphenoid
   E. None of the above

A B C D E

3. WHICH OF THE FOLLOWING IS NOT A COMPONENT OF THE MANDIBLE?
   A. Angle
   B. Ramus
   C. Condyle
   D. Coracoid
   E. Mental protuberance

A B C D E

4. THE MASTOID FONTANEL IS INDICATED BY ____________.

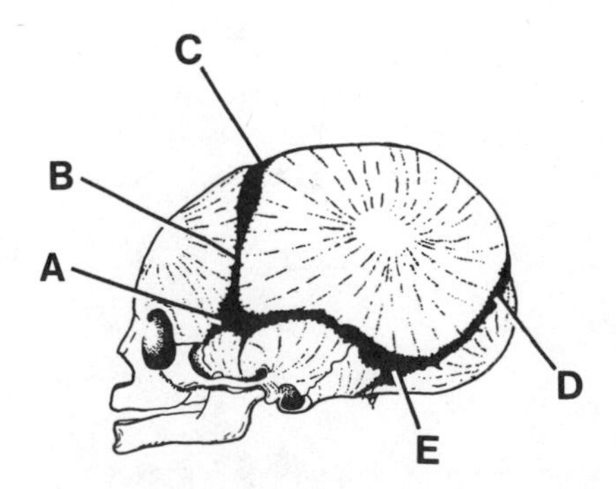

A B C D E

SELECT THE SINGLE BEST ANSWER

5. WHICH OF THE FOLLOWING BONES IS NOT CONSIDERED A CRANIAL BONE?
   A. Frontal
   B. Temporal
   C. Ethmoid
   D. Maxillary
   E. Sphenoid

A B C D E

6. THE SPHENOID BONE IS INDICATED BY __________.

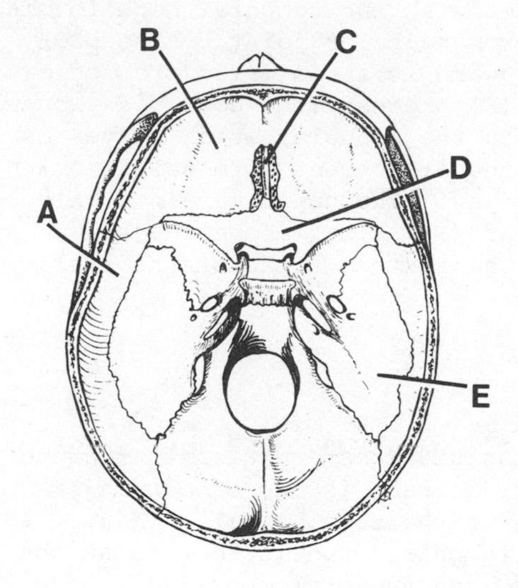

A B C D E

------------------------ ANSWERS, NOTES AND EXPLANATIONS ------------------------

1. B. The optic foramen (B in the diagram) is the opening by which the optic (second cranial) nerve and the ophthalmic artery reach the eye from the floor of the cranial cavity. The foramen is found in the lesser wing of the sphenoid bone, on the posteromedial wall of the orbit. The superior (A) and inferior orbital fissures (D) are seen in the diagram. The sphenoid (E) and ethmoid bones (C) are also labelled.

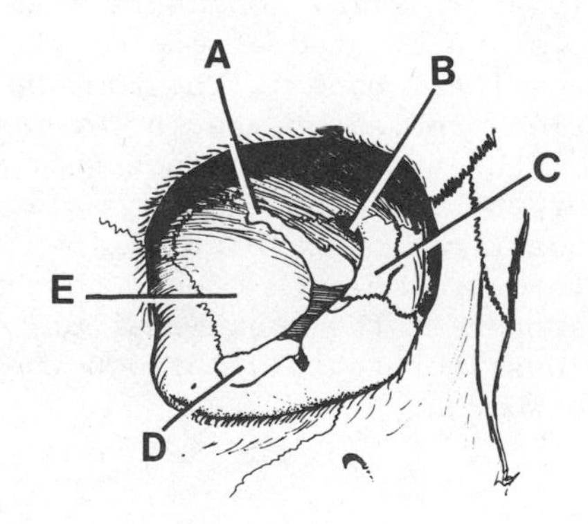

2. C. The skull is made up of 28 bones and can be divided into a cranial portion, or those bones forming the cranium, and a facial portion, or those bones forming the face. Of the bones listed only the palatine bones go to make up part of the face. Other paired bones of the face are the nasal, maxillary, zygomatic, lacrimal, inferior conchae, and the ossicles of the middle ear (malleus, stapes, and incus). The vomer and mandible are single bones of the face. The cranium consists of the paired parietal and temporal bones and the single frontal, occipital, sphenoid and ethmoid bones.

3. D. The mandible (lower jaw) is a U-shaped bone of the face. It gives the face much of its individual character. The mandible consists of a body (A), an angle (B), and a ramus (C). Projecting upward from the ramus is the condyloid (D) process that articulates with the temporal bone forming the temporomandibular joint. Also projecting upward is an anterior process or coronoid (E) process (the coracoid process is found on the scapula) which serves as a site for the attachment of muscles moving the lower jaw. The upper margin or alveolar process (F), lying above the body, contains the sixteen teeth of the adult lower jaw.

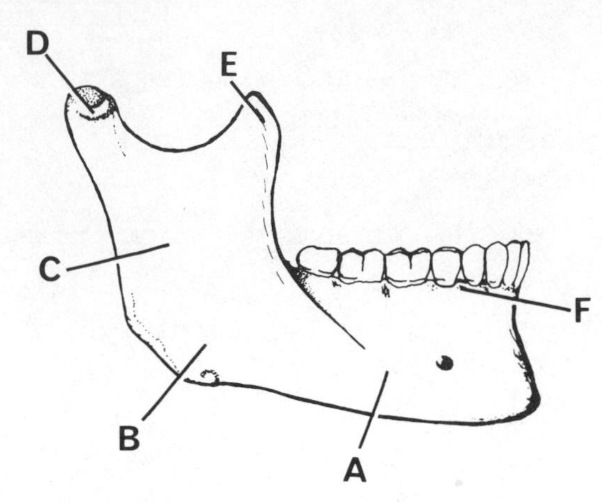

4. E. The mastoid (posterolateral) fontanel (E in the diagram) is one of our major fontanels of the fetal skull. This fontanel is paired and is located at the junction of the temporal, parietal, and occipital bones. Other fontanels are the paired sphenoid (anterolateral) (A), and the single frontal (anterior) (C) and occipital (posterior) (D) fontanels. All of the fontanels are located at one of the angles of the parietal bones. The fontanel is an unossified membranous area between the bones of the skull. The fact that the bones of the skull have not completed their ossification by birth allows these bones to override each other, thus making the diameter of the skull smaller, so that it can pass more easily through the birth canal at birth.

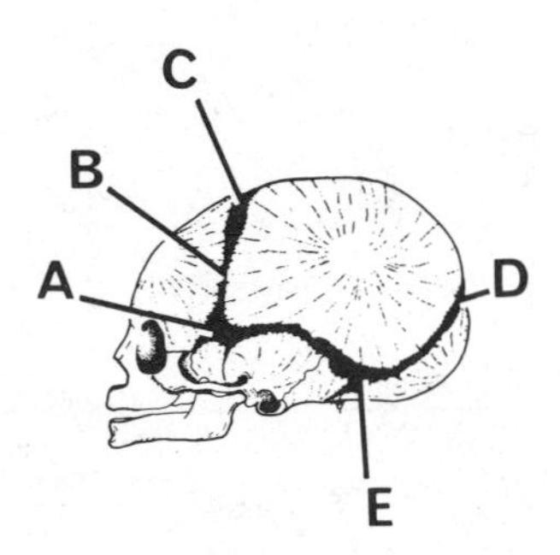

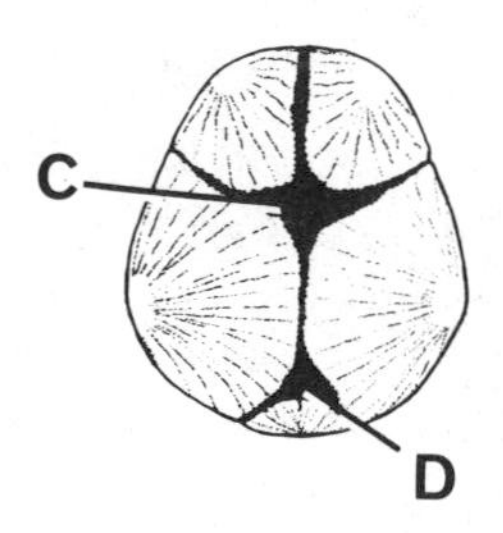

5. D. The maxillary bone (A in the diagram) is a paired bone which forms part of the face of the skull. The cranium is formed by paired parietal (C) and temporal (E) bones and single frontal (B), occipital (D), sphenoid (F), and ethmoid (G) bones.

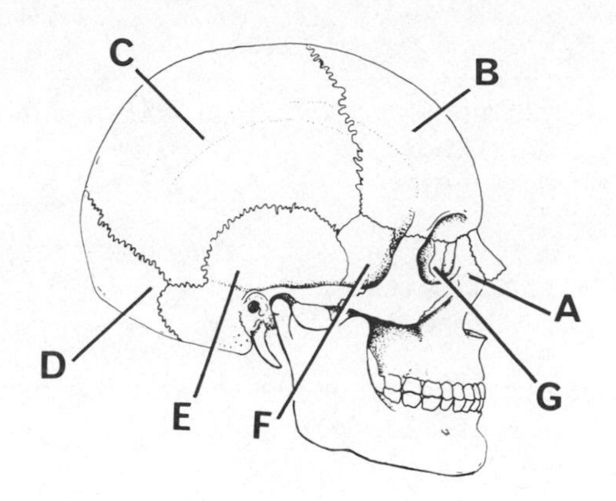

6. D. The sphenoid bone (D) forms the key of the cranial floor of the skull. Other bones forming the floor of the skull are: the frontal (B) and ethmoid (C) bones, anteriorly, and the temporal (E) and occipital (A) bones, posteriorly. The sphenoid bone has paired small and large wings and has been likened to a bat. The sphenoid forms a part of the wall of the orbit, part of the lateral wall of the nasal cavity, and part of the lateral wall of the cranial cavity. The sphenoid also has a small depression in its midline for the hypophysis, called the sella turcica (F). It also contains the sphenoid air sinus, one of the paranasal air sinuses. The occipital bone (A) forms the lateral walls of the cranial cavity.

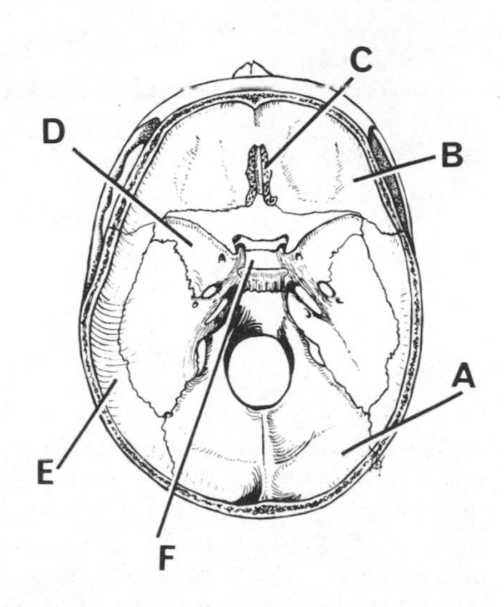

MULTI-COMPLETION QUESTIONS

DIRECTIONS: In each of the following questions or incomplete statements, ONE OR MORE of the completions given is correct. At the lower right of each question, circle A if 1, 2, and 3 are correct; B if 1 and 3 are correct; C if 2 and 4 are correct; D if only 4 is correct; and E if all are correct.

1. WHICH OF THE FOLLOWING BONES HELP TO MAKE UP THE ORBIT?
   1. Frontal
   2. Ethmoid
   3. Maxilla
   4. Nasal

   A B C D E

2. BONES MAKING UP THE FACE INCLUDE THE:
   1. Zygomatic
   2. Ethmoid
   3. Vomer
   4. Frontal

   A B C D E

3. THE ZYGOMATIC BONE ARTICULATES WITH THE:
   1. Frontal
   2. Maxilla
   3. Temporal
   4. Sphenoid

   A B C D E

A	B	C	D	E
1,2,3	1,3	2,4	only 4	all correct

4. THE BONY COMPONENT(S) MAKING UP THE NASAL SEPTUM IS(ARE) THE:
   1. Ethmoid
   2. Nasal
   3. Vomer
   4. Sphenoid

   A B C D E

5. WHICH OF THE FOLLOWING BONES CONTAIN AIR SINUSES?
   1. Occipital
   2. Temporal
   3. Lacrimal
   4. Frontal

   A B C D E

6. THE BONE(S) FORMING THE ROOF OF THE MOUTH IS(ARE) THE:
   1. Inferior concha
   2. Mandible
   3. Vomer
   4. Maxilla

   A B C D E

7. THE HYOID BONE IS:
   1. Part of the axial skeleton
   2. Found below the mandible
   3. U-shaped
   4. Involved with movements of the maxilla

   A B C D E

8. THE BONE(S) IN THE MIDDLE EAR CAVITY OF THE TEMPORAL BONE INCLUDE THE:
   1. Incus
   2. Ilium
   3. Stapes
   4. Sphenoid

   A B C D E

------------------------ ANSWERS, NOTES AND EXPLANATIONS ------------------------

1. A, 1, 2, and 3 are correct. The orbits of the skull are the cavities of the face that contain and protect the eyes. The bones making up these cavities are the following: frontal (A), sphenoid (G), maxilla (E), zygomatic (F), lacrimal (C), and ethmoid (B). Often a small portion of the palatine bone (D) can also form part of the posteromedial aspect of the orbit.

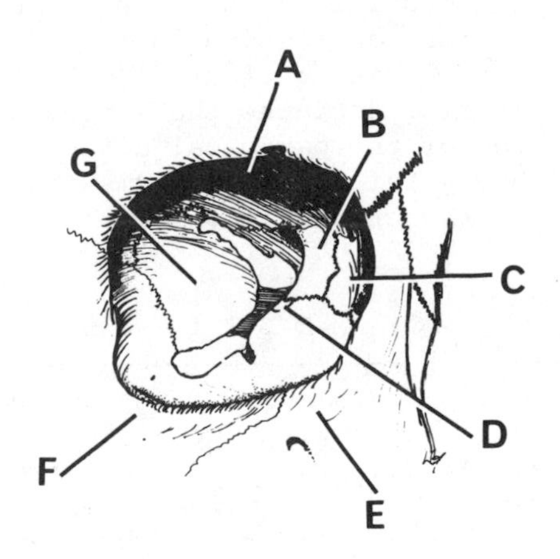

2. B, 1 and 3 are correct. There are 14 paired and single bones making up the face of the skull. The so-called keystone of the face is the maxilla which forms the upper jaw. There are five bones associated with the nasal cavities that make up the face, namely: the vomer, inferior concha, nasal, lacrimal, and palatine bones. The cheeks of the face are made up by the zygomatic (malar) bones and the lower jaw is formed by the mandible.

3. E, All are correct. The zygomatic (malar) bone forms part of the lateral wall and floor of the orbit. It articulates with four other bones of the face, i.e., the temporal, frontal, maxilla, and sphenoid. The cheek of the face is shaped by the underlying zygomatic bone.

4. B, 1 and 3 are correct. The tip of the nasal septum is composed of cartilage and its base is formed by bone. The bone of the nasal septum is formed by the perpendicular plate of the ethmoid and the vomer bones. The plate of the ethmoid forms the upper and anterior part of the septum, whereas the vomer bone extends from below the sphenoid bone down to the hard palate forming the posterior and lower part of the septum.

5. C, 2 and 4 are correct. There are five bones of the skull that contain air cells (sinuses): the frontal, maxilla, ethmoid, sphenoid, and temporal bones. The first four are paranasal air sinuses as they all have openings into the nasal cavities. The other group of air cells is found in the mastoid process of the temporal bone. These cells are associated with the middle ear. The exact function of the air cells is unclear. In the case of the paranasal sinuses some feel that they lighten the skull and may give some resonance to the voice.

6. D, Only 4 is correct. The roof of the mouth or that portion forming the hard palate is formed by two bones: the maxilla and palatine bones. The anterior two-thirds of the hard palate is formed by the palatine process of the maxilla (A); and the posterior one-third is formed by the horizontal plate of the palatine (B) bones. These bones along with the soft palate separate the nasal and oral cavities.

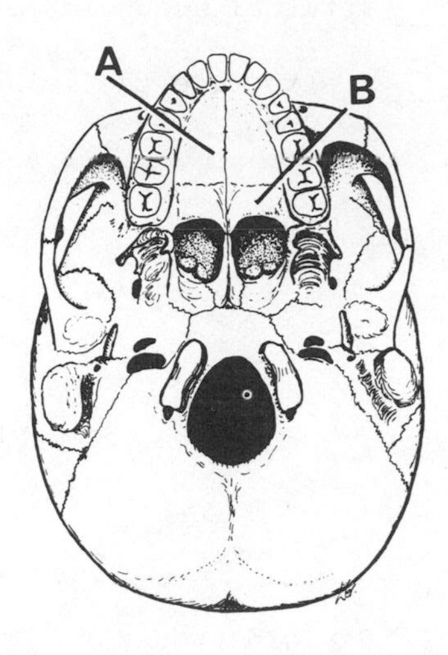

7. A, 1, 2, and 3 are correct. The hyoid bone is a single bone belonging to the axial skeleton. It is an isolated bone, articulating with no other bone. The hyoid is found in the midline of the neck below the mandible. It is U-shaped and consists of a body and paired greater and lesser cornua. The hyoid is suspended by ligaments and through its muscular attachments is involved in speech, swallowing, and movements of the tongue.

8. B, 1, and 3 are correct. The tympanic cavity of the middle ear contains a chain of three movable ossicles (bones), the malleus, stapes, and incus. These three small bones transmit sound waves from the tympanic membrane to the inner ear.

FIVE-CHOICE ASSOCIATION QUESTIONS

DIRECTIONS: Each group of questions below consists of a numbered list of descriptive words or phrases accompanied by a diagram with certain parts indicated by letters, or by a list of lettered headings. For each numbered word or phrase, SELECT THE LETTERED PART OR HEADING that matches it correctly. Then insert the letter in the space to the right of the appropriate number. Sometimes more than one numbered word may be correctly matched to the lettered part or heading.

1. _____ Ethmoid bone
2. _____ Foramen for the ophthalmic artery
3. _____ Sella turcica
4. _____ Internal acoustic foramen
5. _____ A structure the spinal cord passes through

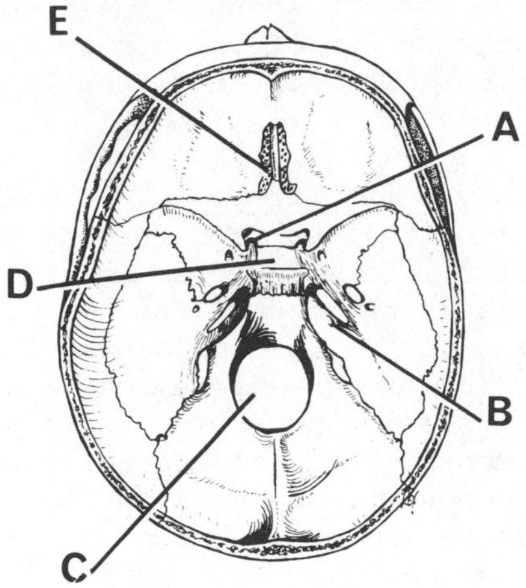

6. _____ Styloid process
7. _____ Condyles
8. _____ Alveolar process
9. _____ External auditory meatus
10. _____ Mental foramen
11. _____ Contains a sinus
12. _____ Keystone of the cranium

A. Occipital
B. Maxilla
C. Temporal
D. Parietal
E. None of the above

------------------------ ANSWERS, NOTES AND EXPLANATIONS ------------------------

1. E. The ethmoid bone, with the upward projecting cristae galli and its perforated cribiform plate, lies between the orbital cavities and forms the roof of the nasal cavities. A downward projection of the ethmoid bone or the perpendicular plate forms the upper part of the nasal septum. The lateral masses of this bone contain ethmoid air cells and also form the superior and middle conchae (turbinates) of the lateral nasal wall.

2. A. The ophthalmic artery, a branch of the internal carotid artery, passes through the optic foramen along with the optic nerve. Both of these structures are intimately involved with the retina of the eye: the ophthalmic artery

through its central artery provides blood to the retina and the optic nerve carries impulses from the retina to the brain.

3. D. The sella turcica is a small depression found in the midline of the sphenoid bone. This bony structure contains the hypophysis (pituitary) of the midbrain.

4. B. The internal acoustic meatus is an opening on the medial aspect of the petrous portion of the temporal bone. Through this foramen the facial and acoustic nerves pass from the inner ear, the organ of hearing and position sense, to the brain.

5. C. The foramen magnum marks the anatomical site between the medulla oblongata, the most caudal portion of the brain, and the upper cervical segments of the spinal cord. The foramen magnum is located in the posterior cranial fossa and is located between the occipital condyles of the occipital bone.

6. C. The styloid process is a downward projection of the temporal bone (petrous portion). To it are attached various ligaments and muscles that are involved in the neck and the floor of the mouth.

7. A. The occipital condyles are articular surfaces found on either side of the foramen magnum. These articular surfaces form paired joints with the superior articulating facets of the atlas, the first cervical vertebrae. This atlanto-occipital joint primarily provides for flexion and extension (forward and backward nodding) of the head.

8. B. Alveolar process is that part of the maxilla and mandible that bears the teeth. Each tooth is embedded in a bony socket or alveolus. The alveolar processes are covered by the mucous membrane of the mouth.

9. C. The external auditory meatus is the bony canal of the external ear. It is located within the temporal bone of the skull and ends internally at the tympanic membrane.

10. E. The mental foramina are openings on the lateral surface of the mandible below the second premolar teeth. Through this foramen the mental nerve and vessels reach the skin and muscles of the chin.

11. B. There are five bones of the skull that contain air sinuses. They are: frontal, maxilla, sphenoid, ethmoid, and temporal bones. The first four are air sinuses related to the nasal cavity and are called paranasal sinuses. The sinuses or air cells found in the temporal bone are associated with the middle ear cavity, and are called the mastoid air cells as they are located in the mastoid process.

12. D. The parietal bones of the skull articulate with all the other bones forming the cranium except for the ethmoid. These bones articulate at immovable joints called sutures. The bones articulating with the parietal bones are: the frontal, occipital, temporal, and sphenoid.

FIVE-CHOICE COMPLETION QUESTIONS

DIRECTIONS: Each of the following questions or incomplete statements is followed by five suggested answers or completions. SELECT THE SINGLE BEST ANSWER in each case and then circle the appropriate letter at the lower right of each question.

1. DEVELOPMENTALLY THE PEDICLES AND LAMINAE OF A VERTEBRA FUSE TO FORM THE ______.
   A. Costal cartilage
   B. Bridge of the nose
   C. Cheek bone
   D. Neural arch
   E. Sternum

   A B C D E

2. THE NUMBER OF VERTEBRAE FOUND IN THE CERVICAL REGION OF THE VERTEBRAL COLUMN IS ______.
   A. 3
   B. 4
   C. 5
   D. 6
   E. None of the above

   A B C D E

3. THE ODONTOID PROCESS IS FOUND ON THE ________.
   A. Ulna
   B. Hyoid
   C. Atlas
   D. Sacrum
   E. None of the above

   A B C D E

4. WHICH OF THE FOLLOWING IS NOT TRUE FOR THE THORACIC VERTEBRAE?
   A. Spinous processes are short
   B. Have facets for the head of the rib
   C. There are 12 in number
   D. Spinous processes are directed downward
   E. Have facets for the tubercle of the rib

   A B C D E

5. WHICH OF THE FOLLOWING IS A CHARACTERISTIC OF THE LUMBAR VERTEBRAE?
   A. Have no transverse processes
   B. Articulate with the ribs
   C. Possess short bifid spinous processes
   D. Have the largest bodies
   E. Form a primary curve

   A B C D E

6. THERE ARE _____ PAIRS OF FALSE RIBS MAKING UP THE RIB CAGE.
   A. 2
   B. 3
   C. 4
   D. 5
   E. None of the above

   A B C D E

7. THE ARROW INDICATES THE ______ OF THE RIBS
   A. Angle
   B. Head
   C. Neck
   D. Tubercle
   E. Shaft

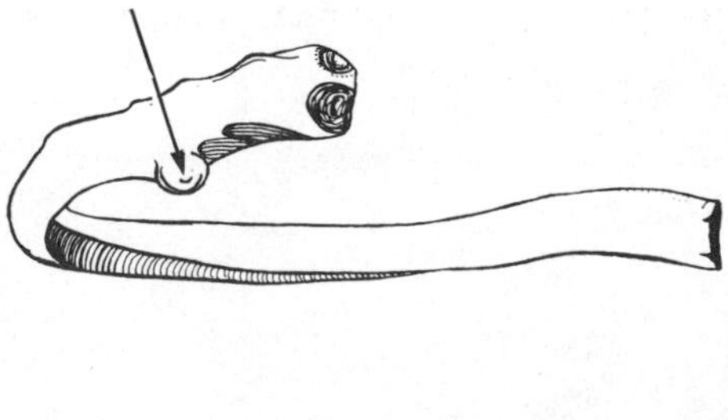

   A B C D E

SELECT THE SINGLE BEST ANSWER

8. AN EXAGGERATED LATERAL CURVE OF THE VERTEBRAL COLUMN IS CALLED:
   A. Kyphosis
   B. Lordosis
   C. Mitosis
   D. Osteoporosis
   E. Scoliosis

A B C D E

------------------------ ANSWERS, NOTES AND EXPLANATIONS ------------------------

1. D. During the development of a vertebra the pedicles (A in the diagram) and the laminae (B) fuse to form the neural (vertebral) arch. The neural arch serves to protect the spinal cord lying within the vertebral foramen (C). The neural arch has a single spinous process (D) and paired transverse processes (E) projecting from it. The massive body of the vertebra is indicated by (F). The broken circles represent the articulating surfaces.

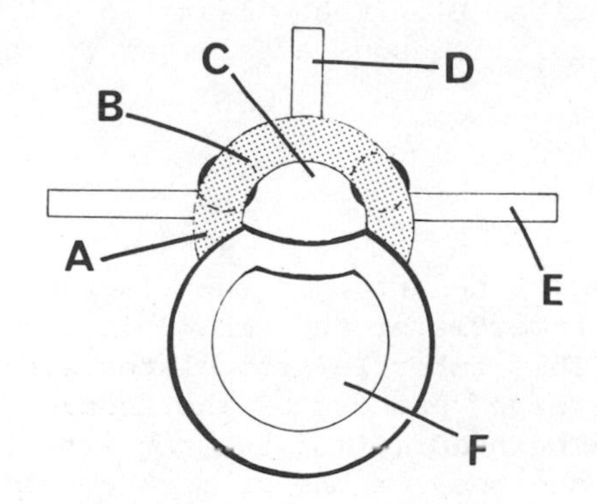

2. E. The number of cervical vertebrae found in the vertebral column is seven. The top two cervical vertebrae are called the atlas and the axis. The atlas articulates with the two occipital condyles of the occipital bone of the skull above and with the two superior articulating surfaces of the axis below.

3. E. The odontoid process (dens) is located as a superior extension of the body of the axis. It represents the body of the first cervical vertebra (atlas) that becomes fused to the axis during development. This arrangement allows the atlas to pivot on the axis. Consequently the atlas has no body.

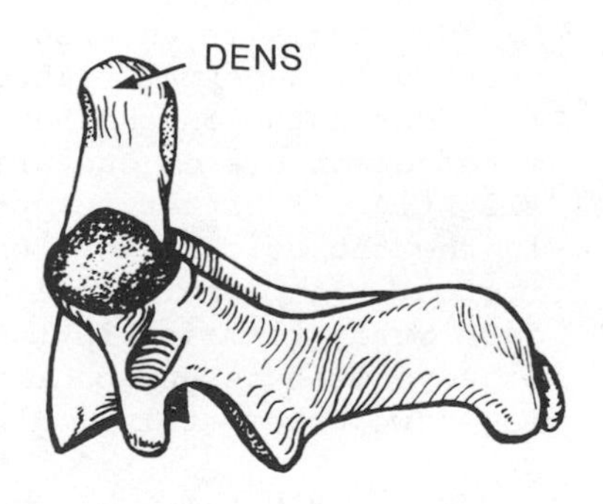

4. A. There are 12 thoracic vertebrae, each of which articulate with a pair of ribs. The ribs articulate by their heads and facets to the articulating facets of the body and transverse processes, respectively. The heads of the second through to the ninth ribs also articulate with the body of the vertebrae above. The spinous processes of the thoracic vertebrae are long and slender, projecting downward and often end at or below the body of the vertebrae below.

5. D. The lumbar vertebrae have the largest bodies of all the vertebrae, as they form the base of the vertebral column. Their spinous and transverse processes are short and thick, serving as sites for attachment of the back muscles. The spinous processes of these vertebrae project horizontally. The superior (A in the diagram) and inferior (B) articulating surfaces project inward and outward, respectively.

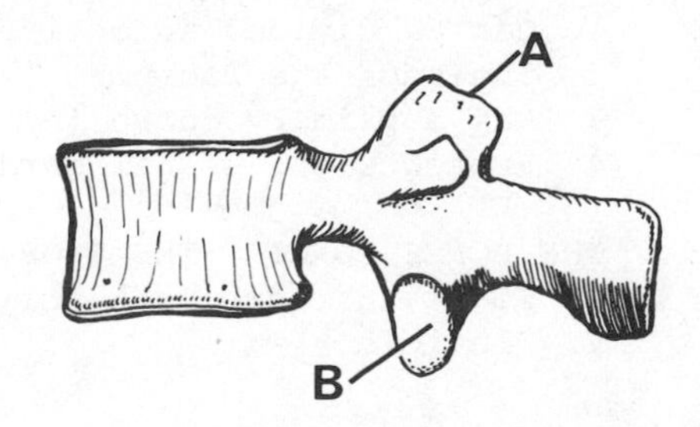

6. B. There are three pairs of false ribs (A in the diagram) involved in the thoracic cage. The eighth, ninth and tenth ribs are considered to be false because they do not articulate with the sternum by their own costal cartilage. Rather their cartilages fuse with the costal cartilage of the seventh rib. The first seven pairs of ribs are considered to be true ribs, whereas the eleventh and twelfth pairs are called floating ribs (B). These last two pairs of ribs have no cartilagenous attachment to the sternum.

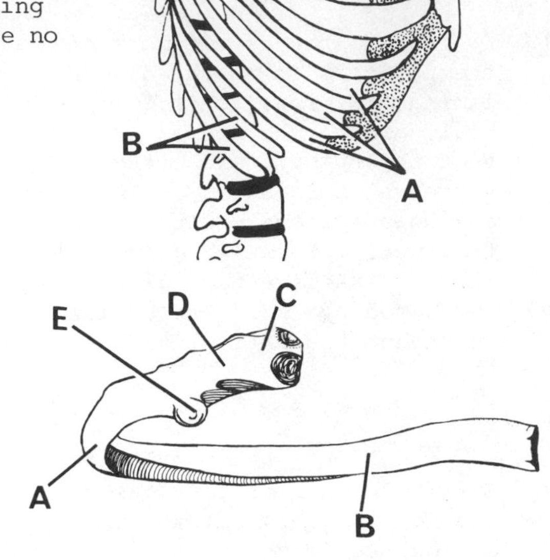

7. D. The arrow in the diagram indicates the tubercle of the rib (E in the diagram). This tubercle articulates with the transverse process of the thoracic vertebrae. The head (C) of the rib articulates with the body of one or more vertebrae. Other named parts of a rib are the neck (D), the angle (A), and the shaft (B).

8. E. Scoliosis is an exaggerated lateral curve of the vertebral column. There usually is a normal slight curve in the vertebral column either to the right or left depending on whether the person is right or left handed. This is due to a more common use of one arm and shoulder over the other and this pulls the column slightly off center. Kyphosis is an exaggerated curve of the vertebral column in the thoracic region, whereas lordosis is an exaggerated curve in the lumbar region. In pathological conditions kyphosis and lordosis quite often offset each other to maintain the center of gravity of the body over the sacrum. Osteoporosis is an abnormal condition of bone in which the osteoblasts fail to lay down bone matrix. Mitosis is the division of the somatic cells of the body.

MULTI-COMPLETION QUESTIONS

DIRECTIONS: In each of the following questions or incomplete statements, ONE OR MORE of the questions given is correct. At the lower right of each question, circle A if 1, 2, and 3 are correct; B if 1 and 3 are correct; C if 2 and 4 are correct; D if only 4 is correct; and E if all are correct.

1. THE VERTEBRAL COLUMN:
   1. Can be divided into five segments
   2. Includes the sacrum
   3. Has a primary curve in the thoracic segment
   4. Protects the spinal cord

   A B C D E

2. THE BONES FORMING THE THORACIC INLET INCLUDE THE:
   1. First thoracic vertebra
   2. Clavicle
   3. Manubrium
   4. Hyoid bone

   A B C D E

A	B	C	D	E
1,2,3	1,3	2,4	only 4	all correct

3. THE INTERVERTEBRAL DISC:
   1. Consists of a soft central core
   2. Allows for some movement
   3. Has an outer portion called the anulus fibrosus
   4. Is found between the atlas and axis

   A B C D E

4. THE ATLAS:
   1. Has no spinous process
   2. Forms a synchondrosis with the axis
   3. Articulates with the skull
   4. Has no transverse foramen

   A B C D E

5. THE SACRUM:
   1. Articulates with the ilia
   2. Has a promontory
   3. Is formed by five fused vertebrae
   4. Has structural sex differences

   A B C D E

6. THE STERNUM:
   1. Consists of three parts
   2. Forms a synchondrosis with the true ribs
   3. Articulates with the clavicle
   4. Is considered an irregular bone

   A B C D E

7. WHICH OF THE FOLLOWING IS(ARE) CORRECT FOR THE RIBS?
   1. Seven of the total are "true" ribs
   2. Their angle forms the lateral boundary of the thoracic cage
   3. Articulate with the bodies of the thoracic vertebrae
   4. Are all attached by cartilage to the sternum

   A B C D E

------------------------ ANSWERS, NOTES AND EXPLANATIONS ------------------------

1. E, All are correct. The vertebral column has five segments: cervical (7 vertebrae); thoracic (12 vertebrae); lumbar (5 vertebrae); sacral (5 fused vertebrae); and coccyx (4 or 5 fused vertebrae). This column both supports the body as a whole and protects the spinal cord which lies within the vertebral foramen of each vertebra. The vertebral column has two primary and two secondary curves. The thoracic (A in the diagram) and sacral (B) segments are considered to be primary curves as they retain the original fetal curve. The cervical (D) and lumbar (C) segments reverse their curves after birth to facilitate the normal posture of the head and the erect position needed during walking.

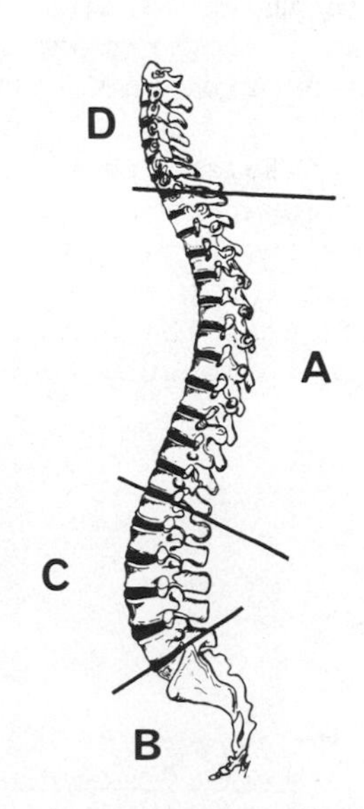

2. B, 1 and 3 are correct. The thoracic inlet (superior thoracic aperture) is formed by the anterior surface of the body of the first thoracic vertebra, the medial edge of the first rib and its cartilage, and the posterior surface of the manubrium. This inlet only measures about 5 cm. from front to back,

and about 10 cm. from side to side. The inlet is occupied by the apices of the lungs and pleura, and all other structures that pass between the neck and thorax. Neither the clavicle nor the hyoid bone form a boundary of the thoracic inlet.

3. A, 1, 2, and 3 are correct. The intervertebral discs are located between all vertebrae from the axis (second cervical) to the sacrum. These discs allow only a small amount of movement between any two vertebral bodies but allow a considerable amount of movement when considering the vertebral column as a whole. The discs consist of a fluid central portion, the nucleus pulposus (A in the diagram) and an outer laminated layer of fibrous connective tissue, the anulus fibrosus (B).

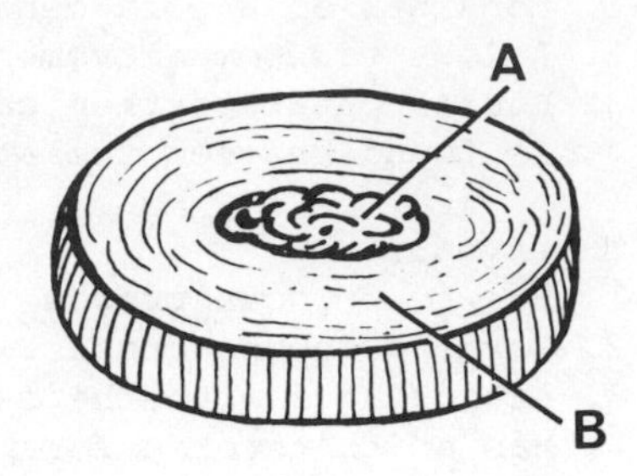

4. B, 1 and 3 are correct. The atlas or the first cervical vertebra lacks both a body and a spinous process. The atlas pivots upon the axis with the ends of the axis forming the pivot. This articulation is a synovial joint. The superior articulating facets of the atlas form concave discs for the condyles of the (skull) occipital bone. The movement of this latter joint is flexion and extension (forward and backward nodding) of the head.

5. E, All are correct. The sacrum of the vertebral column is formed by the fusion of five sacral vertebrae. The base of the sacrum articulates with the fifth lumbar vertebra and is separated from it by a disc. The lateral edges of the sacrum articulate with the ilia forming a combined fibrous and synovial joint (sacroiliac joint). The anterior edge of the superior fused sacral vertebra forms a promontory that projects into the birth canal and may narrow it considerably. The sacrum of a female is shorter, wider, and less curved than in the male.

6. A, 1, 2, and 3 are correct. The sternum is a flat bone consisting of an upper part called the manubrium (A in the diagram), a middle part called the body (B), and a lower part called the xiphoid process (C). The manubrium forms the sternoclavicular (synovial) joint with the medial end of the clavicle. The ribs (true and false) that attach themselves to the sternum do so by hyaline cartilage and therefore are called a synchondrosis articulation (joint). The first rib and the clavicle articulate with the manubrium, whereas the next nine ribs attach themselves directly or indirectly to the body of the sternum.

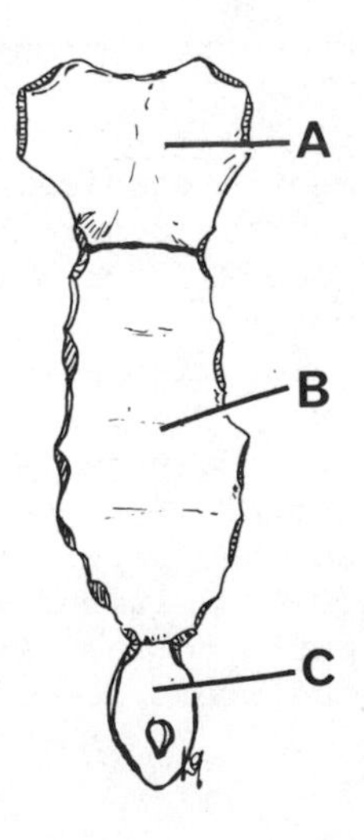

7. A, 1, 2, and 3 are correct. There are a total of 12 pairs of ribs; the first seven (from the top) are called true ribs because they each attach to the sternum by a costal cartilage. The next three pairs of ribs are called false ribs because they do not attach to the sternum directly by a costal cartilage but rather fuse with the costal cartilage of the seventh rib. The last two pairs

of ribs do not articulate anteriorly at all and are called floating ribs. All ribs articulate by their head and facets to the articulating facets of the vertebral body and transverse processes, respectively. The heads of the second through to the ninth ribs also articulate with the body of the vertebrae above. The ribs course anteriorly and downward from their vertebral column attachments and the angle of each rib forms the lateral boundary of the thoracic cage.

FIVE-CHOICE ASSOCIATION QUESTIONS

DIRECTIONS: Each group of questions below consists of a numbered list of descriptive words or phrases accompanied by a diagram with certain parts indicated by letters, or by a list of lettered headings. For each numbered word or phrase, SELECT THE LETTERED PART OR HEADING that matches it correctly. Then insert the letter in the space to the right of the appropriate number. Sometimes more than one numbered word or phrase may be correctly matched to the same lettered part or heading.

1. _____ A ligament connecting the lamina
2. _____ A site where a spinal nerve exits the vertebral canal
3. _____ The posterior longitudinal ligament
4. _____ The supraspinous ligament
5. _____ The anterior longitudinal ligament

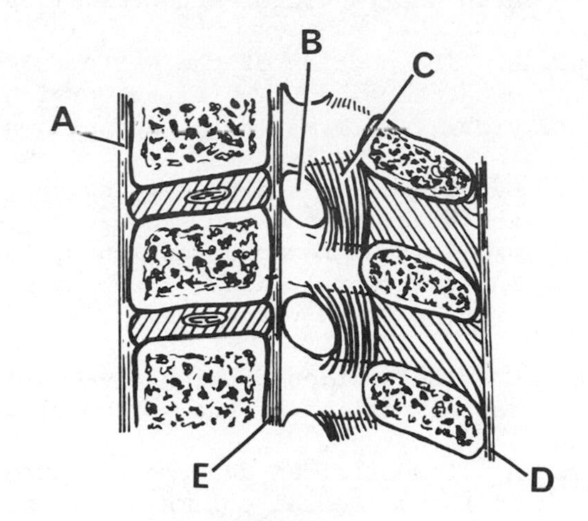

6. _____ A section which contains a pivot joint
7. _____ The section forming the longest primary curve
8. _____ A section forming a wedge-shaped bone
9. _____ The section where lordosis occurs
10. _____ A section forming a part of the upper portion of the birth canal

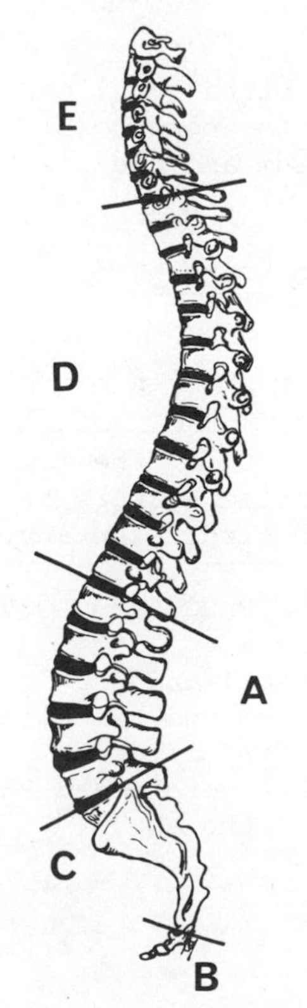

1. C. The ligament found within the vertebral canal connecting the lamina of one vertebra with the lamina of another is the ligamentum flavum.

2. B. The paired spinal nerves from each segment of the spinal cord exit the vertebral canal via the intervertebral foramen.

3. E. The posterior longitudinal ligament lies within the vertebral canal, extending over the bodies of the vertebrae and their intervening discs.

4. D. The supraspinous ligament runs from the tips of the spinous processes of the vertebrae all the way from the base of the skull to the sacrum.

5. A. The anterior longitudinal ligament supports the vertebral column as its anterior aspect. It covers the anterior surfaces of the bodies of the vertebrae and the intervertebral discs.

6. E. The section of the vertebral column containing a pivot joint is the cervical section. The first cervical vertebra (atlas) pivots on the second cervical vertebra (axis) and allows the head to turn from side to side.

7. D. There are two primary curves of the vertebral column, the thoracic (D in the diagram) and the sacro-coccygeal (B and C) sections. The longer of the two is the thoracic section containing twelve thoracic vertebrae and their intervertebral discs. The sacro-coccygeal section consists primarily of eight or nine fused vertebrae.

8. C. The section forming a wedge is the four fused vertebrae that forms the sacral portion of the vertebral column.

9. A. Lordosis, an abnormal or exaggerated curve, is found in the lumbar region of the vertebral column. Kyphosis is an exaggerated curve in the thoracic (D) region.

10. C. The birth canal passing through the pelvis is formed in part by the sacrum (C) and the coccyx (B). The upper part of the canal is bounded posteriorly by the sacrum and the lower part is bounded posteriorly by the coccyx.

THE UPPER AND LOWER LIMBS

FIVE-CHOICE COMPLETION QUESTIONS

DIRECTIONS: Each of the following questions or incomplete statements is followed by five suggested answers or completions. SELECT THE SINGLE BEST ANSWER in each case and then circle the appropriate letter at the lower right of each question.

1. THE INTERTUBERCULAR GROOVE OF THE HUMERUS IS A DEPRESSION LYING BETWEEN THE:
   A. Anatomical and surgical necks
   B. Capitulum and trochlea
   C. Medial and lateral epicondyles
   D. Lesser tubercle and the head
   E. None of the above

A B C D E

SELECT THE SINGLE BEST ANSWER

2. THE ULNA IS INDICATED BY _______.

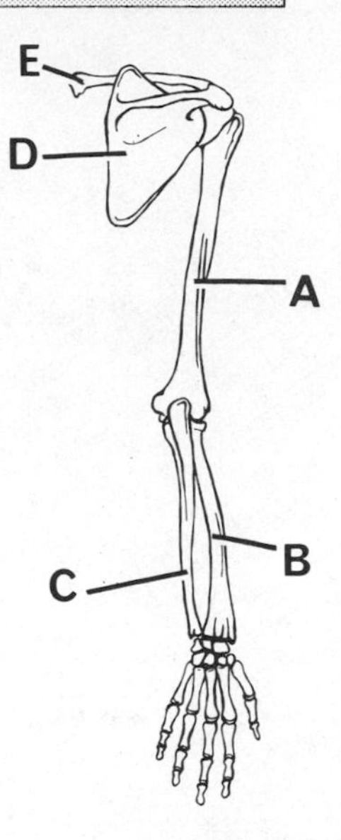

A B C D E

3. THE DISTAL END OF THE FEMUR ENTERING INTO THE FORMATION OF THE KNEE JOINT IS CALLED A ________.
   A. Crest
   B. Condyle
   C. Head
   D. Tuberosity
   E. Tubercle

A B C D E

4. THE ACROMION PROCESS IS INDICATED BY:

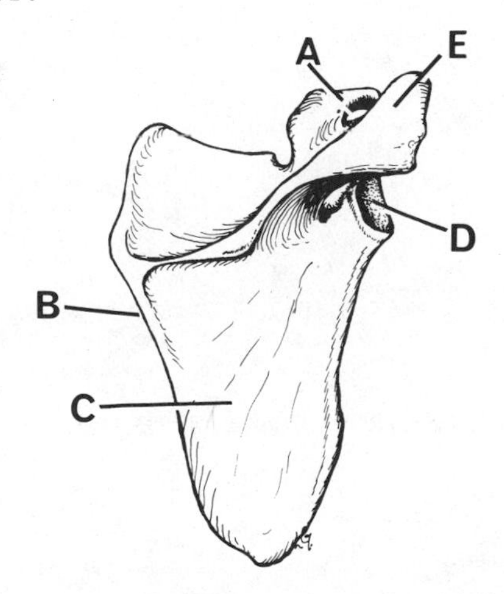

A B C D E

5. THE MEDIAL MALLEOLUS IS PART OF THE _______.
   A. Femur
   B. Fibula
   C. Navicular
   D. Tibia
   E. Talus

A B C D E

6. THE DISTAL END OF THE RADIUS ARTICULATES WITH THE:
   A. Lunate
   B. Triquetrum
   C. Pisiform
   D. Styloid process of the ulna
   E. None of the above

A B C D E

SELECT THE SINGLE BEST ANSWER

7. THE ARROW INDICATES THE:
   A. Deltoid tubercle
   B. Capitulum
   C. Medial epicondyle
   D. Lesser tubercle
   E. None of the above

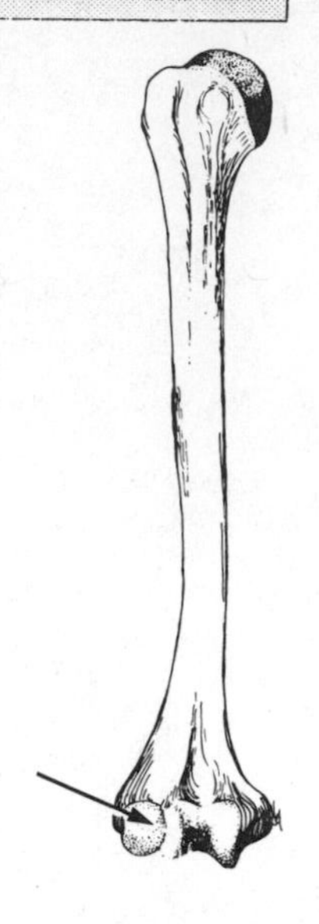

A B C D E

8. THE ISCHIAL TUBEROSITY IS INDICATED BY __________.

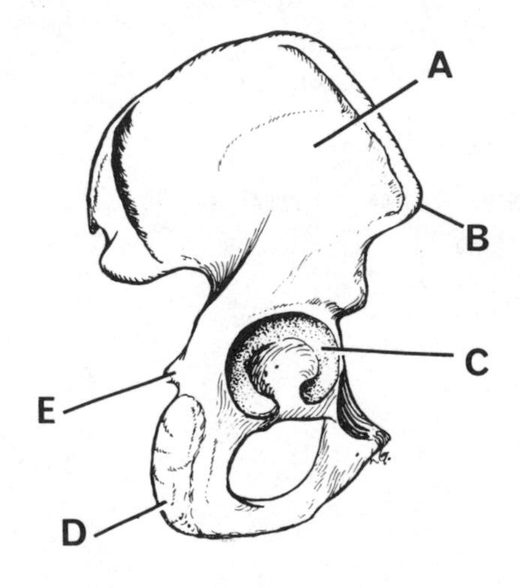

A B C D E

9. WHICH BONE DOES NOT ARTICULATE WITH ONE OF THE METATARSALS OF THE FOOT?
   A. Calcaneus
   B. Cuboid
   C. Cuneiform one
   D. Cuneiform two
   E. Cuneiform three

A B C D E

10. WHICH OF THE FOLLOWING IS NOT A COMPONENT OF THE SCAPULA?
   A. Spine
   B. Coracoid process
   C. Acetabulum
   D. Supraspinous fossa
   E. None of the above

A B C D E

SELECT THE SINGLE BEST ANSWER

11. THE TIBIA IS INDICATED BY ________.

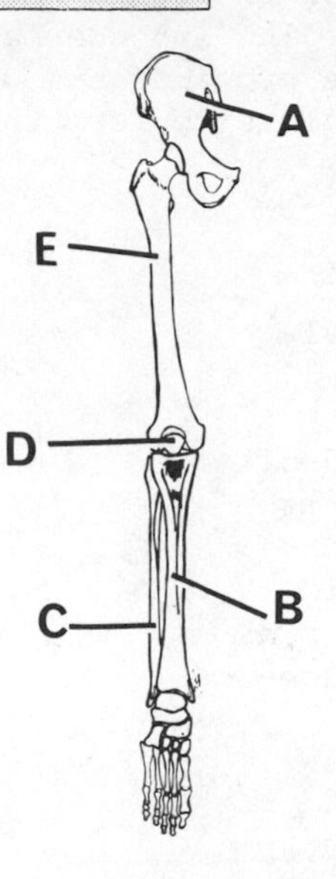

A B C D E

------------------------- ANSWERS, NOTES AND EXPLANATIONS -------------------------

1. E. The intertubercular groove (A in the diagram) lies on the anterior surface of the humerus between the lesser (B) and greater (C) tubercles. This groove accommodates the long head of the biceps brachii muscle of the arm.

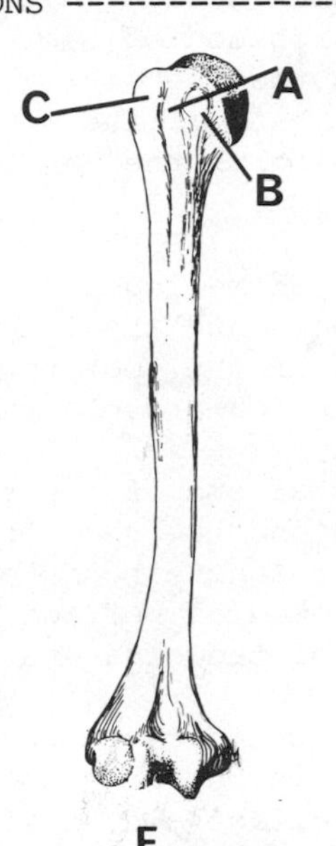

2. C. In this diagram of a posterior view of the limb the ulna is indicated by C. The ulna, a long bone of the forearm, lies medial to the radius (B). The humerus (A) is the long bone of the arm. The shoulder girdle consists of the scapula (D) on the posterior surface of the trunk and the clavicle (E) lying anterior to the trunk.

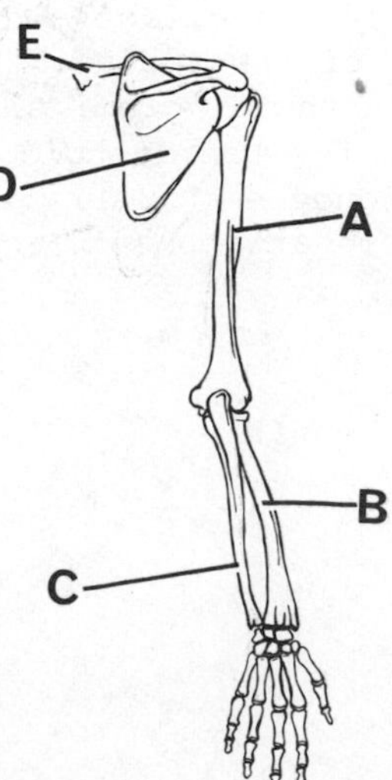

3. B. The distal end of the femur has two bony processes called condyles. These two condyles, one medial and one lateral, articulate with the medial and lateral condyles of the proximal end of the tibia. They are covered by hyaline cartilage and articulate to form the knee joint.

4. E. The acromion process of the scapula is labelled E in the posterior and lateral diagrams. This process articulates with the lateral flat end of the clavicle and forms the acromioclavicular joints. The coracoid process (A) serves as a site for muscle attachments. The medial border and the infrascapular fossa are indicated by B and C, respectively. The glenoid fossa (D) is a shallow depression of the scapula that accommodates the head of the humerus and forms the shoulder joint.

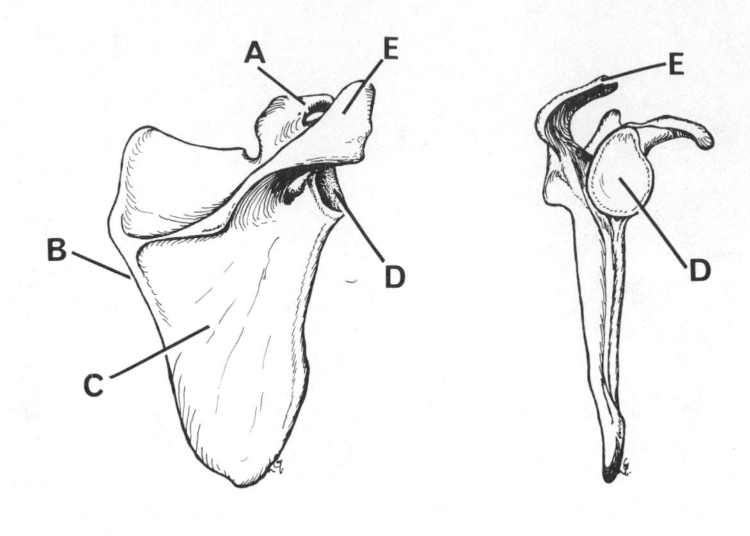

5. D. The medial malleolus is the medial projection of the distal end of the tibia. This malleolus and the lateral malleolus of the fibula grasp the body of the tarsal bone of the foot to form the hinge joint of the ankle. This joint allows for dorsiflexion and plantar flexion of the foot.

6. A. The distal end of the radius articulates with the lunate and scaphoid bones of the hand and the head of the ulna. The distal end of the radius and the scaphoid and lunate bones of the hand form the lateral portion of the wrist joint. The medial portion of this joint is formed by a disc lying between the distal end of the ulna and the triquetrum of the hand.

7. B. The arrow indicates the capitulum of the humerus. This rounded process is covered by hyaline cartilage and articulates with the head of the radius. The deltoid tuberosity (A in the diagram) serves as a site of attachment for the deltoid muscle. The medial epicondyle (B) serves as a site for muscle attachment.

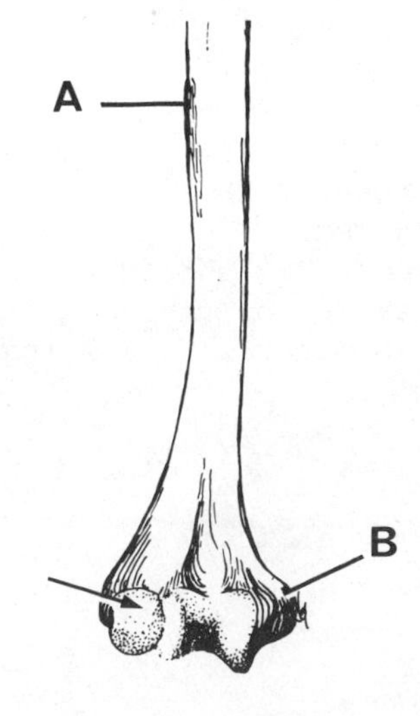

8. D. The ischial tuberosity (D in the diagram), is a large process on the inferior aspect of the ischium. It serves as a site for muscle attachments and in the erect sitting position supports the weight of the body. The ilium (A) represents the upper flaring portion of the hip bone. The anterior superior iliac spines (B) are prominent projections of the iliac crests. The head of the femur articulates with the acetabulum (C) of the hip bone. The ischial spine (E) lies just above the ischial tuberosity.

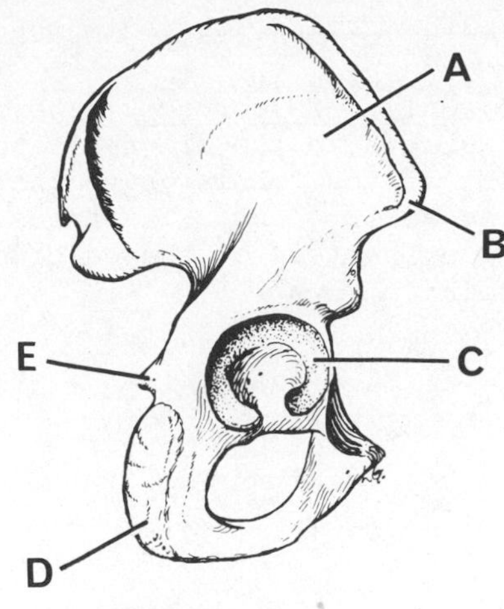

9. A. The three cuneiform bones (A in the diagram) along with the cuboid bone (B) form a row of bones that articulate with the five metatarsals. The calcaneus (C), talus (D), and navicular (E) bones do not articulate with any of the metatarsals (F).

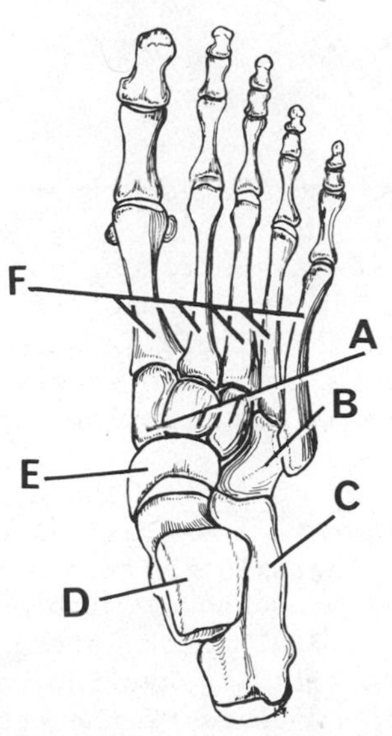

10. C. The acetabulum is not a part of the scapula but is a cup-like structure of the hip bone. This structure accommodates the head of the femur to form the hip joint. The spine, acromion, coracoid process, and the supraspinous fossa are all named structures of the scapula.

11. B. The tibia (B), a long bone of the leg, lies medial to the fibula (C). The femur (E) articulates by a proximal head with the hip bone (A) or os coxa to form the hip joint. The patella (D), a sesamoid bone, lies anterior to the distal end of the femur and articulates with the anterior surfaces of the medial and lateral condyles of the femur.

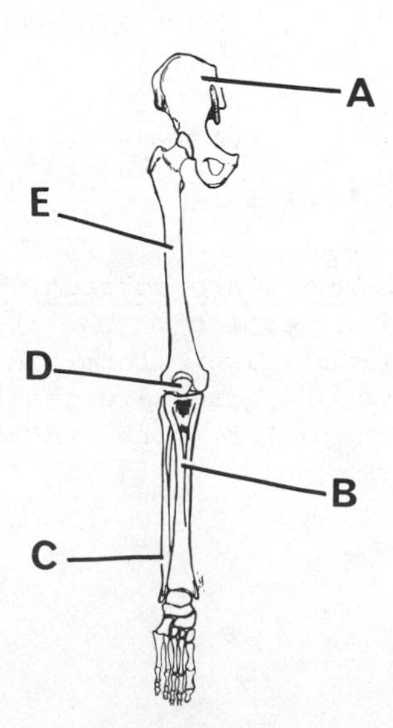

1. D, Only 4 is correct. The medial malleolus of the tibia (A in the diagram) and the lateral malleolus (B) of the fibula grasp the body of the talus to form the ankle joint. The talus articulates with the upper surface of the calcaneus and forms the subtalar joint.

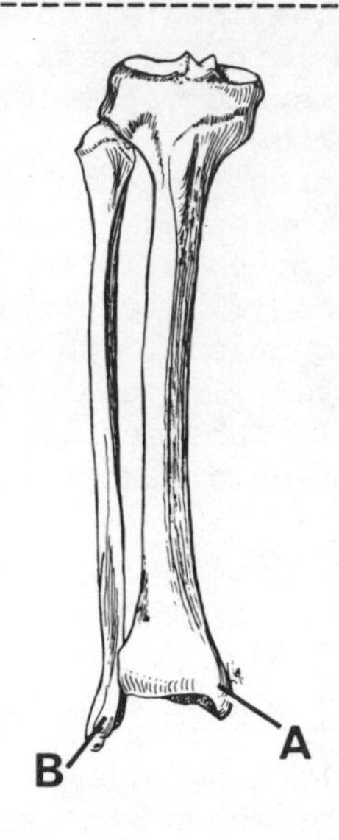

2. A, 1, 2, and 3 are correct. The patella is a sesamoid bone that forms the anterior part of the knee joint. The posterior surface of the patella has two articular surfaces, covered by hyaline cartilage, that articulate with the patellar surface of the femur. The patella does not articulate with the tibia or the fibula. The patella provides the site of insertion for the quadriceps femoris muscle from above and for the ligamentum patella below. The patella also provides protection for the knee joint.

3. E, All are correct. The ulna is the medial bone of the forearm. Its proximal end has the olecranon (A in the diagram), the trochlear notch (B), and the coronoid process (C) with a tuberosity medially (D) and the radial notch laterally (E). The trochlear notch is lined by hyaline cartilage and articulates with the trochlea of the humerus. The distal end of the ulna has a head which bears an articular surface and a styloid process (F). The ulna articulates with the radius laterally at both its distal and proximal ends.

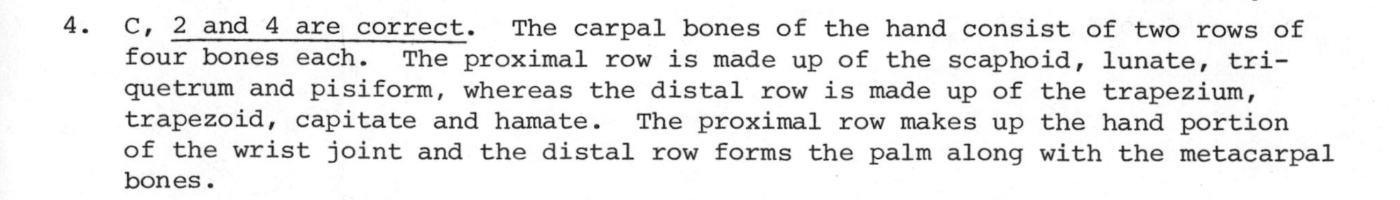

4. C, 2 and 4 are correct. The carpal bones of the hand consist of two rows of four bones each. The proximal row is made up of the scaphoid, lunate, triquetrum and pisiform, whereas the distal row is made up of the trapezium, trapezoid, capitate and hamate. The proximal row makes up the hand portion of the wrist joint and the distal row forms the palm along with the metacarpal bones.

5. B, 1 and 3 are correct. There are 27 bones in the hand: eight carpal, five metacarpal, and 14 phalanges. The eight carpal bones are classified as short bones and are arranged in two rows of four each. The proximal row consists of the scaphoid (A), lunate (B), triquetrum (C), and pisiform (D); whereas the distal row consists of the hamate (E), capitate (F), trapezoid (G), and trapezium (H). Each finger consists of a metacarpal bone and with the exception of the thumb each has three phalanges (proximal, middle, and distal). The thumb has only a distal and proximal phalanges. All the bones of the fingers are considered to be long bones, even though the distal phalanges are extremely short.

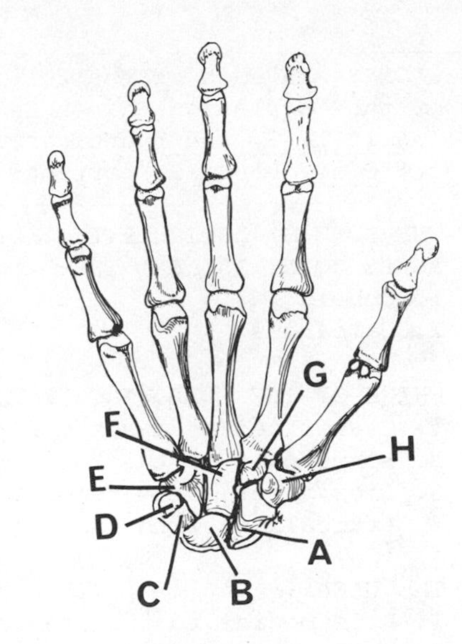

6. E, All are correct. There are five bones of the foot forming the medial longitudinal arch of the foot. From posterior to anterior they are the calcaneus (A), talus (B), navicular (C), first cuneiform (D), and the first metatarsal bone (E). The lateral longitudinal arch consists of three bones: the fifth metatarsal, cuboid and calcaneus. The arches of the foot along with its ligaments and muscles enable man to walk with ease.

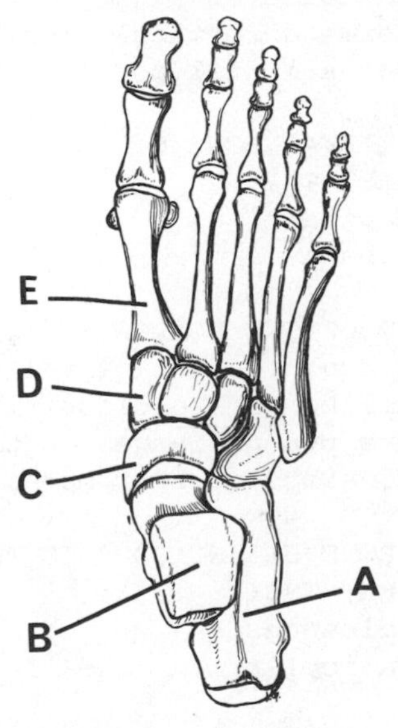

7. A, 1, 2, and 3 are correct. The hip bone (os coxa) is formed by the union of three bones: the ischium, the ilium and the pubis. The right and left hip bones articulate anteriorly to form the symphysis pubis, and posteriorly fuse with the lateral surfaces of the sacrum to form the sacroiliac joints.

8. C, 2 and 4 are correct. The clavicle, along with the scapula, forms the pectoral (shoulder) girdle of the upper limb. The clavicle, a long bone, has two curves in a horizontal plane. It forms a synovial type joint, at its medial end, with the manubrium of the sternum and with the acromion process of the scapula at its lateral end. The medial end of the clavicle is round and the lateral end is flat. The clavicle serves to attach the upper limb by the sternoclavicular joint to the trunk of the body. It also functions as a strut for the upper limb.

MULTI-COMPLETION QUESTIONS

DIRECTIONS: In each of the following questions or incomplete statements, ONE OR MORE of the completions given is correct. At the lower right of each question, circle A if 1, 2, and 3 are correct; B if 1 and 3 are correct; C if 2 and 4 are correct; D if only 4 is correct; and E if all are correct.

1. THE BONE(S) ARTICULATING WITH THE FIBULA AND TIBIA TO FORM THE ANKLE JOINT IS(ARE) THE:
   1. Cuboid
   2. Calcaneus
   3. Navicular
   4. Talus

   A B C D E

2. WHICH OF THE FOLLOWING STATEMENTS IS(ARE) TRUE FOR THE PATELLA?
   1. Classified as a sesamoid bone
   2. Forms part of the knee
   3. Lies anterior to femoral condyles
   4. Articulates with the tibia

   A B C D E

3. THE ULNA:
   1. Lies medial to the radius
   2. Articulates with the trochlea of the humerus
   3. Forms an articulation with the radius
   4. Has a distal portion called a head

   A B C D E

4. THE CARPAL BONE(S) INCLUDED IN THE PROXIMAL ROW OF THE HAND IS(ARE) THE:
   1. Capitate
   2. Lunate
   3. Hamate
   4. Pisiform

   A B C D E

5. WHICH OF THE FOLLOWING IS(ARE) TRUE FOR THE HAND?
   1. Contains eight carpal bones
   2. All bones are classified as short bones
   3. Has two phalanges in the thumb
   4. Consists of four metacarpals

   A B C D E

6. THE BONE(S) INCLUDED IN THE MEDIAL LONGITUDINAL ARCH OF THE FOOT IS(ARE) THE:
   1. Calcaneus
   2. Navicular
   3. First cuneiform
   4. Talus

   A B C D E

7. THE HIP BONE IS FORMED BY THE UNION OF THE:
   1. Ischium
   2. Ilium
   3. Pubis
   4. Sacrum

   A B C D E

8. THE CLAVICLE:
   1. Is an important part of the pelvic girdle
   2. Can be classified as a long bone
   3. Forms a synovial joint with the body of the sternum
   4. Has two curves in a horizontal plane

   A B C D E

9. THE BONE(S) FORMING A COMPONENT OF THE PECTORAL GIRDLE IS(ARE) THE:
   1. First rib
   2. Humerus
   3. Manubrium
   4. Clavicle

   A B C D E

10. WHICH OF THE FOLLOWING IS(ARE) NOT A NAMED PART OF THE FEMUR?
    1. Head
    2. Greater trochanter
    3. Intertrochanteric crest
    4. Intercondylar eminence

    A B C D E

9. D, Only 4 is correct. The shoulder or pectoral girdle is formed by the scapula and the clavicle and serves to attach the upper limb to the trunk. The head of the humerus lies in the glenoid fossa of the scapula and the acromial process of the scapula articulates with the lateral end of the clavicle forming the acromioclavicular joint. The medial end of the clavicle forms the sternoclavicular joint with the manubrium of the sternum. This joint forms the only direct articulation with the trunk. The clavicle acts as a strut for the upper limb. The scapula attaches the upper limb to the trunk by numerous muscles.

10. D, Only 4 is correct. The head (A), greater trochanter (B), and the intertrochanteric crest (C) are all components of the femur. The intercondylar eminence (D) is a small projection on the proximal surface of the tibia lying between the medial (E) and lateral (F) condyles. This eminence serves as a site of attachment for some of the ligaments of the knee joint.

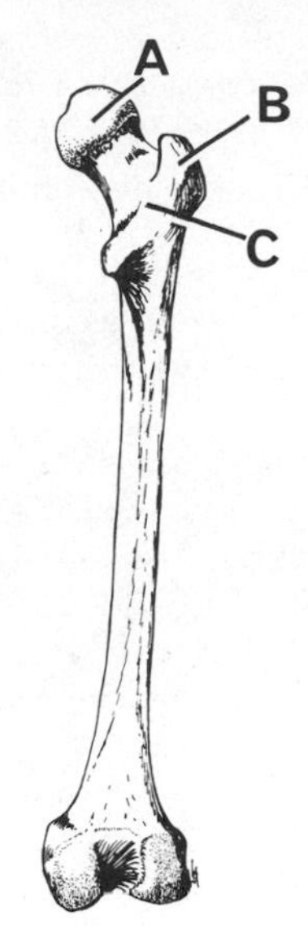

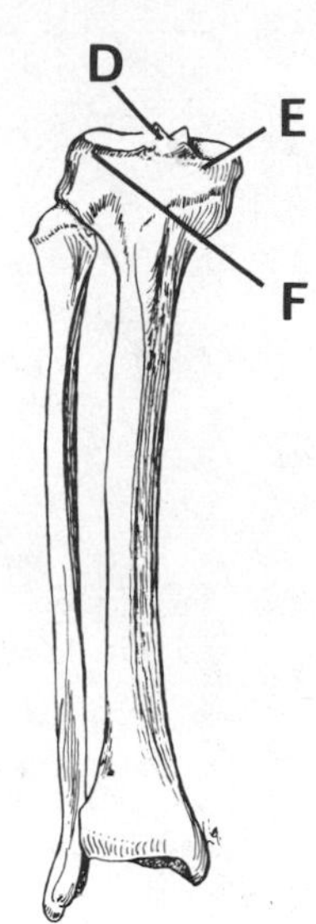

F I V E - C H O I C E   A S S O C I A T I O N   Q U E S T I O N S

DIRECTIONS: Each group of questions below consists of a numbered list of descriptive words or phrases accompanied by a diagram with certain parts indicated by letters, or by a list of lettered headings. For each numbered word or phrase, SELECT THE LETTERED PART OR HEADING that matches it correctly. Then insert the letter in the space to the right of the appropriate number. Sometimes more than one numbered word or phrase may be correctly matched to the same lettered part or heading.

1. _____ Lesser tubercle
2. _____ Acetabulum
3. _____ Acromion process
4. _____ Olecranon process
5. _____ Coronoid fossa
6. _____ Medial malleolus
7. _____ Infraspinous fossa
8. _____ Anatomical neck
9. _____ Intertrochanteric line
10. _____ Iliac crest

A. Scapula

B. Tibia

C. Hip bone

D. Humerus

E. None of the above

------------------------ ANSWERS, NOTES AND EXPLANATIONS ------------------------

1. D. The lesser tubercle is a small process, slightly anterior and lateral to the head of the humerus. It is separated laterally from the greater tubercle by the intertubercular (bicipital) groove which descends down the shaft. This groove contains the long head of the biceps brachii muscle, whereas the tubercles serve as sites for muscle attachments.

2. C. The acetabulum is a round cup-like structure on the lateral surface of the hip bone. It is formed by the union of the ischium, ilium and pubis bones which combine to form the hip bone (os coxa) proper. The acetabulum is the socket portion of the hip joint and accommodates the head of the femur.

3. A. The acromion process is the lateral terminal portion of the spine of the scapula. This process articulates with the lateral end of the clavicle and forms the acromioclavicular joint.

4. E. The olecranon process is the proximal process of the ulna. The semilunar notch of the ulna articulates with the trochlea of the humerus to form the medial half of the elbow joint. The lateral half is formed between the capitulum of the humerus and the fovea of the radial head. This joint is of the hinge variety. When the elbow joint is extended the olecranon process lies in the olecranon fossa on the distal posterior aspect of the humerus.

5. D. The coronoid fossa lies on the anterior surface at the distal end of the humerus just above the trochlea. It accommodates the coronoid process of the ulna when the elbow joint is flexed.

6. B. The tibia forms the medial bone of the leg, whereas the fibula forms the lateral bony component of the leg. The medial malleolus of the tibia is a short process on the distal medial aspect of the tibia. It articulates with the medial side of the talus of the foot and along with the lateral malleolus of the fibula forms the ankle joint.

7. A. The infraspinous fossa is the area below the spine on the dorsal aspect of the scapula. It provides a site for attachment of muscles involved in movements of the humerus.

8. D. The anatomical neck is a slight constriction immediately below the head of the humerus.

9. E. The intertrochanteric line extends between the greater and lesser trochanters of the femur and marks the anterior boundary between the neck and the shaft.

10. C. The iliac crest (A in the diagram) is the expanded upper end of the ilium of the hip bone. It can be palpated in a patient.

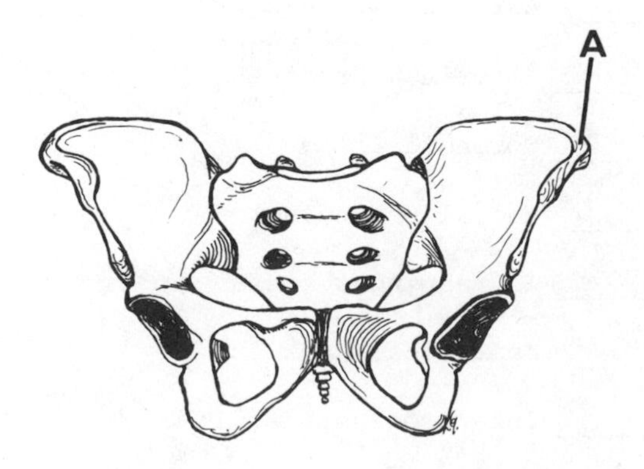

6. JOINTS AND THEIR MOVEMENTS

LEARNING OBJECTIVES

Be Able To:

- Define the term joint (articulation) and describe the classification of joints based on: 1) their structural components, and 2) the degree of movement they allow.
- Give examples of the different types of fibrous and cartilagenous joints.
- Draw and label the components of a 'typical' synovial joint.
- Describe the types of synovial joints based on the shape of their articulating surfaces, giving examples of each.
- Define the following movements of synovial joints: flexion, extension, abduction, adduction, rotation, and circumduction.
- Describe the following special movements and indicate where they take place: inversion, eversion, retraction, supination, pronation, dorsiflexion, and plantar flexion.
- Briefly describe the following disorders of joints as to their cause and site of debilitation: rheumatoid arthritis, osteoarthritis, gout, bursitis, tendinitis, and displaced intervertebral disc.

RELEVANT READINGS

Anthony & Kolthof - Chapters 5 and 6.
Chaffee & Greisheimer - Chapter 3.
Crouch & McClintic - Chapter 8.
Jacob, Francone & Lossow - Chapter 6.
Landau - Chapter 5.
Tortora - Chapter 8.

FIVE-CHOICE COMPLETION QUESTIONS

DIRECTIONS: Each of the following questions or incomplete statements is followed by five suggested answers or completions. SELECT THE SINGLE BEST ANSWER in each case and then circle the appropriate letter at the lower right of each question.

1. IN THE ANATOMICAL POSITION THE FOREARM IS:
   A. Flexed
   B. Rotated
   C. Abducted
   D. Supinated
   E. None of the above

   A B C D E

2. AN EXAMPLE OF A SADDLE JOINT IS THE ________ ARTICULATION.
   A. Elbow
   B. Thumb
   C. Radioulnar
   D. Intercarpal
   E. Acromioclavicular

   A B C D E

3. THE SYNOVIAL MEMBRANE IS INDICATED BY ________.

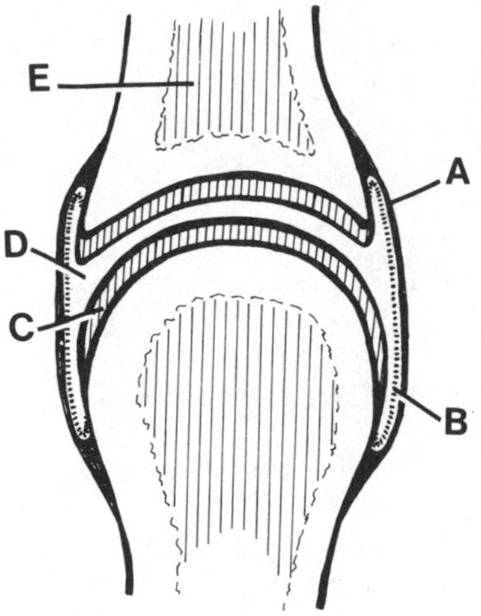

   A B C D E

4. WHICH OF THE FOLLOWING IS AN EXAMPLE OF AN ELLIPSOID JOINT?
   A. Hip
   B. Ankle
   C. Thumb
   D. Wrist
   E. Sacroiliac

   A B C D E

5. THE MOVEMENT WHICH MOVES A BONE AWAY FROM THE MEDIAN PLANE OF THE BODY IS CALLED ________.
   A. Circumduction
   B. Extension
   C. Adduction
   D. Rotation
   E. None of the above

   A B C D E

6. THE SHOULDER JOINT CAN BE CLASSIFIED AS A _____ TYPE OF JOINT.
   A. Ball and socket
   B. Gliding
   C. Hinge
   D. Saddle
   E. None of the above

   A B C D E

7. THE STERNOCLAVICULAR JOINT IS AN EXAMPLE OF A _____ ARTICULATION.
   A. Pivot
   B. Hinge
   C. Saddle
   D. Gliding
   E. Ball and socket

   A B C D E

SELECT THE SINGLE BEST ANSWER

8. WHICH OF THE FOLLOWING CANNOT BE CLASSIFIED AS A HINGE TYPE JOINT?
   A. Ankle
   B. Elbow
   C. Interphalangeal
   D. Knee
   E. Wrist

A B C D E

------------------------ ANSWERS, NOTES AND EXPLANATIONS ------------------------

1. D. In the anatomical position the forearm is supinated with the palm facing forward. In this position the two bones of the forearm, the radius and ulna, lie parallel to each other. In pronation, where the palm faces posteriorly, the radius crosses the ulna just below its midpoint. The elbow joint would be extended in the anatomical position.

2. B. The major saddle joint of the body is the thumb joint, or more correctly, the articulation between the first metacarpal and the trapezium (a carpal bone). The best way to describe this joint is to imagine two saddles placed at right angles to each other, with one inverted on the other. This joint is very mobile, permitting angular movements in any plane, but is restricted in the amount of axial rotation it allows. Its range of motion is exceeded only by a ball and socket type of joint. Abduction of the thumb can be described as moving the thumb away from the hand in a plane which is at a right angle to the palm. The movement that carries the thumb laterally away from the fingers is called extension.

3. B. The synovial membrane (B in the diagram) lines the fibrous capsule (A) of the synovial joint. This membrane is attached to bone at the periphery of the articular cartilages. It secretes a small amount of fluid which lubricates the joint surfaces to reduce friction. The synovial membrane is also thought to be involved in transporting nutrients to the avascular articular (hyaline) cartilage (C). The joint space (D) of a synovial joint normally contains only a small amount of fluid. The compact bone lying beneath the articular cartilage is indicated by E. Some synovial joints (e.g., sternoclavicular) may have an articular disc intervening between the two articular surfaces. This disc may completely divide the joint space into two separate cavities or may form discs (e.g., knee joint) that only partially separate the articular cartilage. The function of the disc may be to help restrain the joint, to allow two movements to take place simultaneously, or to accommodate the shapes of the articulating surfaces.

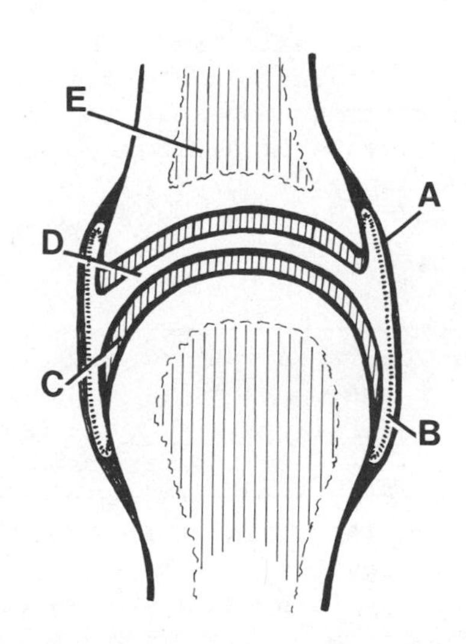

4. D. The wrist (radiocarpal) is an ellipsoid (condyloid) joint and permits movement in two planes (biaxial). The articulating surfaces of the wrist include: the radius and the disc, proximally, and the scaphoid, lunate, and triquetrum bones, distally. The wrist joint allows flexion, extension, abduction, and adduction. The hip joint is a ball and socket type of joint, whereas the ankle is a simple hinge type. The thumb is a saddle type joint and the sacroiliac joint is a combination of a synovial and fibrous type of joints.

5. E. Abduction is the movement which carries a bone away from the median plane of the body. The opposite movement to abduction is adduction. Circumduction is the movement of a structure so as to describe a circle, and is a combination of flexion, extension, abduction, and adduction. Rotation is the movement of a bone about its long axis and it may be further described as being either medial or lateral rotation, depending on the direction it takes.

6. A. The shoulder is an excellent example of a ball and socket type synovial joint. In this joint the large head of the humerus fits into the shallow glenoid fossa of the scapula, and is supported almost entirely by a muscular cuff. The shoulder joint permits flexion, extension, abduction, adduction, rotation, and circumduction of the upper limb. This joint is one of the most freely movable joints in the body and is therefore highly susceptible to dislocation. In contrast, the hip joint, another ball and socket type, is more stable and less susceptible to dislocation. The greater stability of the hip joint is because the head of the femur fits snugly into the deep acetabulum of the hip bone (os coxa). This deep socket plus its ligamentous support allows the same types of movement as permitted at the shoulder joint, but they are somewhat restricted by bony, ligamentous, and soft tissue structures.

7. D. The sternoclavicular joint is a gliding type synovial joint. This joint contains a disc which not only separates the medial end of the clavicle from the sternum but is also attached to the first costal cartilage. The sternoclavicular joint is the only joint holding the shoulder girdle to the thoracic cage. This joint is classified as a multiaxial joint.

8. E. The ankle, knee, elbow, and interphalangeal joints are all classified as hinge joints. In these joints a convex surface fits into a concave structure permitting movement in one plane about a single axis (uniaxial). The typical movements of a hinge joint are flexion and extension. The wrist is a condyloid (ellipsoid) joint between the scaphoid, lunate, and triquetrum bones of the hand and the head of the radius and the articular disc.

MULTI-COMPLETION QUESTIONS

DIRECTIONS: In each of the following questions or incomplete statements, ONE OR MORE of the completions given is correct. At the lower right of each question, circle A if 1, 2, and 3 are correct; B if 1 and 3 are correct; C if 2 and 4 are correct; D if only 4 is correct; and E if all are correct.

1. SYNOVIAL FLUID IN THE JOINT CAVITY:
   1. Provides a cushioning effect
   2. Consists mainly of water
   3. Contains hyaluronic acid
   4. Lubricates the articulating surfaces

   A B C D E

2. FACTORS INVOLVED IN DETERMINING THE DEGREE OF MOVEMENT ALLOWED AT A SYNOVIAL JOINT INCLUDE THE:
   1. Presence of restraining ligaments
   2. Interference by large soft tissue masses
   3. Shape of the articulating bony surfaces
   4. Snugness of the joint capsule

   A B C D E

3. AN AMPHIARTHROSIS IS LOCATED BETWEEN THE:
   1. Parietal and occipital bones of the skull
   2. Adjoining surfaces of the tibia and fibula
   3. Clavicle and the manubrium of the sternum
   4. Two pubic bones

   A B C D E

A	B	C	D	E
1,2,3	1,3	2,4	only 4	all correct

4. DISTINGUISHING FEATURES OF A SYNOVIAL JOINT INCLUDE A(AN):
   1. Potential space
   2. Articular cartilage
   3. Fibrous capsule
   4. Synovial membrane

   A B C D E

5. A HINGE JOINT ALLOWS:
   1. Gliding
   2. Circumduction
   3. Rotation
   4. Extension

   A B C D E

6. A SYNARTHROSIS JOINT:
   1. Contains a small but functional joint space
   2. May have cartilage intervening between its articulating surfaces
   3. Is relatively a weak type of attachment
   4. Permits no movement

   A B C D E

7. WHICH OF THE FOLLOWING IS(ARE) TRUE FOR BURSAE?
   1. Aids the gliding of tendons over a bony surface
   2. Contains a fluid-secreting membrane
   3. Often involved with synovial joints
   4. May become inflamed

   A B C D E

8. A CARTILAGENOUS JOINT IS FORMED BETWEEN THE:
   1. Bodies of vertebrae
   2. Distal epiphysis and the metaphysis of the femur
   3. Sternum and the medial ends of the ribs
   4. Adjoining surfaces of the radius and ulna

   A B C D E

------------------------ ANSWERS, NOTES AND EXPLANATIONS ------------------------

1. E, All are correct. The synovial fluid acts both as a cushion and as as lubricant for the articulating surfaces of a synovial joint. The main component of synovial fluid is water (95 per cent), some protein (1-2 per cent), and mucopolysaccharides (i.e., hyaluronic acid). The hyaluronic acid is thought to provide for a certain amount of viscosity in the fluid.

2. E, All are correct. Synovial joints (diarthroses) are the most mobile type of joint in the body, but they have varying degrees of freedom in the movement they allow. Some of the factors involved are: 1) the shape of the articular bony surfaces which may restrict movement in one or more planes; 2) the snugness of the joint capsule to some degree (it may inhibit some movements if it is too tight); 3) the supporting ligaments play an important role in controlling the movements allowed at any particular joint; and 4) the large soft tissue masses (i.e., muscle, fat, and others) may restrict movement of a joint (e.g., flexion of the hip joint in a large or obese individual).

3. C, 2 and 4 are correct. The term amphiarthroses denotes slight movement between two bony structures joined together by either cartilaginous or fibrous tissues. An example of a cartilaginous type of amphiarthroses is the symphysis pubis which is joined by fibrocartilage. Two examples of this type of joint, joined by fibrous connective tissue, are between the tibia and fibula of the leg and the radius and ulna of the forearm. The parietal and occipital bones of the skull are immovable joints or synarthroses. The articulations between the clavicles and the manubrium are synovial joints (diarthroses).

4. E, All are correct. The synovial joint (diarthrosis) is by far the most numerous type of joint in the body. It consists of two or more surfaces of bone covered by an articular (hyaline) cartilage, contained within a fibrous tissue capsule. The capsule is lined by a thin synovial membrane that attaches to the bone at the periphery of the articular cartilage. This membrane secretes a serous fluid that both reduces the friction between the articulating surfaces and also provides nourishment to the avascular cartilage. This fluid is found in space (joint space) located between the two articulating surfaces.

5. D, Only 4 is correct. A hinge joint is a uniaxial joint and therefore allows action in only one plane. Hinge joints such as the knee, elbow, and ankle allow both flexion and extension. Gliding action is possible in all directions and a joint of this type is considered a multiaxial joint. Circumduction is exemplified by swinging the arm in a circular fashion, and in doing so flexion, extension, abduction, and adduction are all carried out during some phase of circumduction. Rotation is the movement around the long axis of the bone, taking place between the atlas and axis of the cervical vertebrae and in the radius during pronation and supination.

6. C, 2 and 4 are correct. A synarthrosis joint is an immovable joint in which the articulating bones are united by small amounts of fibrous connective tissue or hyaline cartilage. The two best examples of this type of joint are the sutures of the skull, which are united by fibrous connective tissue, and the synchondrosis of long bones, which are temporary joints of hyaline cartilage. This latter type of synarthrosis is considered temporary because as the growth of the long bones approaches completion the cartilage bands (epiphyseal plates) will eventually disappear to be replaced by bone. These immovable joints are strong.

7. E, All are correct. Bursae are closed sacs of connective tissue lined by a synovial membrane. They are located near and are involved with synovial joints and some directly communicate with the synovial joint cavity. Typically, bursae are usually located beneath muscle tendons that cross over a bony structure or a series of ligaments. They function in aiding the gliding movement of the tendon by secreting a small amount of fluid into the sac to reduce friction. Bursae may become inflamed (bursitis) because of an increase in the stress or tension placed on them or by a local or systemic inflammatory process. Bursitis is usually quite painful.

8. A, 1, 2, and 3 are correct. Cartilaginous joints (synchrondrosis) are nothing more than fibrocartilage or hyaline cartilage joining two bones. Examples of this type of joint, where fibrocartilage is the binding material, are at the symphysis pubis and between the bodies of adjacent vertebrae (except between the atlas and axis). Hyaline cartilage forms a joint between the medial ends of the upper ten ribs and the sternum. Another example where hyaline cartilage forms a synchondrosis is in the developing long bone where the epiphyseal plates join the epiphysis and metaphysis. However, this cartilagenous joint is a temporary one and disappears by the end of puberty.

FIVE-CHOICE ASSOCIATION QUESTIONS

DIRECTIONS: Each group of questions below consists of a numbered list of descriptive words or phrases accompanied by a diagram with certain parts indicated by letters, or by a list of lettered headings. For each numbered word or phrase, SELECT THE LETTERED PART OR HEADING that matches it correctly. Then insert the letter in the space to the right of the appropriate number. Sometimes more than one numbered word may be correctly matched to the same lettered part of heading.

ASSOCIATION QUESTIONS

1. _____ The radius lies obliquely across the ulna.
2. _____ The elbow joint is straight.
3. _____ The lower jaw is pulled back.
4. _____ Movement about the long axis of a bone.
5. _____ The sole of the foot is turned inward.
6. _____ Movement away from the midline.

A. Eversion
B. Extension
C. Rotation
D. Retraction
E. None of the above

7. _____ A disturbance of purine metabolism.
8. _____ A chronic degenerative disorder of weight-bearing joints.
9. _____ Inflammation of a synovial bursa.
10. _____ Antibody reacts to chemicals produced by an inflamed synovial membrane.

A. Gout
B. Tendinitus
C. Osteoarthritis
D. Rheumatoid arthritis
E. None of the above

------------------------ ANSWERS, NOTES AND EXPLANATIONS ------------------------

1. E. When the radius lies obliquely across the ulna the forearm is said to be pronated. In this position the palm of the hand would be facing posteriorly. The forearm in the anatomical position is said to be supinated. In supination the radius and ulna of the forearm lie parallel to each other with the palm of the hand facing anteriorly.

2. B. When the elbow is straight, or the bones of the forearm and the arm lie in the same plane, the elbow is said to be extended. In the anatomical position all joints of the body are considered to be extended. This movement is opposite to flexion. In extension of the foot at the ankle joint we talk of plantar flexion (A in the diagram). Here the ball of the foot drops to lie below the level of the heel; the reverse of this movement is called dorsiflexion (B) of the ankle joint.

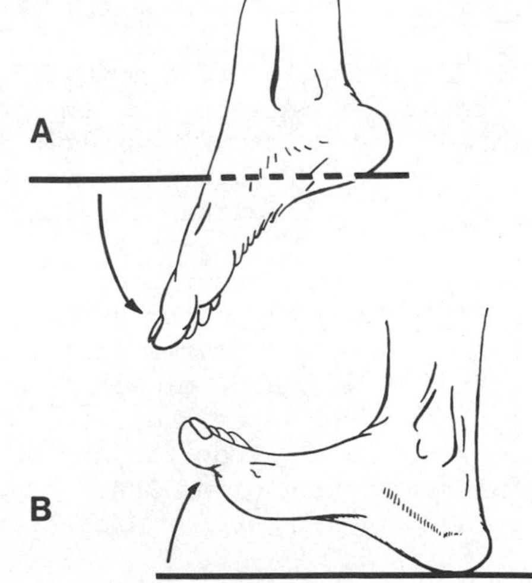

3. D. Retraction is where a part of the body is pulled back such as in the pulling back of the lower jaw (mandible). The opposite to this movement is protraction or a pushing out of a part such as the lower jaw.

4. C. Movement about the long axis of a bone is called rotation. The best way to demonstrate this action is to place the flexed elbow at the side of the trunk. By moving the hand towards the midline of the body we demonstrate

medial rotation of the humerus. By returning to the starting position we would be producing lateral rotation of the humerus. Similarly, the head and the atlas rotate upon the axis of the vertebral column in moving the head from the right to left (or left to right).

5. E. Inversion is a special movement of the ankle joint that turns the sole of the foot inward (A). Eversion of the ankle joint would leave the sole of the foot facing outward (B).

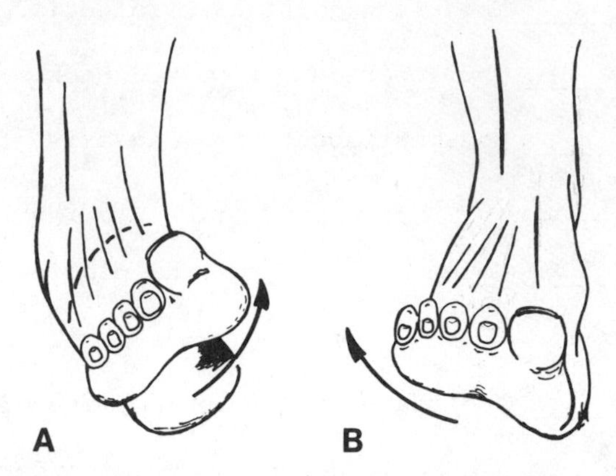

6. E. Movement of a bone away from the midline is called abduction, whereas movement towards the midline or returning to the anatomical position is called adduction.

7. A. Gout or gouty arthritis is attributed to an error in purine metabolism which produces an increase in the concentration of uric acid in the blood. Uric acid crystals become deposited in joints such as the knee and swelling and pain ensue.

8. C. Osteoarthritis, a chronic degenerative disease of weight-bearing joints, is characterized by a degeneration of the articular cartilage, a small amount of inflammation, an overgrowth of bone at the margins of the joint, and often twisting and enlargement of the affected appendage. It usually occurs after 40 years of age and seldom cripples unless the hip joint is seriously affected.

9. E. Inflammation of a synovial bursa is called bursitis. Bursitis may occur as a result of an injury to the joint, rheumatoid inflammation, gout, or bacterial infections. Chronic bursitis may eventually result in the deposition of calcium in the sac and may produce severe pain on movement. A common form of bursitis is olecranon bursitis or tennis elbow.

10. D. Rheumatoid arthritis is an inflammation of the synovial membrane of unknown origin which may occur at any age. The inflamed membrane is infiltrated by lymphocytes and plasma cells, producing a rheumatoid factor that acts as an antibody to the chemicals produced by the inflamed membrane. This antibody further aggravates the situation and the lysosomes of dying cells are released and tend to destroy the articular cartilage and the surrounding joint tissues. This is a very painful and progressively debilitating disease which may be treated by surgery or drug therapy.

7. MUSCLE TISSUE AND ITS PHYSIOLOGY

LEARNING OBJECTIVES

STRUCTURE AND FUNCTION

Be Able To:

- Describe the general functions of muscle tissue.
- Give the distinguishing features of skeletal, cardiac, and smooth muscle, paying particular attention to their: location in the body, microscopic structure, gross arrangement, and innervation.
- Distinguish between the terms somatic, visceral, efferent (motor), and afferent (sensory) as they pertain to nerve fibers.
- Describe the fundamental property of excitability (irritability) for muscle tissue, including the following topics: resting membrane potential, action potential, depolarization, repolarization, propagation of membrane potentials, rhythmicity, stimuli, neuromuscular transmission, and the motor unit.
- Discuss the mechanism of muscle contraction, paying particular attention to molecular changes, energy sources, strength, and the resultant production of heat.
- Define the following terms as they pertain to muscle: isotonic, isometric, and tetanic contractions, hypertrophy, atrophy, fatigue, fasciculation, fibrillation, and rigor mortis.
- Describe the additional fundamental properties of tonus, extensibility, and elasticity as they pertain to the three types of muscle.

SKELETAL MUSCLES

Be Able To:

- Define the following terms related to skeletal muscle: origin, insertion, belly, tendon, aponeurosis, agonist, antagonist, fixator, and synergist.
- Describe the attachments and action of the following muscles of the face: frontalis, orbicularis oculi, zygomaticus, levator palpebrae superioris, buccinator, masseter, temporalis, and platysma.

- Give the attachments and function of the muscles of mastication and the extrinsic muscles of the tongue.

- Give the cranial nerve(s) involved in controlling the following groups of muscles of the head: mastication, facial expression, and extrinsic muscles of the tongue.

- Describe the major function and attachments of the following muscles of the neck: trapezius, sternocleidomastoid, scalenes, stylohyoid, digastric, mylohyoid, geniohyoid, omohyoid, sternohyoid, sternothyroid, and thyrohyoid.

- Determine from the following list which muscles form the abdominal and thoracic walls and the floor of the pelvis, giving their general attachments and functions: diaphragm, external intercostal, external oblique, external anal sphincter, internal oblique, internal intercostal, levator ani, quadratus lumborum, rectus abdominis, and transversus abdominis.

- Describe the location and functions of the erector spinae and transversospinalis muscles of the back.

- Describe the general attachments and action of the following muscles involved in the movement of the shoulder and the arm: deltoid, infraspinatus, latissimus dorsi, levator scapulae, pectoralis major, pectoralis minor, rhomboids, serratus anterior, trapezius, teres major, teres minor, and supraspinatus.

- Describe each of the following muscles or functioning muscle groups of the arm, forearm, wrist, and hand by listing their attachments, locations, and general actions: biceps brachii, brachialis, triceps brachii, pronator and supinator groups, flexor and extensor groups, and thenar and hypothenar groups.

- Describe each of the following muscles or functioning muscle groups of the thigh and leg by listing their attachment, location, and general actions: gluteal group, psoas major and iliacus (iliopsoas), adductor group, sartorius, quadriceps femoris and hamstring group.

- Describe the location, general attachments and actions of the following muscles as they pertain to the leg and foot: gastrocnemius, soleus, tibialis anterior and posterior, flexor and extensor digitorum longus, flexor and extensor hallucis longus, and peroneus brevis and longus.

RELEVANT READINGS

Anthony & Kolthof - Chapter 6.
Chaffee & Greisheimer - Chapter 4.
Crouch & McClintic - Chapter 9.
Jacob, Francone & Lossow - Chapter 7.
Landau - Chapters 6 and 7.
Tortora - Chapters 9 and 10.

FIVE-CHOICE COMPLETION QUESTIONS

DIRECTIONS: Each of the following questions or incomplete statements is followed by five suggested answers or completions. SELECT THE SINGLE BEST ANSWER in each case and then circle the appropriate letter at the lower right of each question.

1. WHICH OF THE FOLLOWING IS NOT CORRECT FOR CARDIAC MUSCLE CELLS?
   A. Contain peripherally placed nuclei
   B. Form a functional syncytium
   C. Have striations similar to skeletal muscle
   D. Interdigitate at intercalated discs
   E. Contract automatically without a stimulus

   A B C D E

2. SMOOTH MUSCLE IS NOT FOUND IN THE WALL OF THE:
   A. Aorta
   B. Gallbladder
   C. Parotid duct
   D. Heart
   E. Stomach

   A B C D E

3. SKELETAL MUSCLE CELLS ARE STIMULATED BY:
   A. Distention
   B. Nerve impulses
   C. Mechanical pressure
   D. Changes in temperature
   E. None of the above

   A B C D E

4. AT THE NEUROMUSCULAR JUNCTION, THE END PLATE OF THE NERVE FIBER RELEASES A CHEMICAL SUBSTANCE THAT EXCITES THE SKELETAL MUSCLE FIBER. THIS SUBSTANCE IS CALLED __________.
   A. Acetoacetic acid
   B. Acetylcholine
   C. Cholinesterase
   D. Norepinephrine
   E. None of the above

   A B C D E

5. NERVE IMPULSES REACH SKELETAL MUSCLE CELLS BY THE ______ NERVE FIBERS.
   A. Somatic efferent
   B. Somatic afferent
   C. Visceral efferent
   D. Visceral afferent
   E. None of the above

   A B C D E

6. WHICH OF THE FOLLOWING IS INCORRECT FOR A MYOFIBRIL?
   A. Forms a component of a skeletal muscle fiber
   B. Are longitudinally arranged in the cell
   C. Contains three types of parallel myofilaments
   D. Is striated because of the arrangement of the myofilaments
   E. Cross-bridges extend between the different types of myofilaments

   A B C D E

7. THE MATRIX SURROUNDING THE MYOFILAMENTS OF A MUSCLE CELL IS CALLED THE ______.
   A. Sarcolemma
   B. Sarcomere
   C. Sarcoplasm
   D. Syncytium
   E. None of the above

   A B C D E

8. THE CONNECTIVE TISSUE SURROUNDING AN INDIVIDUAL MUSCLE FIBER IS CALLED THE ______.
   A. Endomysium
   B. Epimysium
   C. Perimysium
   D. Sarcolemma
   E. None of the above

   A B C D E

9. THE SUBSTANCE PRODUCED BY THE METABOLIC PROCESSES OF SKELETAL MUSCLE IN THE ABSENCE OF A SUFFICIENT OXYGEN SUPPLY IS CALLED:
   A. Glucose
   B. Lactic acid
   C. Acetylcholine
   D. Ethyl alcohol
   E. Carbon dioxide

   A B C D E

------------------------ ANSWERS, NOTES AND EXPLANATIONS ------------------------

1. A. The cardiac muscle fiber cell is long and cylindrical with a centrally placed nucleus. The individual fibers interdigitate at sites of union called the intercalated discs. Since the cells function as if they were one single sheet or syncytium they are said to form a 'functional' syncytium. The cardiac muscle fibers are striated, similar to skeletal muscle but function more like smooth muscle cells in that they are involuntary. Cardiac muscle cells contract independent of any external stimulus and the nerves found in the heart are solely involved in controlling the rate and force of contraction.

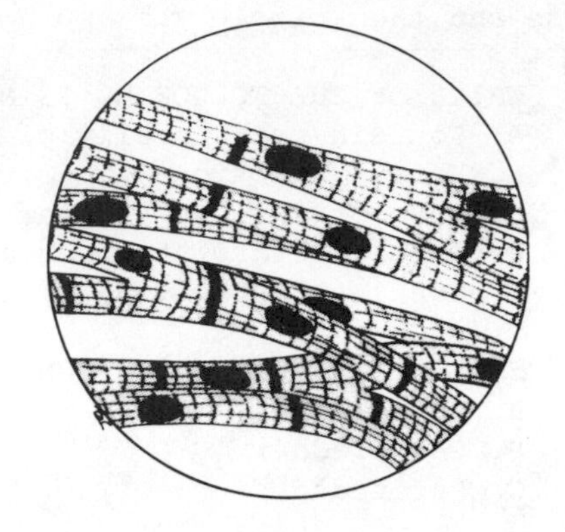

2. D. The heart wall consists of cardiac muscle, not smooth muscle. Smooth or visceral muscle is found in the walls of the larger blood vessels, including the aorta; the walls of the gallbladder, stomach and other tubular components of the gastrointestinal tract; and in the walls of the ducts of the larger glands of the body.

3. B. All three types of muscle will respond to nerve impulses, but nerve impulses are the only stimulus to which skeletal muscle will respond. The failure of nerve impulses to reach skeletal muscle renders it inactive and prolonged inactivity will cause it to undergo atrophy. In addition to nerve impulses, smooth muscle will also respond to distention, temperature changes, and to chemical changes in the blood, related to hormones, drugs, salts, and other substances. Smooth muscle, found lining the gut wall, has a certain degree of rhythmicity or self-excitatory property which produces peristalsis. Cardiac muscle on the other hand has a well developed inherent rhythmicity whose rate and force of contraction may be altered by a nerve impulse. Electrical, thermal, chemical, and mechanical stimuli may also moderate the rhythmicity of cardiac muscle.

4. B. The chemical substance released by the motor end plate of a nerve to stimulate skeletal muscle contraction is acetylcholine. This substance initiates contraction of the muscle fiber. Another substance, cholinesterase, is released within a fraction of a second after the acetylcholine has stimulated the muscle. Cholinesterase quickly destroys acetylcholine and prevents re-excitation of the muscle fiber before another impulse can be transmitted from the nerve. A similar situation occurs in the stimulation of both smooth and cardiac muscle.

5. A. Nerve impulses reach skeletal muscle cells by the somatic efferent (motor) nerve fibers. These motor fibers, conducting impulses away from the central nervous system (CNS), stimulate the skeletal muscle cells to contract. Afferent or sensory fibers conduct impulses towards the CNS; keeping it informed about the degree of contraction of all types of muscles found in the body. The term visceral refers to muscle (smooth or cardiac) found in the viscera or body organs. Afferent (sensory) visceral fibers would therefore relay information to the CNS from the smooth and cardiac muscle cells. Often excessive contraction or stretching of smooth muscle cells will produce pain that reaches the

brain by these afferent visceral fibers. The visceral efferent or autonomic nerve fibers relay messages from the CNS to the smooth or cardiac muscle cells. These messages modulate the automatic (rhythmic) contractions that are inherent in these two types of muscle.

6. C. The myofibrils are longitudinally arranged fibers located within the skeletal muscle cell. These myofibrils are made up of two types of parallel strands of myofilaments and are arranged to give a striated effect within a segmental unit called a sarcomere. The myofilaments are of two types: one thick, composed of the protein myosin and the other thin, composed of the protein actin. The myosin myofilament is arranged so that it is surrounded by six thinner actin myofilaments and similarly a thin actin myofilament is surrounded by three myosin myofilaments (as seen in the diagram). Chemical evidence, produced during contraction, suggests that there are cross-bridges extending between the two different types of myofilaments.

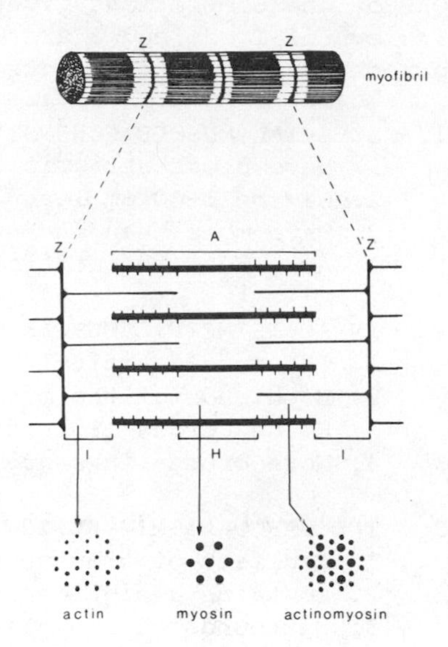

7. C. The cytoplasmic matrix surrounding the myofilaments of the muscle cell is called the sarcoplasm. This sarcoplasm contains the usual organelles and inclusions that one would typically find in other cells of the body.

8. A. Each muscle fiber is enclosed within a delicate connective tissue sheath or endomysium (endo = within). A large number of muscle fibers form a fasciculus (bundle) which is surrounded by a connective tissue sleeve called the perimysium (peri = around). Many of these fasciculi unite to form a complete skeletal muscle which itself is surrounded by a connective tissue layer or epimysium (epi = upon). These connective tissue elements surrounding the components of a skeletal muscle, provide support and nourishment, and help to segmentalize the pull of the individual fibers.

9. B. The aerobic and anaerobic (lack of molecular oxygen) breakdown of carbohydrates, the splitting of adenosine triphosphate (ATP) to adenosine diphosphate (ADP), the hydrolysis of phosphocreatine, and the oxidation of free fatty acids, all provide sources of energy for muscular contraction under differing conditions. During moderate muscular activity sufficient oxygen is made available from breathing, so that oxygen consumption is proportionate to the energy expended and all the energy needs are met by the aerobic processes. However, during excessive muscle activity when there is a lack of sufficient quantities of oxygen, the aerobic resynthesis of the energy stores does not keep pace with their utilization. During this period of insufficient oxygen, pyruvic acid formed from the breakdown of glucose does not enter the TCA cycle but instead is converted to lactic acid. In this state of oxygen debt, the excessive accumulation of lactic acid decreases the activity of the muscle cells by depressing both their irritability and contractility. This state of muscle, where lactic acid accumulates and inhibits contraction, is called muscle fatigue. After a period of rest and rapid breathing, lactic acid is oxidized back to pyruvic acid and glycogen.

MULTI-COMPLETION QUESTIONS

DIRECTIONS: In each of the following questions or incomplete statements, ONE OR MORE of the completions given is correct. At the lower right of each question, circle A if 1, 2, and 3 are correct; B if 1 and 3 are correct; C if 2 and 4 are correct; D if only 4 is correct; and E if all are correct.

1. SKELETAL MUSCLE FIBERS:
   1. Have cross striations
   2. May be 1-40 mm long
   3. Have more than one nucleus
   4. Are under the control of the will

   A B C D E

2. IN AN ISOMETRIC MUSCLE CONTRACTION:
   1. The muscle actively shortens against a load
   2. Both ends of the muscle are fixed
   3. No new cross-links are formed
   4. More cross-links are formed

   A B C D E

3. THE SARCOPLASMIC RETICULUM:
   1. Consists of transverse sarcotubules
   2. Contains a high concentration of calcium ions
   3. Surrounds each myofibril
   4. Includes attached ribosomes

   A B C D E

4. THE FUNDAMENTAL PHYSIOLOGICAL PROPERTIES OF MUSCLE INCLUDE:
   1. Contractility
   2. Elasticity
   3. Excitability
   4. Extensibility

   A B C D E

5. IN CONTRACTION OF SKELETAL MUSCLE WHICH OF THE FOLLOWING STATEMENTS IS(ARE) TRUE?
   1. The A band remains constant
   2. The H band increases in length
   3. The I band decreases in length
   4. The distance between the Z bands remains constant

   A B C D E

6. WHICH OF THE FOLLOWING IS(ARE) CONSIDERED A FUNCTION OF SKELETAL MUSCLE?
   1. Major producer of heat
   2. Serves as an endocrine organ
   3. Contracts to produce movement
   4. Protects vital organs

   A B C D E

7. CHEMICAL CONSTITUENTS OF MUSCLE TISSUE INCLUDE:
   1. 20 per cent protein
   2. Mostly water
   3. Small amounts of glucose and glycogen
   4. Adenosine triphosphate (ATP)

   A B C D E

8. SMOOTH MUSCLE IS:
   1. Found in the walls of viscera
   2. Considered involuntary
   3. Nonstriated
   4. Arranged in sheets

   A B C D E

9. THE MAIN SOURCE(S) OF ENERGY IN SKELETAL MUSCLE CONTRACTION IS(ARE) FROM:
   1. ATP
   2. Phosphocreatine
   3. Free fatty acids
   4. $NADP^+$

   A B C D E

A	B	C	D	E
1,2,3	1,3	2,4	only 4	all correct

10. WHEN A MUSCLE FIBER IS FIRST STIMULATED THERE IS A(AN):
    1. Movement of potassium out of the cell
    2. Rapid movement of sodium ions into the cell
    3. Immediate repolarization of the action potential
    4. Change in the electrical potential of its membrane

A B C D E

------------------------ ANSWERS, NOTES AND EXPLANATIONS ------------------------

1. E, <u>All are correct</u>. Skeletal muscle (cells) fibers are elongated cylinders with transverse (cross) striations. The multinucleated cells range from 1-40 mm long with their nuclei peripherally arranged. Skeletal muscle is considered to be voluntary or under the control of the will.

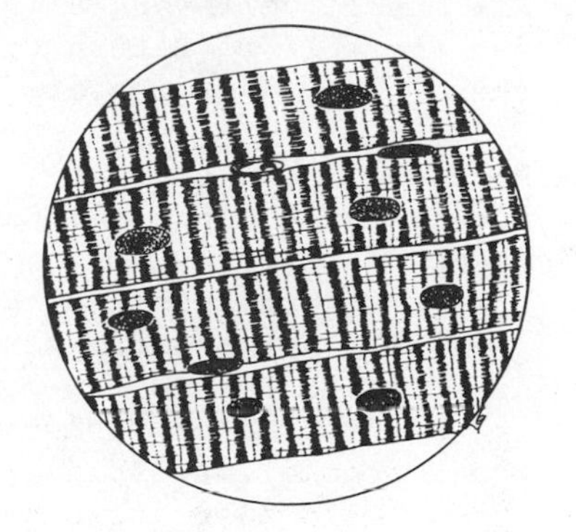

2. C, <u>2 and 4 are correct</u>. In muscle contractions of any type there is an increase in the number of cross-links formed. However, since muscle tissue has elastic and viscous elements which are in series with the contractile mechanism it is possible for contraction to occur without an appreciable decrease in the length of the whole muscle. This type of contraction is called isometric (same length) contraction. An isotonic (same tension) contraction is where there is contraction against a constant load and the ends of the muscle come closer together.

3. <u>A, 1, 2, and 3 are correct.</u> The sarcoplasmic reticulum is an organelle found within the sarcoplasm of striated (skeletal and cardiac) muscle cells. This reticulum consists of a series of tubules, not unlike the endoplasmic reticulum of other cells but they are not associated with ribosomes. The longitudinally oriented sarcotubules (A in the diagram) anastomose in the H band and are continuous with the paired transversely running terminal cisternae (B). On top of these cisternae lies the T-tubule (C). These three transverse tubules are collectively called the triad (D). The T-tubule does not communicate with the cisternae but opens into the extracellular space surrounding the cell. The triad is located at the junction of the A and I bands of the striated muscle. The function of the T-tubule is the rapid transmission of the action potential to all of the fibrils in the muscle. It triggers the release of calcium ions

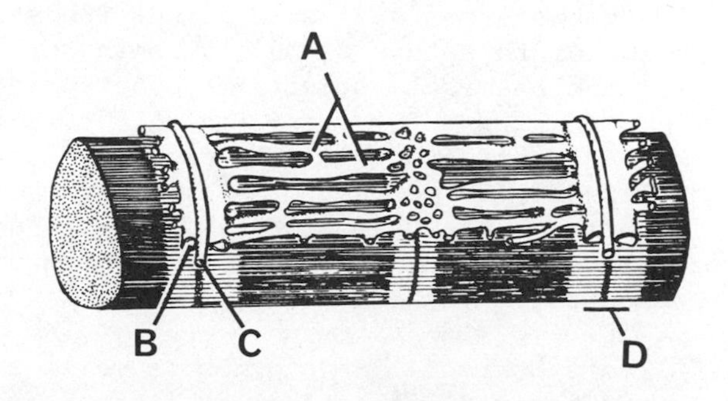

from the terminal cisternae into the sarcoplasm. This sudden increase of calcium activates the contractile mechanism of the myofibrils and the sarcomere shortens.

4. E, All are correct. All types of muscle tissue have the following fundamental properties of contractility, elasticity, excitability, extensibility, and tonus. When properly stimulated, all muscle cells have the capacity to contract and the very fact that these cells can respond to a stimulus is referred to as excitability. Normal muscle cells are said to always be in a steady state of partial contraction to some degree and this property is called tonus. All muscle cells are also capable of being stretched when force is applied, and are able to return to their original state when the force is released; these two properties are called extensibility and elasticity, respectively.

5. B, 1 and 3 are correct. In contraction of skeletal muscle the A band remains constant in its length (compare figures (1) and (2)). The Z lines move closer together and so the sarcomere becomes contracted. The H and I bands both become smaller, in fact the H band may completely disappear. In the stretching of skeletal muscle the sarcomere would become longer, and therefore the I and H bands would also increase. However, the A band remains the same.

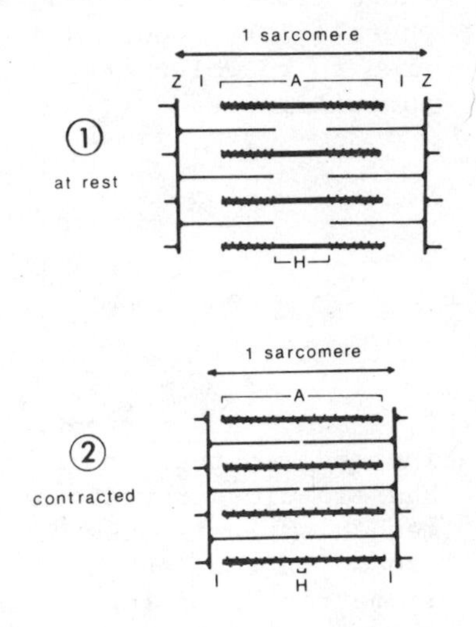

6. B, 1 and 3 are correct. An important function of skeletal muscle is to produce movement of the body as a whole or its parts. Such movements are vital for a normal active life. Skeletal muscle contraction also allows the body to assume various postures such as one finds in standing or sitting. The third function of this type of muscle is the production of great quantities of heat during the chemical activity of contraction.

7. E, All are correct. The main constituents of muscle are water (75 per cent) and proteins (20 per cent). Actin and myosin are the main contractile proteins of muscle. Carbohydrates (glucose and glycogen), lipids, inorganic salts and the nonprotein nitrogenous compounds of adenosine triphosphate (ATP) and phosphocreatine make up the remaining five per cent.

8. E, All are correct. Smooth muscle cells are long and tapered without any striations as found in the other types of muscle. Smooth muscle is located in the wall of the hollow organs of the gastrointestinal tract and the bladder and therefore is often called visceral muscle. It is also found in the walls of the larger blood vessels. The smooth muscle of many hollow structures is found in sheets or layers encircling the lumen. Here the smooth muscle is involved in moving substances along these hollow structures. Smooth

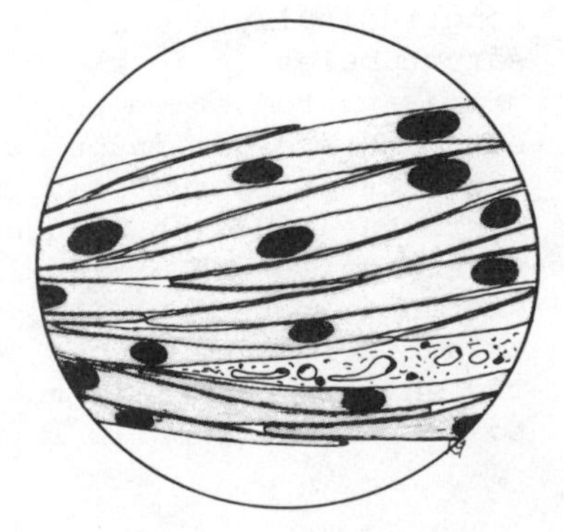

muscle cells are considered to be involuntary or not subject to control by the conscious will.

9. A, <u>1, 2, and 3 are correct</u>. The hydrolysis of ATP and phosphocreatine are major sources of energy for muscle contractions. Skeletal muscle also takes up some free fatty acids from the blood and oxidizes them to produce CO_2, H_2O, and ATP. The major sources of ATP production for muscle contraction comes from the aerobic and anaerobic breakdown of glucose. $NADP^+$ is a coenzyme which serves as a hydrogen receptor in some oxidation processes.

10. C, <u>2 and 4 are correct</u>. When a muscle cell is first stimulated there is a change in the permeability of its cell membrane to sodium. Sodium ions flow into the cells and this changes the electrical potential of its membrane from negative to positive on its internal surface and from positive to negative on its external surface. This change in the action potential is called depolarization. Within a split second of depolarization the membrane becomes relatively impermeable to sodium ions and there is a rapid movement of potassium out of the cell, which re-establishes the original electrical (resting) potential. This re-establishment of the resting membrane potential is called repolarization. Once a muscle cell is stimulated this depolarization and repolarization of the cell membrane cycles down the length of the membrane and propagates an impulse for the contraction of the muscle fiber.

FIVE-CHOICE ASSOCIATION QUESTIONS

DIRECTIONS: Each group of questions below consists of a numbered list of descriptive words or phrases accompanied by a diagram with certain parts indicated by letters, or by a list of lettered headings. For each numbered word or phrase, SELECT THE LETTERED PART OR HEADING that matches it correctly. Then insert the letter in the space to the right of the appropriate number. Sometimes more than one numbered word may be correctly matched to the same lettered part or heading.

1. _____ Stimulates skeletal muscle
2. _____ Skeletal muscle wasting
3. _____ An increase in muscle fiber size
4. _____ A repetitive discharge of action potentials
5. _____ A resistance to stretching
6. _____ Uncoordinated contraction of muscle fibers
7. _____ Sustained contractions with no apparent relaxation
8. _____ The accumulation of excessive amounts of lactic acid produces

A. Tetanic contraction
B. Rhythmicity
C. Hypertrophy
D. Motor unit(s)
E. None of the above

------------------------- ANSWERS, NOTES AND EXPLANATIONS -------------------------

1. D. The term motor unit refers to the muscle fibers innervated by the axon of a single motor neuron. The axon of a neuron branches to supply many muscle fibers; the exact number varies from neuron to neuron. In muscles involved in fine movements (eye or hand) there are 3-6 muscle fibers per motor unit, whereas those muscles involved in gross movement, such as the back, contain as many as 160 fibers per motor unit.

2. E. Atrophic shrinkage, death and disappearance of muscle cells occurs under a variety of circumstances. Localized muscle atrophy results from the interference with the individual motor units supplying the muscle. Minute local involvement of fibers often does not produce an appreciable muscle loss, since adjacent unaffected fibers will undergo compensatory hypertrophy. In generalized involvements, large numbers of muscle cells die and the affected muscles become grossly shrunken and flabby.

3. C. Muscle cells may adapt to an overwork situation by increasing their size. This adaptive process is called hypertrophy and can be seen in both skeletal and cardiac muscle. Increased work of both skeletal (as in the case of an athlete) and cardiac muscle (in a patient with an elevated blood pressure) increases the size of the individual muscle cells without an increase in the number of cells.

4. B. A repetitive discharge of action potentials is called rhythmicity. When the threshold for stimulation is sufficiently low, automatic depolarization is quickly followed by repolarization. This phenomenon can take place in all excitable tissues (i.e., muscle and nerve). These discharges normally take place in cardiac and some smooth muscle cells. Similar discharges may take place in skeletal muscle and nerve cells, but such discharges signify some abnormality.

5. E. Tonus is a property of muscle which can be defined as its resistance to stretch. It is present at all times and is due to the taut connective tissue fibers within and around the muscle cells and also to the elasticity and turgor of the tissue. It is not due to any electrical stimuli by motor units, since a muscle at complete rest has no electrical activity as witnessed by electromyography. In smooth muscle, tonus is highly sensitive to its chemical environment.

6. E. The uncoordinated contraction of individual muscle fibers is called fibrillation and fails to produce contraction of the whole muscle. Fibrillation occurs in skeletal and cardiac muscle, but usually is not observable in skeletal muscle. In cardiac muscle such an uncoordinated contraction of the heart appears as a state of continuous convulsive movement. If the ventricles of the heart are involved death occurs quickly.

7. A. Tetanic (complete) contraction (tetanus) is a result of a continual barrage of closely spaced stimuli, producing a sustained contraction with no relaxation. During a complete tetanus, the tension developed is about four times that of an individual twitch contraction. An incomplete tetanus is when there are periods of complete relaxation between the summated stimuli.

8. E. The accumulation of excessive amounts of lactic acid (oxygen debt) is a key factor in the production of muscle fatigue. Such fatigue causes a marked reduction in the strength of the contraction. The build-up of lactic acid decreases the irritability of the muscle fibers and interferes with their ability

to contract. Rest and rapid breathing allows the lactic acid to be converted back to pyruvic acid, thus removing the oxygen debt.

SKELETAL MUSCLES

F I V E - C H O I C E   C O M P L E T I O N   Q U E S T I O N S

DIRECTIONS: Each of the following questions or incomplete statements is followed by five suggested answers or completions. SELECT THE SINGLE BEST ANSWER in each case and then circle the appropriate letter at the lower right of each question.

1. A MUSCLE CONTRACTING TO OPPOSE THE ACTIONS OF ANOTHER MUSCLE IS CALLED A(AN) ______.
   A. Agonist
   B. Antagonist
   C. Synergist
   D. Fixator
   E. None of the above

   A B C D E

2. THE EXTRINSIC MUSCLES OF THE TONGUE ARE INNERVATED BY THE __________ NERVE.
   A. Facial
   B. Hypoglossal
   C. Oculomotor
   D. Trigeminal
   E. Trochlear

   A B C D E

3. THE FIBERS OF THE OUTER LAYER OF THE INTERCOSTAL MUSCLES ARE ORIENTED IN THE SAME DIRECTION AS THE _______ MUSCLE.
   A. External oblique
   B. Internal oblique
   C. Rectus abdominis
   D. Transversus abdominis
   E. None of the above

   A B C D E

4. THE OMOHYOID MUSCLE IS INDICATED BY ______.

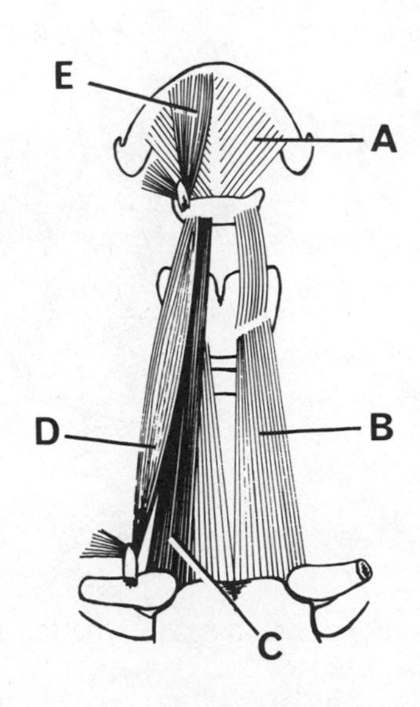

   A B C D E

5. THE MUSCLE ATTACHING TO THE HUMERUS, CLAVICLE, AND STERNUM IS THE:
   A. Deltoid
   B. Biceps brachii
   C. Brachialis
   D. Pectoralis major
   E. Subclavius

   A B C D E

SELECT THE SINGLE BEST ANSWER

6. AN EXTENSOR OF THE ELBOW JOINT IS THE _______ MUSCLE.
   A. Biceps brachii
   B. Brachialis
   C. Coracobrachialis
   D. Teres major
   E. Triceps brachii

A B C D E

7. THE MUSCLE LYING LATERAL TO THE RAMUS OF THE MANDIBLE IS THE ____.
   A. Buccinator
   B. Medial pterygoid
   C. Lateral pterygoid
   D. Temporalis
   E. None of the above

A B C D E

8. A MUSCLE FORMING A MAJOR PORTION OF THE POSTERIOR ABDOMINAL WALL IS THE:
   A. Diaphragm
   B. Internal oblique
   C. Quadratus lumborum
   D. Rectus abdominis
   E. None of the above

A B C D E

9. THE INSERTION OF A LOWER LIMB MUSCLE CAN BE DEFINED AS THE:
   A. Fleshy portion of the muscle
   B. Distal attachment of the muscle
   C. Proximal attachment of the muscle
   D. Area where connective tissue blends with the muscle
   E. None of the above

A B C D E

10. THE MUSCLE THAT IS BOTH A SUPINATOR OF THE FOREARM AND A FLEXOR OF THE ELBOW JOINT IS THE:
   A. Biceps brachii
   B. Brachialis
   C. Brachioradialis
   D. Supinator
   E. Triceps brachii

A B C D E

11. THE TEMPORALIS MUSCLE IS INDICATED BY ______.

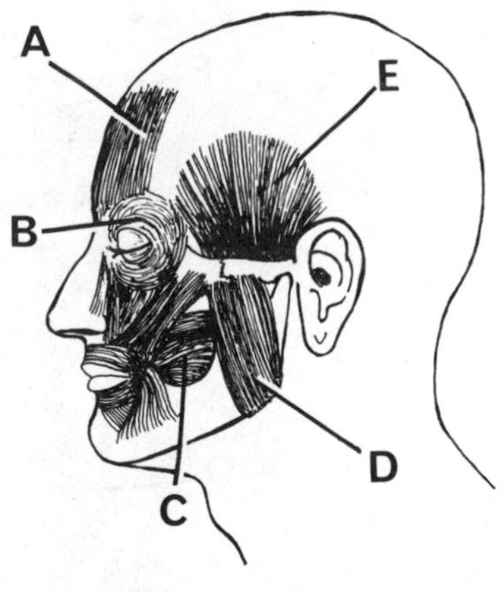

A B C D E

12. THE MUSCLE GROUP WHICH DORSIFLEXES THE ANKLE JOINT IS LOCATED:
   A. Anterolateral to the tibia
   B. On the dorsum of the foot
   C. Posterior to the tibia and fibula
   D. On the lateral side of the fibula
   E. None of the above

A B C D E

SELECT THE SINGLE BEST ANSWER

13. THE MUSCLE LIFTING THE UPPER EYELID IS THE ______.
    A. Frontalis
    B. Orbicularis oculi
    C. Temporalis
    D. Zygomaticus
    E. None of the above

    A B C D E

14. THE ANTERIOR COMPARTMENT OF THE FOREARM CONTAINS THE:
    A. Extensor muscles of the ulna
    B. Rotator muscles of the radius
    C. Abductor muscles of the humerus
    D. Adductor muscles of the middle finger
    E. Flexor muscles of the wrist and finger joints

    A B C D E

15. A MEDIAL ROTATOR OF THE HIP IS THE _____ MUSCLE.
    A. Biceps femoris
    B. Gluteus maximus
    C. Gluteus medius
    D. Sartorius
    E. None of the above

    A B C D E

16. WHICH OF THE FOLLOWING IS NOT A MUSCLE OF MASTICATION?
    A. Buccinator
    B. Medial pterygoid
    C. Lateral pterygoid
    D. Masseter
    E. Temporalis

    A B C D E

17. THE GROUP OF LARGE MUSCLES ON THE ANTERIOR THIGH IS THE:
    A. Quadriceps femoris
    B. Hamstrings
    C. Abductors
    D. Triceps femoris
    E. None of the above

    A B C D E

------------------------ ANSWERS, NOTES AND EXPLANATIONS ------------------------

1. B. An antagonist (opponent) is a muscle or a group of muscles that contracts to oppose and in doing so regulates the speed and power of another muscle or group of muscles. Flexor muscles often oppose extensor muscles and in this way control their movements. A prime mover (agonist) is a muscle responsible for a particular desired movement. A muscle contracting to eliminate some undesired movement of a prime mover is considered to be a synergist. Fixator muscles have little to do with the main action but rather stabilize a bone or fix its position. Postural muscles are considered to be fixators.

2. B. The extrinsic muscles of the tongue (hyoglossus, styloglossus and genioglossus) are innervated by the hypoglossal (XII) nerve. If the palatoglossus muscle is included (some feel it is a muscle of the soft palate) as an extrinsic muscle of the tongue then the pharyngeal plexus (CN X and XI) would also be included. The extrinsic muscles of the tongue determine its position. The genioglossus muscle is mainly a depressor of the tongue, but it also protrudes the tongue. The hyoglossus, styloglossus and the posterior part of the genioglossus muscles retract the tongue. The palatoglossus muscle lifts the tongue.

3. A. The external intercostal muscle fibers run downward and anteromedially from an upper rib to the rib below. The fibers of the external oblique muscle of the anterolateral abdominal wall run in the same direction as the external intercostal muscle. The fibers of the internal intercostal and internal oblique muscles run at right angles to the appropriate external muscles, respectively. The transversus abdominis muscle fibers run in a transverse plane around the abdominal wall, deep to the internal oblique muscle. The rectus abdominis is a flat strap-like muscle extending up from the pubic bone to the costal cartilage of the fifth, sixth and seventh ribs. The external and internal oblique

and the transversus abdominis muscles compress the abdominal contents and depress the ribs. The rectus abdominis muscle also compresses the abdomen and like the external and internal oblique (abdominal) muscles is also involved in flexing the vertebral column.

4. D. The omohyoid muscle has two bellies. Its superior belly arises from the hyoid bone and its central tendon passes through a sling attached to the clavicle. Its inferior belly passes posteriorly and attaches to the scapula. The omohyoid muscle depresses the hyoid bone. The mylohyoid (A in the diagram) and digastric (anterior belly)(E) muscles raise the hyoid, whereas the sternohyoid muscle (C) depresses the hyoid bone. The sternothyroid (B) muscle depresses the thyroid cartilage.

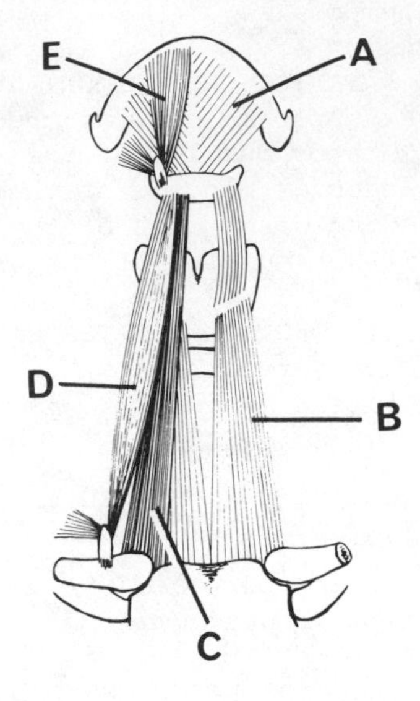

5. D. The pectoralis major muscle has a clavicular head extending from the front of the medial half of the clavicle and a sternal head arising from the sternum and the costal cartilages of the upper six ribs. Both heads insert into the bicipital groove of the humerus. The pectoralis major muscle adducts the arm. Its clavicular head also flexes and rotates the arm medially; whereas the sternal head depresses the arm and the shoulder.

6. E. The triceps brachii muscle is the major extensor of the elbow jont. It has three heads, one arising from the scapula (long head) and the other two arising from the posterior surface of the humerus (lateral and medial heads). The three heads converge and subsequently form a broad aponeurosis in the lower arm, which then inserts onto the olecranon process of the ulna. The long head of the triceps brachii is also involved in extension of the shoulder joint.

7. E. The muscle lying lateral of the ramus of the mandible is the masseter (A in the diagram). This muscle belongs to a group of four muscles that moves the lower jaw (muscles of mastication). The masseter, along with the temporalis (B) and medial pterygoid muscles, are involved in raising the lower jaw as in biting. The masseter arises from the zygomatic arch and inserts onto the outer surface of the ramus of the mandible.

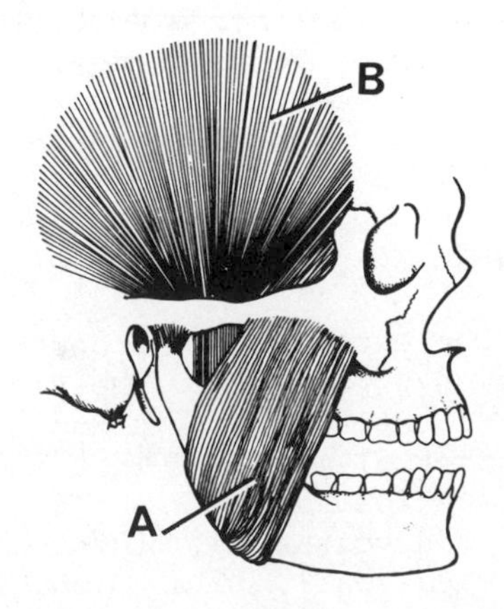

8. C. The muscles forming the posterior abdominal wall are the quadratus lumborum, psoas major, iliacus, and the lower part of the diaphragm. The major contributor to this wall is the quadratus lumborum muscle which extends from the last rib down to the iliac crest of the ilium. The bodies of the lumbar vertebrae, their intervening discs and the ilium of the hip bone also contribute to the formation of the posterior abdominal wall.

9. B. The insertion of a muscle in the upper or lower limbs is considered to be the distal attachment of the muscle. Some define the "moving end" of a muscle as the insertion, but use of this definition is poor since the moving end may change depending upon which bony structure is stabilized. The origin of a muscle in the limbs is considered to be the proximal end.

10. A. The biceps brachii originates by a long head from an area just above the glenoid fossa and by a short head attached to the coracoid process of the scapula. The muscle crosses the elbow joint anteriorly and inserts into the back of the tuberosity of the radius. When the forearm is pronated, the tendon of the biceps is wrapped behind the radius, and on contraction this tendon uncoils and supinates the hand or forearm. The biceps brachii can also help to flex the elbow joint against resistance. However, the main flexor of the elbow joint is the brachialis muscle.

11. E. The temporalis muscle (E in the diagram) is one of the muscles of mastication. The other muscles of mastication are the masseter (D) and the medial and lateral pterygoids. The frontalis (A), buccinator (C) and the orbicularis oculi (B) muscles are all considered muscles of facial expression.

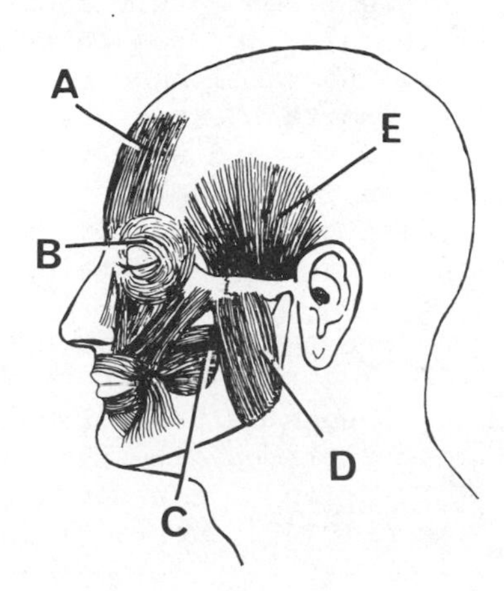

12. A. The muscle group that dorsiflexes the ankle joint is located anterolateral to the tibia. The major muscles in this group are the tibialis anterior, extensor digitorum longus, and permeus tertius. The tibialis anterior is the strongest dorisflexor of the ankle joint with the other two muscles assisting. The tibialis anterior is also an important invertor of the foot. The extensor digitorum longus muscle extends the lateral four toes, whereas the extensor hallucis longus extends the big toe.

13. E. The muscle responsible for raising the upper eyelid is the levator palpebrae superioris. This muscle arises from the roof of the orbit and by a broad aponeurosis inserts into the skin of the upper eyelid. It is innervated not by the facial nerve, but the oculomotor nerve located in the orbit.

14. E. The anterior compartment of the forearm contains the flexors of the wrist and fingers, and the pronators of the forearm. The six flexors of the forearm are arranged in three layers. The superficial layer consists of the flexor carpi radialis, palmaris longus, and flexor carpi ulnaris muscles acting on the wrist and carpal joints. The flexor digitorum superficialis muscle

forms the middle layer. It inserts onto the middle phalanges of the medial four fingers and acts on all of the intermediate joints in passing. The deep layer, formed by the flexor digitorum profundus and flexor pollicis longus muscle, inserts onto the terminal phalanges and also acts on all the intermediate joints. The pronator teres and pronator quadratus muscles are also deeply placed muscles and they pronate the forearm.

15. C. The gluteus medius muscle is a medial rotator and an abductor of the hip. The biceps femoris muscle is an extensor of the hip, whereas the gluteus maximus is an extensor and a lateral rotator of the hip. The sartorius is a weak abductor of the hip. Other medial rotators of the hip include the gluteus minimus and the tensor fascia lata.

16. A. The buccinator muscle is not a muscle of mastication, but helps to maintain the food between the molar teeth during mastication. It also helps move food out of the vestibule, the space found between the cheek and the lateral surfaces of the upper and lower teeth. The muscles of mastication include the masseter, temporalis, and the lateral and medial pterygoids.

17. A. The quadriceps femoris, of the anterior thigh, is a complex of four muscles which are not completely separated from each other. This complex includes the vastus medialis, vastus lateralis, vastus intermedius and rectus femoris muscles. The vasti muscles originate from the anterior surface of the femur and are important extensors of the knee joint. The rectus femoris originates from the pelvis and with the vasti inserts by tendons onto the patella of the knee. The rectus femoris is not only an extensor of the knee joint, but is also an important flexor of the hip joint. Another muscle found on the anterior surface of the thigh is the strap-like sartorius muscle.

MULTI-COMPLETION QUESTIONS

DIRECTIONS: In each of the following questions or incomplete statements, ONE OR MORE of the completions given is correct. At the lower right of each question, circle A if 1, 2, and 3 are correct; B if 1 and 3 are correct; C if 2 and 4 are correct; D if only 4 is correct; and E if all are correct.

1. SYNERGISTIC MUSCLES, WHEN THEY CONTRACT, ARE:
   1. Responsible for the desired movement
   2. Involved in positioning of the body
   3. Responsible for regulating the prime mover
   4. Eliminators of undesired movements

   A B C D E

2. THE ORBICULARIS ORIS IS:
   1. A muscle of the pharynx
   2. A muscle of facial expression
   3. An elevator of the hyoid bone
   4. Innervated by the seventh cranial nerve

   A B C D E

3. THE EXTRINSIC MUSCLES OF THE TONGUE INCLUDE THE:
   1. Hyoglossus
   2. Genioglossus
   3. Palatoglossus
   4. Styloglossus

   A B C D E

4. THE MUSCLES THAT "GUARD" THE SHOULDER JOINT INCLUDE THE:
   1. Infraspinatus
   2. Subscapularis
   3. Supraspinatus
   4. Teres minor

   A B C D E

A	B	C	D	E
1,2,3	1,3	2,4	only 4	all correct

5. THE ILIOPSOAS MUSCLE:
   1. Lies partly within the pelvis
   2. Consists of two separate muscles
   3. Flexes the hip joint
   4. Attaches to the lesser trochanter

   A B C D E

6. THE PERONEI MUSCLES OF THE LEG ARE:
   1. Located lateral to the fibula
   2. Plantar flexors of the foot
   3. Evertors of the foot
   4. Inserted into the calcaneus

   A B C D E

7. WHICH OF THE FOLLOWING STATEMENTS IS(ARE) CORRECT FOR THE MYLOHYOID MUSCLE?
   1. Attaches to the mandible
   2. Depresses the hyoid bone
   3. Forms the floor of the mouth
   4. Has a fusiform shapc

   A B C D E

8. THE HAMSTRING MUSCLES INCLUDE THE:
   1. Biceps femoris
   2. Semimembranosus
   3. Semitendinosus
   4. Adductor magnus

   A B C D E

9. THE RECTUS ABDOMINIS MUSCLE:
   1. Arises from the front of the fourth to sixth costal cartilages
   2. Helps compress the abdominal contents
   3. Attaches to the front of the pubis
   4. Is contained within a sheath

   A B C D E

10. THE STERNOCLEIDOMASTOID MUSCLE:
    1. Contracting bilaterally, flexes the head and neck
    2. Attaches to the temporal bone
    3. Inserts into the clavicle
    4. Is considered a strap muscle

    A B C D E

11. THE ORBICULARIS ORIS AND BUCCINATOR MUSCLES ARE INNERVATED BY WHICH OF THE FOLLOWING NERVES?
    1. Oculomotor
    2. Hypoglossal
    3. Trigeminal
    4. Facial

    A B C D E

12. THE MUSCLES OF THE ANTEROLATERAL ABDOMINAL WALL:
    1. Protect the viscera
    2. Flex the vertebral column
    3. Aid in all expulsive acts
    4. Are important in forced expiration

    A B C D E

13. THE DELTOID MUSCLE FUNCTIONS AS A(AN) _______ OF THE SHOULDER JOINT.
    1. Abductor
    2. Extensor
    3. Flexor
    4. Adductor

    A B C D E

14. THE GASTROCNEMIUS MUSCLE:
    1. Is a superficial muscle
    2. Has two heads
    3. Crosses two joints
    4. Plantar flexes the ankle joint

    A B C D E

A	B	C	D	E
1,2,3	1,3	2,4	only 4	all correct

15. WHICH OF THE FOLLOWING STATEMENTS IS(ARE) TRUE FOR THE ERECTOR SPINAE GROUP OF MUSCLES?
    1. Form a superficial group
    2. Consists of three longitudinal muscles
    3. Can produce lateral bending of the vertebral column
    4. Includes the semispinalis muscles

    A B C D E

16. THE HYPOTHENAR MUSCLES ARE:
    1. Short muscles of the hand
    2. Extrinsic to the hand
    3. Responsible for moving the little finger
    4. Found in the middle of the palm

    A B C D E

17. MUSCLES THAT ROTATE THE GLENOID CAVITY OF THE SCAPULA DOWNWARDS ARE THE:
    1. Subscapularis
    2. Trapezius
    3. Serratus anterior
    4. Levator scapulae

    A B C D E

18. THE MUSCLES OF THE POSTERIOR COMPARTMENT OF THE FOREARM:
    1. Are primarily flexors of the wrist or fingers
    2. Form deep and superficial layers
    3. Consist primarily of short strong muscles
    4. Originate in part from the lateral epicondyle of the humerus

    A B C D E

19. THE MEDIAL PTERYGOID MUSCLE:
    1. Inserts into the ramus of the mandible
    2. Has fibers running parallel to the zygomatic arch
    3. Raises the jaw
    4. Originates from the zygomatic arch

    A B C D E

------------------------ ANSWERS, NOTES AND EXPLANATIONS ------------------------

1. D, Only 4 is correct. A synergist is a muscle or group of muscles that contract to eliminate the undesired movements of an agonist or prime mover. An agonist is responsible for a desired movement. A muscle whose contractions are involved in positioning the body is called a fixator, whereas an antagonist is a muscle that regulates the speed and power of the prime mover.

2. C, 2 and 4 are correct. The orbicularis oris is a circular muscle lying within the upper and lower lips and encircling the mouth. This muscle contracts and constricts the orifice to the oral cavity. It is innervated by the facial nerve (cranial nerve VII) and is considered one of the muscles of facial expression.

3. E, All are correct. The extrinsic muscles of the tongue, those arising from structures outside the tongue, include the: hyoglossus, genioglossus, palatoglossus, and styloglossus muscles. Some consider the palatoglossus muscle to be more involved with the soft palate than the tongue and it is innervated by the pharyngeal plexus (CN X and XI). The other extrinsic and intrinsic muscles of the tongue are innervated by the hypoglossal nerve (CN XII). The intrinsic (skeletal) muscles of the tongue are involved in changing its shape and run in all directions and are called longitudinal, transversus, and verticalis muscles.

4. E, All are correct. The shoulder joint is guarded by four muscles that form the 'rotator cuff'. They are the supraspinatus (A in the diagram), subscapularis (B), teres minor (C), and infraspinatus (D) muscles. These four muscles provide a great deal of stability to the shoulder joint.

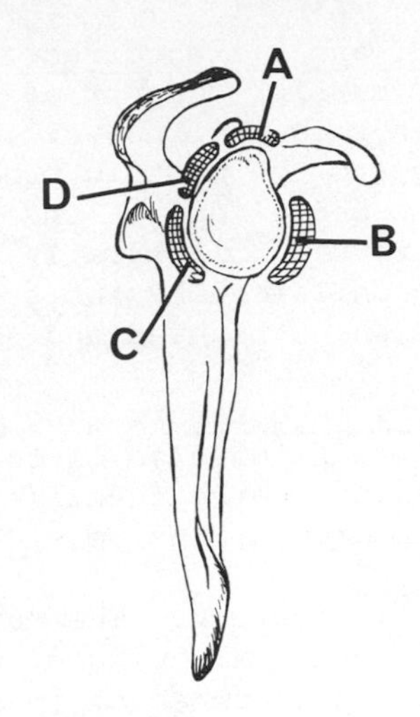

5. E, All are correct. The iliopsoas muscle consists of two muscles that share a common tendon of insertion and a common action. The psoas major and iliacus muscles form the iliopsoas. The iliacus muscle arises from the false pelvis, whereas the psoas major muscle arises from the lumbar vertebrae and their intervening discs. They both leave the pelvis and insert onto the lesser trochanter of the femur. They function as strong flexors of the hip joint.

6. B, 1 and 3 are correct. There are three peronei muscles of the leg: peroneus longus, brevis and tertius. Peroneus tertius is considered a part of the extensor digitorum longus muscle. The peroneus longus and brevis muscles lie lateral to the fibula and are the evertors of the foot. They both originate from the fibula. The peroneus brevis inserts onto the base of the fifth metatarsal and the peroneus longus inserts onto the lateral side of the base of the first metatarsal after having crossed beneath the arches of the foot.

7. B, 1 and 3 are correct. The mylohyoid, a flat muscle, attaches to the inner surface of the mandible and passes inferomedially to attach to the body of the hyoid bone. The mylohyoid muscle from each side unites to form a median raphe that extends from the hyoid bone to the mandible. This muscle forms the floor of the mouth and probably raises it during other movements involving the structures of the oral cavity.

8. A, 1, 2, and 3 are correct. The group of muscles lying behind the adductor magnus muscle of the posterior thigh are called the hamstrings. These muscles extend the hip or flex the knee or both. The muscles included in this group are the biceps femoris, semitendinosus, and semimembranosus. These muscles are extremely important in walking and running.

9. E, All are correct. The rectus abdominis is a paired muscle and extends from the front of the fourth, fifth, and sixth costal cartilages to the front of the pelvis. It is contained within a sheath formed by the aponeuroses of the lateral abdominal muscles. The rectus abdominis muscle helps to compress the abdominal contents and is important in the flexion of the trunk. It also helps to control hyperextension of the vertebral column and can elevate the pelvis upwards.

10. A, 1, 2, and 3 are correct. The sternocleidomastoid muscle is not considered a strap muscle. The strap (infrahyoid) muscles include the thyrohyoid, sternothyroid, sternohyoid, and omohyoid. The sternocleidomastoid muscle attaches to the mastoid process of the temporal bone and inserts by two heads, onto the medial end of the clavicle and the manubrium of the sternum. This muscle working bilaterally, against resistance, flexes the head and neck. Working unilaterally the sternocleidomastoid muscle rotates the chin away and upwards when the head and neck are flexed.

11. D, Only 4 is correct. The orbicularis oris and buccinator muscles are considered to be muscles of facial expression. All the muscles of facial expression receive their (efferent somatic) motor innervation from the facial (seventh cranial) nerve.

12. E, All are correct. The muscles of the anterolateral abdominal wall (external oblique, internal oblique, transversus abdominis, and rectus abdominis) are involved in a number of important functions. They function to protect the viscera from blows and are extremely important in increasing the intra-abdominal pressure in expulsive acts such as parturition, defecation, respiration, and others. Except for the rectus abdominis muscle, which plays little part in breathing, they are the most important muscles of forced expiration. These muscles are also involved in rotation and flexion of the vertebral column, control hyperextension of the trunk, and fix the thoracic cage during movements of the upper limbs.

13. E, All are correct. The deltoid muscle originates from the front of the lateral third of the clavicle, the lateral border of the acromion, and the crest of the spine of the scapula. It inserts into the deltoid tuberosity of the humerus. The deltoid can be split functionally into anterior, middle, and posterior fibers. The anterior fibers flex and medially rotate the shoulder joint, the middle fibers abduct it and the posterior fibers extend, adduct, and laterally rotate the shoulder joint.

14. E, All are correct. The gastrocnemius muscle along with the soleus muscle are important plantar flexors of the ankle joint. The gastrocnemius muscle is a powerful superficial muscle arising by two heads from the femur just above the medial and lateral condyles, respectively. Inferiorly these two heads fuse to form a flat tendon which contributes to the formation of the tendocalcaneus. The soleus and plantaris muscles also contribute to this tendon. The gastrocnemius muscle crosses both the knee and ankle joints. It can flex the knee joint or plantar flex the ankle joint. However, it cannot perform both of these actions at the same time, because of the shortness of its fibers.

15. A, 1, 2, and 3 are correct. The superficial erector spinae and the deep transversospinalis groups of muscles form the major intrinsic muscles of the back. The erector spinae group consists of three longitudinally oriented muscles and from lateral to medial they are the iliocostalis, longissimus, and spinalis muscles. The deeper transversospinalis group includes the semispinalis, multifidus, and rotatores muscles. The muscles of the superficial group are involved in lateral bending and extension of the vertebral column. The short deeper group of muscles are involved in local finite movements of the vertebral column.

16. B, 1 and 3 are correct. Three hypothenar muscles form the ball of the little finger; they include the abductor digiti minimi, opponens digiti minimi and flexor digiti minimi brevis. These three muscles originate from the flexor retinaculum situated in the palm and their bellies form the medial raised area (ball) of the hand, lying just medial to the palm. These muscles are named for the actions upon the little finger.

17. D, Only 4 is correct. The scapula is rotated so that the glenoid cavity faces downward by the levator scapulae (A in the diagram) and the two rhomboids (B). The glenoid cavity is rotated upwards by the upper (C) and lower fibers (D) of the trapezius and the serratus anterior (E) muscles. Rotation of the glenoid cavity upwards is necessary for maximum abduction of the arm.

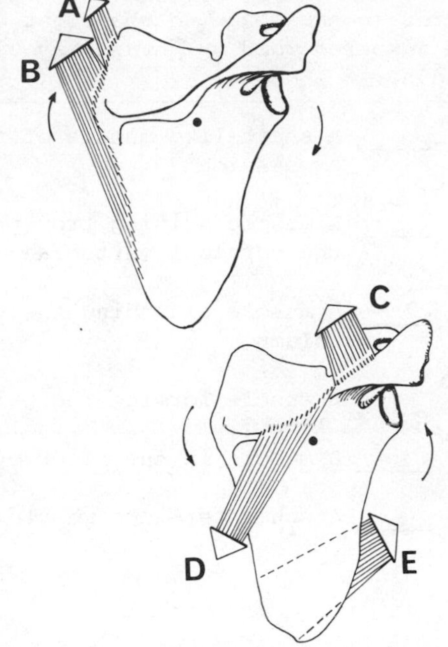

18. C, 2 and 4 are correct. The muscles forming the posterior compartment of the forearm are primarily long muscles, functioning as extensors of the wrist and fingers. The muscles of this compartment are arranged in two layers, one superficial and the other deep. The superficial layer of extensors consists of the extensor carpi radialis brevis and longus, extensor digitorum, extensor digiti minimi, and extensor carpi ulnaris. This superficial group originates for the most part from the lateral epicondyle of the humerus and inserts into the metacarpals of the hand. These extensors are important synergistic stabilizers of the wrist. The deep group of extensor muscles consists of the extensor pollicis longus and brevis, and the extensor indicis. Another deep muscle closely related to this deep group of extensors is the adductor pollicis longus.

19. B, 1 and 3 are correct. The medial pterygoid (A in the diagram), a muscle of mastication, is responsible for raising the lower jaw. It originates from the lateral pterygoid plate and inserts into the ramus of the mandible. The fibers of the medial pterygoid muscle run in the same direction as those of the masseter, whereas the fibers of the lateral pterygoid (B) muscle run parallel to the zygomatic arch.

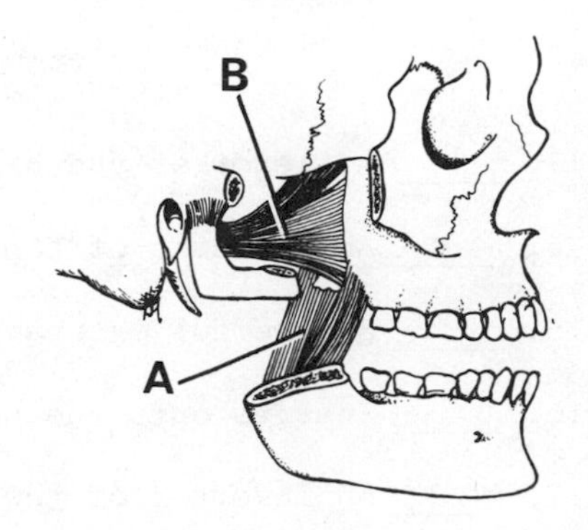

FIVE-CHOICE ASSOCIATION QUESTIONS

DIRECTIONS: Each group of questions below consists of a numbered list of descriptive words or phrases accompanied by a diagram with certain parts indicated by letters, or by a list of lettered headings. For each numbered word or phrase, SELECT THE LETTERED PART OR HEADING that matches it correctly. Then insert the letter in the space to the right of the appropriate number. Sometimes more than one numbered word or phrase may be correctly matched to the same lettered part or heading.

1. _____ A sheet-like muscle of facial expression
2. _____ A muscle arising from processes of the cervical vertebrae
3. _____ A muscle extending the vertebral column
4. _____ A muscle forming the pelvic floor
5. _____ A muscle in the floor of the mouth
6. _____ A sphincter-like muscle

A. Levator ani
B. Geniohyoid
C. Scalenus anterior
D. Platysma
E. None of the above

7. _____ A medial rotator of the shoulder joint
8. _____ A supinator of the forearm
9. _____ An abductor of the arm
10. _____ A flexor of the elbow joint
11. _____ A muscle attaching to the radius and ulna
12. _____ A flexor of the wrist joint

A. Supraspinatus
B. Brachialis
C. Latissimus dorsi
D. Pronator quadratus
E. None of the above

13. _____ A flexor of the hip joint
14. _____ An extensor of the knee joint
15. _____ A powerful extensor of the hip joint
16. _____ Inserts onto the patella
17. _____ An invertor of the foot
18. _____ A muscle that unlocks the knee joint

A. Vastus medialis
B. Gluteus maximus
C. Tibialis posterior
D. Iliopsoas
E. None of the above

ANSWERS, NOTES AND EXPLANATIONS

1. D. The platysma is a sheet-like muscle of facial expression. It lies below the skin and extends from the upper two ribs into the neck and inserts into the margin of the mandible. The platysma, like all the muscles of facial expression, is innervated by the facial nerve.

2. C. The scalenus anterior arises from the front of the transverse processes of the lower five cervical vertebrae. The larger scalenus medius arises from the posterior surfaces of the same cervical transverse processes as the scalenus anterior and from the axis. The scalenus anterior and posterior muscles insert into the top of the first and second ribs, respectively. An important function of these muscles is that they raise the upper two ribs during inspiration.

3. E. The muscles extending the vertebral column belong primarily to the erector spinae and transversospinalis groups.

4. A. The levator ani are paired muscles that form the floor of the pelvic cavity. Their fibers attach to the walls of the pelvis and blend in the midline surrounding the rectum. These muscles support the organs of the pelvis and contribute to the voluntary control of the rectum. They also contract vigorously during expulsive movements of the rectum, as do the abdominal wall muscles.

5. B. The geniohyoid muscle lies immediately above the mylohyoid muscle and thus forms a component of the floor of the oral cavity. This narrow muscle extends from the inner posterior surface of the mandible to the hyoid bone. The geniohyoid, mylohyoid, and the anterior belly of the digastric muscles raise the hyoid bone when the mandible is fixed and help to lower the mandible when the hyoid bone is fixed.

6. E. None of the muscles are sphincter-like muscles. Examples of sphincter muscles are the orbicularis oris, orbicularis oculi and the anal sphincters.

7. C. The medial rotators of the shoulder joint include the latissimus dorsi, pectoralis major, teres major, subscapularis, and the anterior fibers of the deltoid muscles.

8. E. The supinators of the forearm are the supinator and biceps brachii muscles. Both of these muscles wrap themselves around the radius during pronation and on their contraction produce the opposite action or supination.

9. A. The supraspinatus muscle is an important muscle in abduction of the humerus. This muscle is important at the beginning of abduction in that it pulls the head of the humerus into the glenoid fossa allowing the deltoid to exert its force. The deltoid is greatly hindered in abduction when the supraspinatus muscle is paralysed.

10. B. The brachialis muscle extends across the anterior surface of the elbow joint, from the humerus to the ulna. This muscle is the strongest flexor of the elbow joint and is assisted by the biceps brachii, brachioradialis and pronator teres muscles.

11. D. The pronator quadratus muscle runs transversely between the radius and the ulna just proximal to the wrist joint. This deep muscle pulls the radius over the ulna and pronates the hand or forearm. The pronator teres, passing

obliquely from the medial epicondyle region of the humerus to the anterior surface of the shaft of the radius, assists the pronator quadratus in its action.

12. E. Flexors of the wrist joint are the flexor carpi radialis and ulnaris muscles, and the palmaris longus muscle (when it is present).

13. D. The iliopsoas (psoas major and iliacus) muscle is a strong flexor of the hip joint. The psoas major because of its origin from the lumbar vertebrae is also a flexor of the vertebral column when the hip joint is stabilized.

14. A. The vastus medialis is one of four muscles contributing to the quadratus femoris of the anterior thigh. The three vasti (medialis, lateralis and intermedius) originate from the anterior femur and insert into the upper aspect of the patella. The rectus femoris, the last member of this group, originates from the anterior inferior spine of the ilium and inserts into the patella. All four muscles are powerful extensors of the knee joint. The rectus femoris, in addition to its function as an extensor of the knee, also flexes the hip joint.

15. B. The powerful extensor of the hip joint is the gluteus maximus muscle. This function is aided by the hamstring muscles (semitendinosus, semimembranosus, and biceps femoris) and the posterior fibers of the adductor magnus muscle. The other glutei muscles (medius and minimus) are medial rotators of the hip joint and do not extend it.

16. A. The vastus medialis muscle along with the other three muscles forming the quadratus femoris insert onto the upper aspect of the patella.

17. C. The tibialis posterior muscle is an invertor of the foot. It is the deepest muscle of the leg and originates from the posterior aspect of the tibia, fibula, and the interosseous membrane. The tendon of the tibialis posterior passes behind the medial malleolus and inserts by numerous slips into the base of several tarsal and metatarsal bones. The other invertor of the foot is the tibialis anterior.

18. E. The muscle responsible for 'unlocking' the knee is the popliteus. The knee, when it is fully extended, is also slightly laterally rotated or 'locked'. The popliteus originating from the lateral femoral condyle passes inferiorly across the posterior aspects of the knee to insert onto the tibia. It contracts, rotating the knee joint at the beginning of flexion.

REVIEW EXAMINATION OF PART 2

INTRODUCTORY NOTE: This review examination consists of 80 multiple-choice questions based on the learning objectives listed for the following chapters of this Study Guide and Review Manual: 4. THE SKIN AND OTHER MEMBRANES; 5. THE SKELETAL SYSTEM; 6. JOINTS AND THEIR MOVEMENTS; and 7. MUSCLE TISSUE AND ITS PHYSIOLOGY. Before beginning, tear out an answer sheet from the back of the book and read the directions on how to use it. The key to the correct response is on page 339.

FIVE-CHOICE COMPLETION QUESTIONS

DIRECTIONS: Each of the following questions or incomplete statements is followed by five suggested answers or completions. SELECT THE SINGLE BEST ANSWER in each case and then blacken the appropriate space on the answer sheet.

1. TO PRODUCE MUSCLE CONTRACTION WHICH OF THE FOLLOWING REACTANTS ARE REQUIRED?
   A. ADP, Ca^{++}, myofilaments
   B. ATP, Ca^{++}, myofilaments
   C. ADP, myofilamements, impulse
   D. ATP, Ca^{++}, myofilaments, impulse
   E. ADP, Ca^{++}, myofilaments, impulse

2. THE DELTOID TUBEROSITY IS INDICATED BY _______.

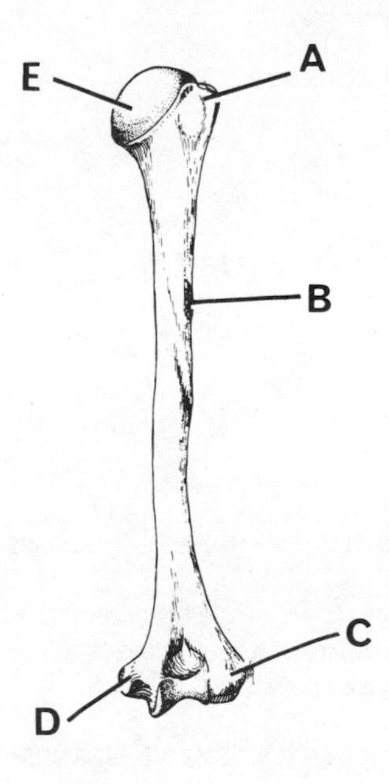

SELECT THE SINGLE BEST ANSWER

3. IN THE LIMBS THE PROXIMAL ATTACHMENT OF A MUSCLE IS CONSIDERED TO BE THE:
   A. Belly
   B. Aponeurosis
   C. Insertion
   D. Origin
   E. None of the above

4. A PLANE JOINT IS LOCATED BETWEEN THE:
   A. Two pubic bones
   B. Carpal bones of the hand
   C. Occipital and parietal bones
   D. Mandible and the temporal bone
   E. Head of the radius and the ulna

5. THE PERIOSTEUM OF THIS CROSS SECTION OF THE DIAPHYSIS OF A LONG BONE IS INDICATED BY _______.

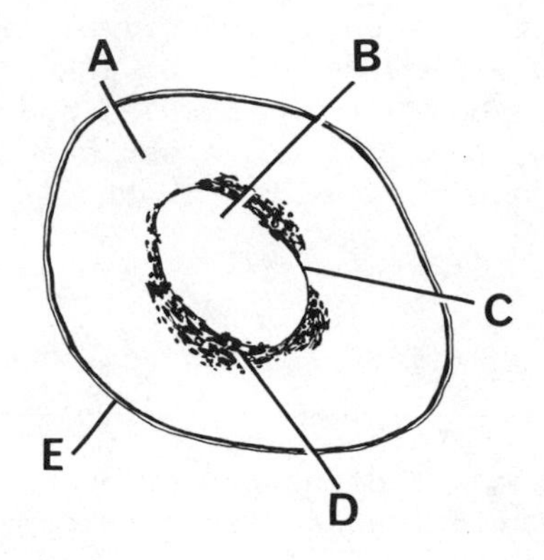

6. THE ARROW INDICATES THE:
   A. Femur
   B. Talus
   C. Tibia
   D. Fibula
   E. Calcaneus

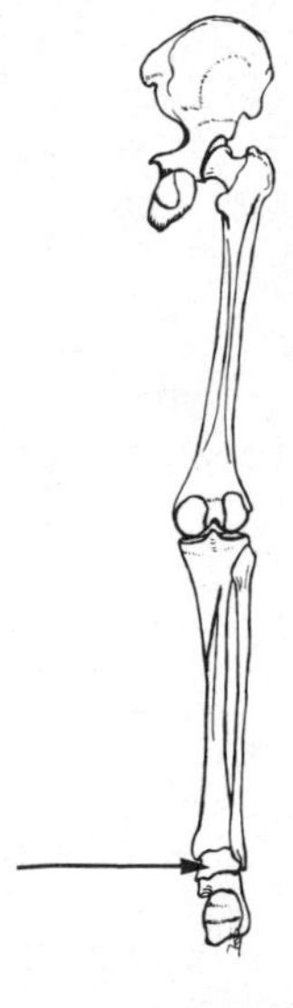

7. A MUSCLE PRIMARILY RESPONSIBLE FOR A DESIRED MOVEMENT IS CONSIDERED A(AN) _______.
   A. Agonist
   B. Antagonist
   C. Synergist
   D. Fixator
   E. None of the above

8. THE MODIFIED SWEAT GLANDS LOCATED IN THE EXTERNAL EAR CANAL ARE CALLED ______ GLANDS.
   A. Serous
   B. Mucous
   C. Mammary
   D. Salivary
   E. Ceruminous

SELECT THE SINGLE BEST ANSWER

9. THE HEAD OF THE RADIUS ARTICULATES WITH THE ______ OF THE HUMERUS.
   A. Head
   B. Capitulum
   C. Medial epicondyle
   D. Trochlea
   E. Lesser trochanter

10. THE BROAD MUSCLE ARISING FROM THE OUTER SURFACES OF THE LOWER EIGHT RIBS IS CALLED THE _______ MUSCLE.
   A. External oblique
   B. Internal oblique
   C. Quadratus lumborum
   D. Rectus abdominis
   E. Transversus abdominis

11. THE TEMPORAL BONE IS INDICATED BY _______.

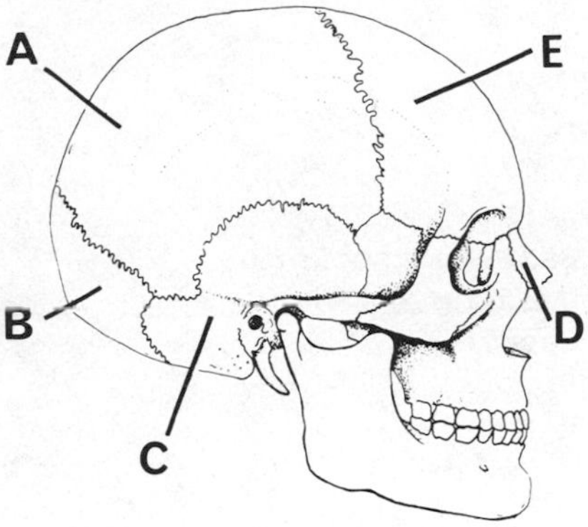

A
E
B
D
C

12. A SKIN CONDITION OF UNKNOWN ORIGIN, PRESENTING NON-INFECTIOUS PATCHES OF DRY SKIN IS CALLED ________.
   A. Acne
   B. Hives
   C. Nevus
   D. Psoriasis
   E. Decubitus ulcer

13. THE ARROW INDICATES WHICH OF THE FOLLOWING SUTURES?
   A. Coronal
   B. Lambdoidal
   C. Sagittal
   D. Squamosal
   E. None of the above

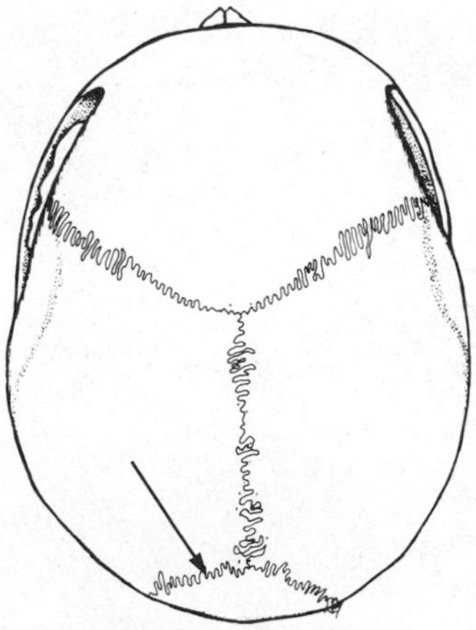

14. THE INFERIOR CONCHA IS A PART OF THE _____ BONE.
   A. Vomer
   B. Palatine
   C. Ethmoid
   D. Sphenoid
   E. None of the above

SELECT THE SINGLE BEST ANSWER

15. THIS IS A DIAGRAM OF THE _____.
    A. Axis
    B. Hyoid bone
    C. Coccyx
    D. Occipital bone
    E. None of the above

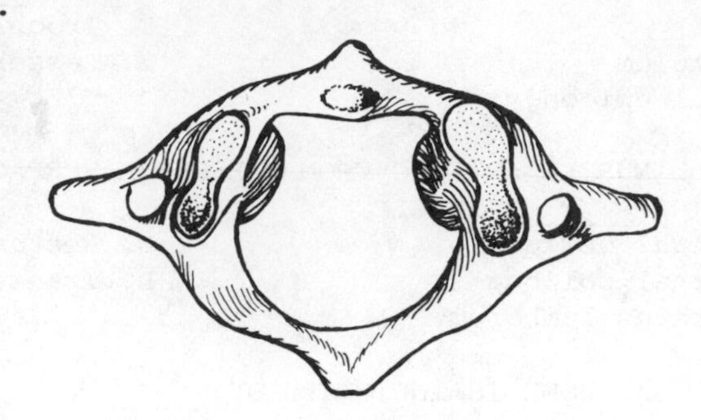

16. THE ARROW INDICATES A ______ JOINT.
    A. Ball and socket
    B. Gliding
    C. Saddle
    D. Pivot
    E. Hinge

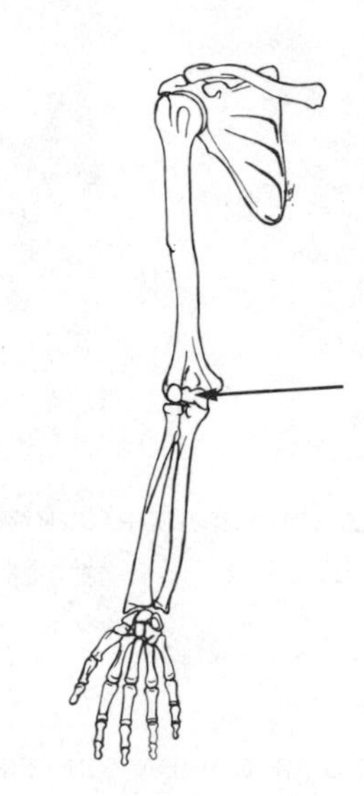

17. THE POSTERIOR BELLY OF THE DIGASTRIC MUSCLE IS INDICATED BY ______.

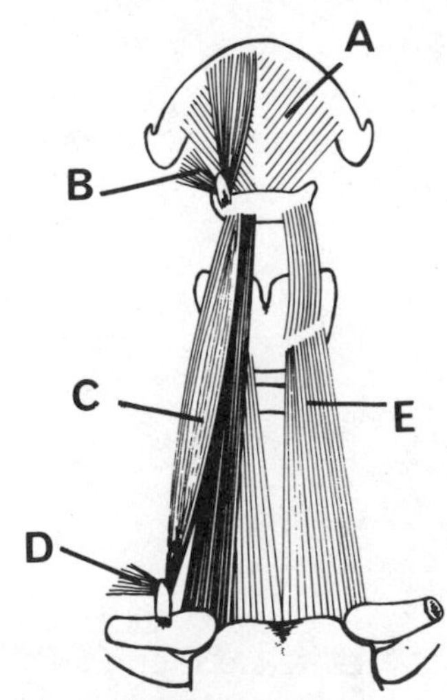

18. A BONE FORMING THE POSTERIOR ASPECT OF THE CRANIAL FLOOR AND WALLS IS THE:
    A. Ethmoid
    B. Sphenoid
    C. Temporal
    D. Occipital
    E. Palatine

SELECT THE SINGLE BEST ANSWER

19. THE ARROW INDICATES THE STRATUM ________ OF THE SKIN.
   A. Lucidum
   B. Corneum
   C. Spinosum
   D. Granulosum
   E. Germinativum

20. THE RECTUS FEMORIS ________ THE HIP.
   A. Medially rotates
   B. Adducts
   C. Extends
   D. Flexes
   E. None of the above

21. THE SELLA TURCICA IS PART OF THE ______ BONE.
   A. Ethmoid
   B. Sphenoid
   C. Maxilla
   D. Frontal
   E. Palatine

22. A DEEP MUSCLE PROTECTING THE SHOULDER JOINT IS THE:
   A. Deltoid
   B. Levator scapulae
   C. Pectoralis major
   D. Teres major
   E. Supraspinatus

23. THE OUTER MEMBRANE COVERING THE CARTILAGE MODEL IN ENDOCHONDRAL OSSIFICATION IS TERMED THE:
   A. Periosteum
   B. Perimysium
   C. Perineurium
   D. Perichondrium
   E. None of the above

24. THE PEDICLE OF THIS VERTEBRA IS INDICATED BY ________.

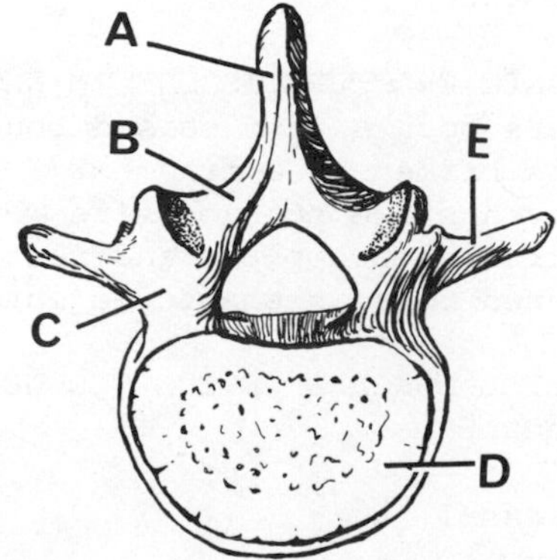

SELECT THE SINGLE BEST ANSWER

25. TEMPERATURE REGULATING CENTERS OF THE SKIN AND THE BODY IN GENERAL ARE LOCATED IN THE ______ OF THE BRAIN.
    A. Cerebrum
    B. Pons
    C. Midbrain
    D. Medulla
    E. Hypothalamus

26. THE SACRAL PROMONTORY IS INDICATED BY ______.

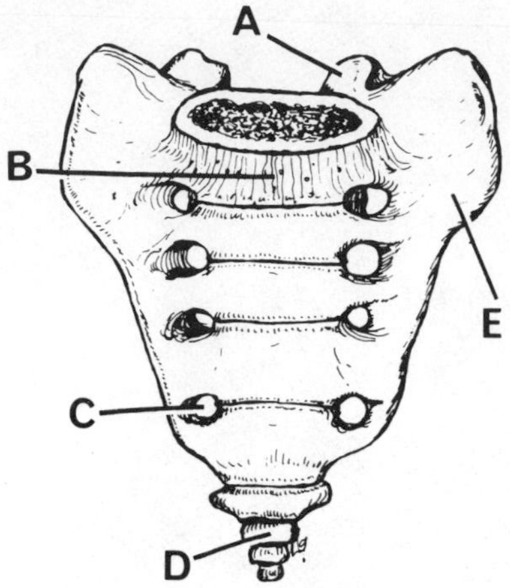

27. ROTATION OF THE SCAPULA SO THAT THE GLENOID CAVITY FACES UPWARD IS DUE TO CONTRACTION OF WHICH OF THE FOLLOWING MUSCLES?
    A. Rhomboids
    B. Levator scapulae
    C. Subscapularis
    D. Trapezius
    E. None of the above

28. WHICH OF THE FOLLOWING IS NOT INVOLVED IN BONE TISSUE HOMEOSTASIS?
    A. Vitamin D
    B. Vitamin B_{12}
    C. Parathyroid
    D. Somatotropin
    E. Normal blood calcium levels

29. THE SARCOLEMMA OF A MUSCLE FIBER IS A:
    A. Network of tubules
    B. Substance similar to actin
    C. Sarcomere
    D. Plasma membrane
    E. Region about its nucleus

30. THE MUSCLES OF FACIAL EXPRESSION ARE INNERVATED BY CRANIAL NERVE _____.
    A. Abducens
    B. Hypoglossal
    C. Oculomotor
    D. Trigeminal
    E. None of the above

31. A PATHOLOGIC FRACTURE IS ONE IN WHICH THE:
    A. Fracture occurs in diseased bone
    B. Skin is broken by a fragment
    C. Bone breaks due to excessive weight loading
    D. Fracture is incomplete and most often occurs in growing bone
    E. Fractured bone is shattered into many small pieces

32. THE SKELETAL MUSCLES OF THE TONGUE ARE INNERVATED BY THE ________ NERVE.
    A. Abducens
    B. Facial
    C. Hypoglossal
    D. Trochlear
    E. None of the above

SELECT THE SINGLE BEST ANSWER

33. WHICH OF THE FOLLOWING CELLS ARE ASSOCIATED WITH BONE RESORPTION?
    A. Fibroblasts
    B. Macrophages
    C. Osteoblasts
    D. Osteoclasts
    E. Osteocytes

34. THE CONNECTIVE TISSUE COMPONENT FOUND ENVELOPING A FASCICULUS OF SKELETAL MUSCLE IS CALLED THE _______.
    A. Sarcolemma
    B. Endomysium
    C. Epimysium
    D. Perimysium
    E. None of the above

35. WHICH OF THE FOLLOWING PAIRS OF BONY STRUCTURES ARE NOT COMPARABLE WITH EACH OTHER?
    A. Femur and humerus
    B. Scapula and ilium
    C. Scaphoid and fibula
    D. Acetabulum and glenoid
    E. Metatarsals and metacarpals

MULTI-COMPLETION QUESTIONS

DIRECTIONS: In each of the following questions or incomplete statements, ONE OR MORE of the completions given is correct. On the answer sheet, blacken the space under A if 1, 2, and 3 are correct; B if 1 and 3 are correct; C if 2 and 4 are correct; D if only 4 is correct; and E if all are correct.

36. THE SUPERFICIAL FLEXORS OF THE FOREARM:
    1. Flex the fingers
    2. Flex the wrist
    3. Insert into the phalanges
    4. Originate from the medial epicondyle

37. THE STABILITY OF A JOINT IS PROVIDED BY THE:
    1. Contours of its articulating surfaces
    2. Tendons crossing the joint
    3. Extracapsular ligaments
    4. Joint capsule

38. WHICH OF THE FOLLOWING STATEMENTS IS(ARE) TRUE FOR ENDOCHONDRAL BONE FORMATION?
    1. Osteoblasts migrate into the area
    2. Calcium phosphate is laid down
    3. Cartilage cells undergo hypertrophy
    4. Somatotropin is relatively unimportant

39. FUNCTIONS OF THE SKIN INCLUDE THE:
    1. Regulation of body heat
    2. Absorption of selected materials
    3. Ability to perceive pain
    4. Protection against water loss

40. WHICH OF THE FOLLOWING IS(ARE) NOT CONSIDERED A FUNCTION OF SKELETAL MUSCLE?
    1. Major producer of heat
    2. Maintains posture
    3. Produces movement
    4. Protects vital organs

A	B	C	D	E
1,2,3	1,3	2,4	only 4	all correct

41. A FIBROUS JOINT IS FORMED BETWEEN THE:
    1. Distal epiphysis and the metaphysis of the femur
    2. Adjacent surfaces of the tibia and fibula
    3. Medial end of the ribs and the sternum
    4. Bodies of the vertebrae

42. THE PERIOSTEUM:
    1. Is located on the articular surface
    2. Contains blood vessels and nerves
    3. Forms the inner membrane of the metaphysis
    4. Has cells involved in the diameter growth of long bones

43. THE DERMIS OF THE SKIN:
    1. Is of ectodermal origin
    2. Has numerous blood vessels
    3. Contains a deep papillary layer
    4. Usually is thicker than the epidermis

44. THE Z-LINE OF SKELETAL MUSCLE IS:
    1. Found in the middle of the I band
    2. Approached by actin filaments
    3. The termination of a sarcomere
    4. Formed by the T-tubule

45. AMONG THE SUPRAHYOID MUSCLES ONE FINDS THE:
    1. Palatoglossus
    2. Sternohyoid
    3. Glossopharyngeus
    4. Stylohyoid

46. WHICH OF THE FOLLOWING PROTEINS ARE FOUND IN SKELETAL MUSCLE TISSUE?
    1. Actin
    2. Tropomyosin
    3. Myosin
    4. Troponin

47. STRUCTURES CONTAINING SYNOVIAL FLUID INCLUDE THE:
    1. Intervertebral discs
    2. Bursae
    3. Symphysis pubis
    4. Ankle joint

48. BONES OF THE APPENDICULAR SKELETON INCLUDE THE:
    1. Atlas
    2. Sacrum
    3. Mandible
    4. Radius

49. IN AN ISOTONIC MUSCLE CONTRACTION:
    1. More cross-links are formed
    2. Both ends of the muscle are fixed
    3. The muscle actively shortens against a load
    4. No new cross-links are formed

50. THE MEDIAL LONGITUDINAL ARCH INCLUDES THE:
    1. Cuboid
    2. Talus
    3. Third cuneiform
    4. Navicular

51. BURNS ARE CLASSIFIED BY WHICH OF THE FOLLOWING:
    1. Skin layers involved
    2. Sensitivity remaining
    3. Regenerative capacity
    4. Character of remnants

A	B	C	D	E
1,2,3	1,3	2,4	only 4	all correct

52. WHICH OF THE FOLLOWING STRUCTURES FORM THE THORACIC OUTLET?
    1. Tenth thoracic vertebrae
    2. Costal arch
    3. Last false rib
    4. Xiphoid process

53. THE MUSCLES OF THE TONGUE ARE:
    1. Mainly innervated by the hypoglossal nerve
    2. Involved in swallowing
    3. Divided into two major groups
    4. Formed by smooth muscle

54. THE MUSCLES OF THE ANTERIOR COMPARTMENT OF THE LEG:
    1. Extend the knee joint
    2. Lie on the medial side of the tibia
    3. Act to maintain stability of the knee
    4. Dorsiflex the foot

55. THE PROXIMAL RADIO-ULNAR JOINT ALLOWS FOR ______.
    1. Flexion
    2. Circumduction
    3. Extension
    4. Rotation

56. THE BONE(S) FORMING A PORTION OF THE CRANIAL FLOOR IS(ARE) THE:
    1. Temporal
    2. Zygomatic
    3. Ethmoid
    4. Maxilla

57. THE 'STRAP' MUSCLES OF THE NECK INCLUDE THE:
    1. Mylohyoid
    2. Sternohyoid
    3. Digastric
    4. Thyrohyoid

58. SEROUS MEMBRANES ARE FOUND:
    1. Lining the pericardial cavity
    2. Forming the inner layer of blood vessels
    3. Covering the surface of the lung
    4. To contain mucous glands

59. WHICH OF THE FOLLOWING STATEMENTS IS(ARE) TRUE FOR THE GLUTEUS MAXIMUS MUSCLE?
    1. Inserts for the most part into the iliotibial tract
    2. May be used as a site for intramuscular injection
    3. Functions when powerful extension is needed
    4. Is a relatively thin muscle

60. THE ARTICULATION BETWEEN THE TEMPORAL AND SPHENOID BONE:
    1. Is considered to be a suture
    2. Contains fibrous connective tissue
    3. Allows no movement
    4. Forms a symphysis

61. THE HUMERUS:
    1. Has an olecranon fossa
    2. On its distal end has a capitulum
    3. Has a fossa proximal to the trochlea
    4. Articulates with the glenoid fossa

62. THE ______ JOINT(S) IS(ARE) CONSIDERED TO BE OF THE BALL AND SOCKET TYPE.
    1. Hip
    2. Elbow
    3. Shoulder
    4. Knee

A	B	C	D	E
1,2,3	1,3	2,4	only 4	all correct

63. WHICH OF THE FOLLOWING STATEMENTS PERTAIN TO THE MUSCLES OF MASTICATION?
    1. There are four in number
    2. Innervated by the trigeminal nerve
    3. Are involved in chewing
    4. Includes the frontalis muscle

64. THE QUADRICEPS FEMORIS GROUP OF MUSCLE:
    1. Extends the knee joint
    2. Is attached to the patella
    3. Consists of 4 separate muscles
    4. Is located anteriorly on the femur

65. WHICH OF THE FOLLOWING STATEMENTS IS(ARE) TRUE FOR THE LONG MUSCLES OF THE THIGH?
    1. The anterior group is a flexor of the knee
    2. The rectus femoris flexes the thigh
    3. The hamstrings flex the hip
    4. The adductors form the medial compartment

66. WHICH OF THE FOLLOWING STRUCTURES IS(ARE) NOT FOUND ON THE HIP BONE?
    1. Anterior superior spine
    2. Sciatic notch
    3. Obturator foramen
    4. Intertrochanteric line

67. THE FLEXORS OF THE ELBOW INCLUDE THE:
    1. Biceps brachii
    2. Brachialis
    3. Brachioradialis
    4. Pronator teres

68. WHICH OF THE FOLLOWING BONES IS(ARE) LOCATED IN THE HAND?
    1. Lunate
    2. Capitate
    3. Scaphoid
    4. Cuboid

69. WHICH OF THE FOLLOWING STATEMENTS CONCERNING THE RADIUS IS(ARE) CORRECT?
    1. Lies lateral to the ulna
    2. Has a distal portion called a head
    3. Forms two articulations with the ulna
    4. Articulates with the trochlea of the humerus

70. THE FLEXORS OF THE WRIST AND FINGERS ARE LOCATED:
    1. On the posterior side of the forearm
    2. On the anterior side of the forearm
    3. Only in the hand
    4. Toward the medial epicondyle of the humerus

DIRECTIONS: Each group of questions below consists of a numbered list of descriptive words or phrases accompanied by a diagram with certain parts indicated by letters, or by a list of lettered headings. For each numbered word or phrase, SELECT THE LETTERED PART OR HEADING that matches it correctly. Then insert the letter in the space to the right of the appropriate number. Sometimes more than one numbered word may be correctly matched to the same lettered part or heading.

71. _____ The matrix of the hair

72. _____ The hair papilla

73. _____ The external root sheath

74. _____ The shaft of the hair

75. _____ The dermal sheath

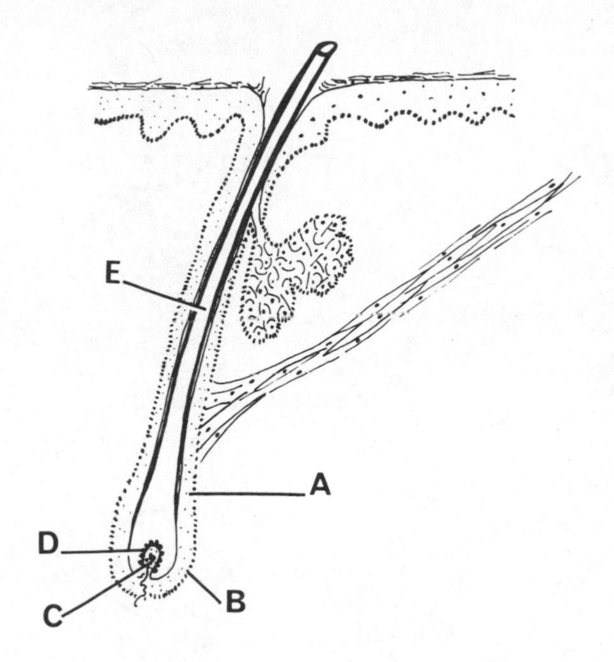

76. _____ An extensor of the elbow joint

77. _____ Belongs to the quadriceps group

78. _____ A flexor of the elbow joint

79. _____ An abductor of the hip joint

80. _____ A supinator of the forearm

A. Rectus femoris

B. Biceps brachii

C. Biceps femoris

D. Triceps brachii

E. None of the above

PART THREE

THE NERVOUS SYSTEM

SPECIAL SENSES

THE ENDOCRINE SYSTEM

THE CARDIOVASCULAR SYSTEM

THE LYMPHATIC SYSTEM

8. THE NERVOUS SYSTEM

LEARNING OBJECTIVES

INTRODUCTION

Be Able To:

- Describe the general functions of the nervous system.
- List the components of the central and peripheral divisions of the nervous system and define the terms somatic and visceral.
- Illustrate, using a diagram, the following components of a 'typical' myelinated motoneuron: cell body (soma), Nissl body, nucleus, nucleoplasm, dendrites, axon, axon hillock, axoplasm, myelin, Schwann cell, node of Ranvier, and telodendria.
- Describe the general appearance of unipolar, bipolar, and multipolar neurons.
- Discuss the following functional components of the reflex arc: motor (efferent) neuron, sensory (afferent) neuron, and internuncial (intercalated) neuron.
- Describe the different types of receptors and the sensations they modulate.
- Describe the types and functions of the neuroglial cells of the central nervous system (CNS).
- Define the following terms: nerve, endoneurium, perineurium, epineurium, gray matter, tract, white matter, ganglion, and nucleus.
- Describe the stimulation and transmission of nerve impulses with reference to the following topics: resting potential, action potential, subthreshold stimulation, all-or-none law, refractory period, and saltatory conduction.
- Describe the anatomy and function of a synapse, paying particular attention to the types of transmitter substances, their synthesis and actions.
- Describe the anatomy of the neuromuscular junction and its transmission, paying particular attention to the transmitter substances, the end plate potential, and the effects of cholinesterase.

THE SOMATIC NERVOUS SYSTEM

Be Able To:

- Describe the arrangement of the meningeal layers of the brain and the spinal cord along with their spaces and extensions.

- Describe the ventricular system of the brain and the formation, circulation, and function of the cerebral spinal fluid (CSF).

- Using simple diagrams illustrate the three and five vesicle stages of brain development, indicating the adult structures derived from the walls and cavities of the forebrain, midbrain, and hindbrain.

- Define the terms fissure, sulcus, and gyrus, and on a diagram of a lateral or medial view of the surface of a cerebral hemisphere identify the following: lateral fissure (Sylvius), central sulcus (Rolando), parieto-occipital fissure, calcarine fissure, precentral gyrus, postcentral gyrus, superior frontal gyrus, frontal, parietal, occipital and temporal lobes, cuneus, lingual gyrus, hippocampal gyrus, uncus, fornix, and corpus callosum.

- Discuss the following topics pertaining to the structure and location of the cerebrum: (a) outer cortical layer (gray matter); (b) cerebral tracts (white matter) including association, commissural, ascending projection, and descending projection fibers; and (c) deep basal ganglia (gray matter).

- Using a diagram of a midsagittal section of the brain locate the following structures: diencephalon (thalamus and hypothalamus), midbrain (cerebral peduncles and corpora quadrigemina), cerebellum, pons, medulla oblongata, choroid plexus, third and fourth ventricles, and the cerebral aqueduct.

- Discuss the functions of the thalamus and the hypothalamus as they pertain to sensation, arousal mechanism, reflex movement, control of autonomic nervous system, relay station, hormonal control, and maintenance of body temperature.

- Define the term brain stem, giving the functions of the brain stem and the cerebellum.

- On a diagram of the ventral or lateral view of the brain identify the 12 pairs of cranial nerves, noting their site of exit from the brain.

- Describe the functions of the cranial nerves along with the location of their nuclei.

- Describe the spinal cord as to its: location, shape (enlargements), extent, number of segments.

- On a representative diagram of a spinal cord segment label the following: dorsal median sulcus; ventral median fissure; dorsal, intermediolateral, and ventral gray columns; dorsal, lateral, and ventral funiculi; and the central canal.

- Describe the spinal nerves as to their total number, and number in each of the following segments: cervical, thoracic, lumbar, sacral, and coccygeal.

- Describe the typical spinal nerve as to its following components: dorsal (sensory) root, dorsal root ganglion, ventral (motor) root, ventral primary ramus and dorsal primary ramus.

- Describe the distribution of a 'typical' segmental spinal nerve and the formation and general distribution of the cervical, brachial, and lumbosacral plexuses.

- For each of the following nerves give the plexus from which they are derived and their general functions: axillary, intercostal, femoral, median, musculocutaneous, phrenic, pudendal, radial, sciatic, and ulnar.

- Describe the physiology of the sensory pathways with reference to the following: location of neurons; sensations conveyed by the lateral spinothalamic, medial lemniscal, spinothalamic, and ventral spinothalamic pathways; and the somatic areas of the cerebral cortex.

- Define the term pain, differentiating between somatic, referred, and visceral pain.

- Describe the components of the somatic motor pathways with special reference to the pyramidal and extrapyramidal pathways.

THE AUTONOMIC NERVOUS SYSTEM

Be Able To:

- Define the autonomic nervous system (ANS) in terms of the tissues and organs innervated, visceral effectors, motoneurons, and its control.

- Describe the anatomy of the sympathetic (thoracolumbar) division of the autonomic nervous system as to its site of origin, preganglionic and postganglionic neurons, rami communicans, sympathetic ganglia, sympathetic trunk, plexuses involved, and collateral ganglia.

- Describe the anatomy of the parasympathetic (craniosacral) nervous system with reference to its origin, preganglionic and postganglionic neurons, nerves it is involved with, associated plexuses and ganglia.

- Compare the chemical transmitters at the preganglionic and postganglionic synapses in the sympathetic and parasympathetic divisions of the ANS.

- Describe the physiological functions of the sympathetic and parasympathetic divisions of the ANS.

RELEVANT READINGS

Anthony & Kolthoff - Chapters 7, 8, and 9.
Chaffee & Greisheimer - Chapters 6, 7, 8, and 9.
Crouch & McClintic - Chapters 11, 12, 13, 14, 15, and 19.
Jacob, Francone & Lossow - Chapter 8.
Landau - Chapters 8, 9, 10, 11, and 13.
Tortora - Chapter 12, 13, 14, and 15.

INTRODUCTION

FIVE-CHOICE COMPLETION QUESTIONS

DIRECTIONS: Each of the following questions or incomplete statements is followed by five suggested answers or completions. SELECT THE SINGLE BEST ANSWER in each case and then circle the appropriate letter at the lower right of each question.

1. THE CELL RESPONSIBLE FOR TRANSMITTING AN IMPULSE IS THE:
   A. Fibrocyte
   B. Neuroglial cell
   C. Neuron
   D. Schwann cell
   E. Satellite cell

   A B C D E

2. A GROUP OF NERVE FIBERS IS SURROUNDED BY THE ________.
   A. Endoneurium
   B. Epineurium
   C. Epimysium
   D. Endosteum
   E. None of the above

   A B C D E

3. WHICH OF THE FOLLOWING IS NOT LOCATED IN THE CENTRAL NERVOUS SYSTEM?
   A. Astrocytes
   B. Ependyma
   C. Microglia
   D. Schwann cell
   E. Oligodendria

   A B C D E

4. THIS DIAGRAM REPRESENTS A ________ NEURON.
   A. Unipolar
   B. Multipolar
   C. Motor
   D. Bipolar
   E. None of the above

   A B C D E

5. WHICH OF THE FOLLOWING IS NOT A COMPONENT OF A SYNAPSE?
   A. Cleft
   B. Terminal
   C. Receptor area
   D. Myelin sheath
   E. Transmitter vesicles

   A B C D E

6. A LOCAL ANESTHETIC, SUCH AS COCAINE, FUNCTIONS BY:
   A. Increasing the extracellular calcium level
   B. Decreasing the excitability of the cell membrane
   C. Inhibiting acetylcholine at the synaptic cleft
   D. Increasing the membrane permeability to potassium
   E. None of the above

   A B C D E

7. BY DEFINITION, THE CELL PROCESS CARRYING THE NERVE IMPULSE TO THE CELL BODY IS THE:
   A. Axon
   B. Dendrite
   C. Node of Ranvier
   D. Axon hillock
   E. None of the above

   A B C D E

8. A CHARACTERISTIC CHANGE IN THE MEMBRANE OF A NERVE CELL BECAUSE OF A STIMULUS IS CALLED A(AN)________.
   A. Divergence
   B. Facilitation
   C. Refractory period
   D. Action potential
   E. None of the above

   A B C D E

SELECT THE SINGLE BEST ANSWER

9. A VISCERAL EFFERENT NERVE FIBER IS ONE WHICH CARRIES IMPULSES:
   A. Towards the CNS from the skeletal muscles
   B. Away from the CNS to the smooth and cardiac muscles and glands
   C. Away from the CNS to the skeletal muscles
   D. Towards the CNS from the smooth and cardiac muscles and glands
   E. None of the above

A B C D E

------------------------ ANSWERS, NOTES AND EXPLANATIONS ------------------------

1. C. The nervous system consists of two types of cells: neurons and neuroglia. The neuron is the cell responsible for carrying an impulse or action potential. These cells are the very essence of the nervous system, as they are the cells which both inform the body of its internal and external environments and provide a stimulus for the body to respond or adjust if need be to a changing environment. The neuroglial cells are found only in the central nervous system and are responsible for a variety of functions related to the support, nutritional, and phagocytic needs of the neuron.

2. E. A group of nerve fibers is surrounded by connective tissue elements called the perineurium (A in the diagram). Each individual neuron (nerve fiber)(B) is separated from other neurons by a fine connective tissue network called the endoneurium (C). The larger groups of nerve fibers, surrounded by the perineurium, form a peripheral nerve which is contained within a connective tissue sleeve or epineurium (D). The connective tissue elements surrounding the different units of a nerve blend with each other and provide a strong support.

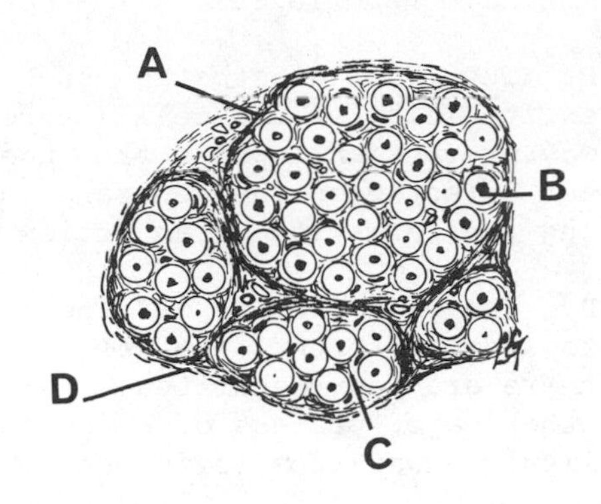

3. D. The Schwann (neurilemmal) cell is not found in the central nervous system (CNS) but surrounds the neurons of a peripheral nerve. These cells play an important role in the myelination of peripheral nerve fibers. However, these cells are also associated with unmyelinated nerve fibers. The neuroglial cells are thought to play a similar role in the CNS. The four types of neuroglial cells found in the CNS are: astrocytes, ependymal, microglia, and oligodendria.

4. A. A unipolar neuron has one process (A in the diagram) which splits into two. One process carries impulses towards the cell body and is therefore a functional dendrite, whereas the other process carries impulses away from the soma and is therefore an axon. Bipolar neurons (B) have two processes, an axon and a dendrite, whereas multipolar neurons (C) have many dendrites and one axon. The dendrites of unipolar or bipolar neurons are either in synaptic contact with specialized receptor cells or the dendrites themselves function as specialized peripheral receptors in the tissues of the body. In contrast, all the nerve cells of the CNS (brain and spinal cord) are of the multipolar variety.

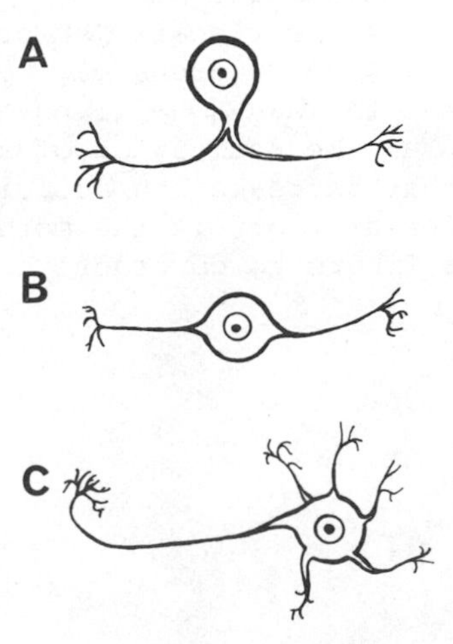

5. D. The myelin sheath plays no part in the synapse between neurons. The presynaptic terminal (knob, telodendria, or bouton; A in the diagram) represents the terminal end of the axon of a presynaptic neuron. This presynaptic terminal lies on the receptor area (B) of either a dendrite or a cell body (soma) of a postsynaptic neuron. Between the presynaptic terminal and the receptor area lies a small space called the synaptic cleft (C). The presynaptic terminal has transmitter vesicles (D) which contain either inhibitory or excitatory transmitter (protein) substances that are released into the synaptic cleft. These substances function either to inhibit or to stimulate the postsynaptic neuron. Most of the receptor areas are located on the dendrites of neurons. The presynaptic terminal (knob) continually processes new transmitter substances utilizing energy provided by the mitochondria (E).

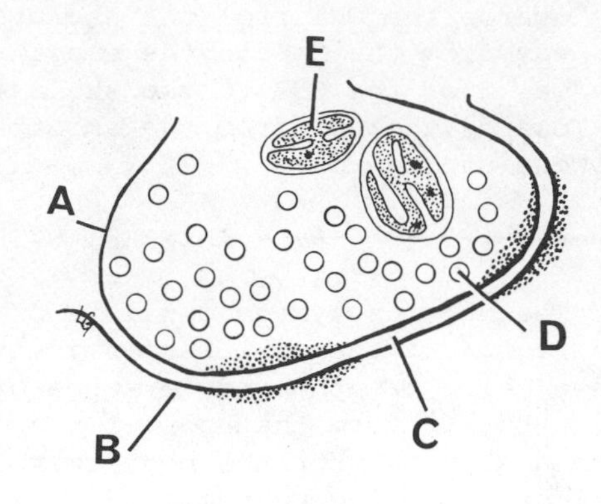

6. B. The local anesthetics such as cocaine and many other drugs reduce the excitability of the cell (neuron) membrane below a critical level and in this way prevent an impulse from passing through the anesthetized area. These drugs act directly on the membrane and decrease its permeability to sodium, preventing the formation of an action potential.

7. B. The cell process of a neuron responsible for carrying a nerve impulse towards (afferent) the cell body (soma) is the dendrite. In multipolar neurons there are many dendrites, whereas in bipolar neurons there is only one. The other major process of a neuron is called the axon and it transports a nerve impulse away from (efferent) the cell body.

8. D. An action potential (nerve impulse) is a characteristic change in the cell membrane of a neuron produced by a stimulus. This impulse is very similar to the one produced in muscle tissue on its excitation. It can be defined as a self-propagating wave of electrical negativity travelling along the surface of the membrane of a neuron. In the case of skeletal muscle such an impulse is responsible for its contraction, however in the case of the neuron such an impulse is passed on across a synaptic junction to another neuron and eventually to an effector.

9. B. A visceral efferent nerve fiber carries an impulse away from the CNS and stimulates the viscera (smooth muscle, cardiac muscle and glands). These visceral efferent neurons may be of three types: 1) secretory, transmitting impulses to induce a gland to secrete; 2) accelerator neurons, those that speed up the activities of smooth and cardiac muscle; and 3) inhibitory neurons that decrease the activities of smooth and cardiac muscle. Another type of efferent fiber is the motor neuron (motoneuron) which stimulates skeletal muscle fibers to contract.

M U L T I - C O M P L E T I O N   Q U E S T I O N S

DIRECTIONS: In each of the following questions or incomplete statements, ONE OR MORE of the completions given is correct. At the lower right of each question, circle A if 1, 2, and 3 are correct; B if 1 and 3 are correct; C if 2 and 4 are correct; D if only 4 is correct; and E if all are correct.

1. WHICH OF THE FOLLOWING STATEMENTS IS(ARE) TRUE FOR SALTATORY CONDUCTION?
   1. Restricted to myelinated nerves
   2. Produces a high velocity of nerve transmission
   3. Conserves energy for the axon
   4. Impulses are conducted continuously along the entire fiber

   A B C D E

2. THE CENTRAL NERVOUS SYSTEM INCLUDES THE:
   1. Brain
   2. Spinal nerves
   3. Spinal cord
   4. Second cranial nerve

   A B C D E

3. THE FUNCTION OF NEUROGLIAL CELLS INCLUDE WHICH OF THE FOLLOWING?
   1. Transportation of chemicals to the neuron
   2. Phagocytosis of cellular debris
   3. Production of a myelin sheath
   4. Structural support to the neuron

   A B C D E

4. WHICH OF THE FOLLOWING STATEMENTS IS(ARE) CORRECT AS A FUNCTION OF THE NERVOUS SYSTEM IN MAN?
   1. Involved in controlling homeostasis
   2. Has the unique ability to reason
   3. Stimulates movements of the body
   4. Carries nutrients throughout the body

   A B C D E

5. WHICH OF THE FOLLOWING IS(ARE) CONSIDERED TO BE EXCITATORY TRANSMITTERS OF THE NERVOUS SYSTEM?
   1. Dopamine
   2. Serotonin
   3. Acetylcholine
   4. Gamma aminobutyric acid

   A B C D E

6. RECEPTORS:
   1. May mediate more than one type of sensation
   2. Detect changes in the environment
   3. Conduct impulses towards the CNS
   4. Are involved in the reflex arc

   A B C D E

7. A GANGLION OF THE NERVOUS SYSTEM IS:
   1. Composed of many axonal processes
   2. An aggregation of nerve cell bodies
   3. Formed by a mass of neuroglial cells
   4. Located outside the CNS

   A B C D E

8. THE NEUROMUSCULAR JUNCTION:
   1. Forms the effector end of the reflex arc
   2. Is the site where acetylcholine is released
   3. Consists of a gap or gutter
   4. Has cholinesterase destroying the chemical stimulator

   A B C D E

9. WHITE MATTER OF THE NERVOUS SYSTEM:
   1. Consists of myelinated axons
   2. May form specific tracts
   3. Is located in the CNS
   4. Contains cell bodies and dendrites

   A B C D E

A	B	C	D	E
1,2,3	1,3	2,4	Only 4	All correct

10. NUCLEI OF THE CENTRAL NERVOUS SYSTEM:
    1. Are cross sections of small collagen fibers
    2. Consist of myelinated axons
    3. Contain the soma of neuroglial cells
    4. Are made up of nerve cell bodies

A B C D E

------------------------ ANSWERS, NOTES AND EXPLANATIONS ------------------------

1. A, <u>1, 2, and 3 are correct</u>. Saltatory conduction is where impulses are conducted from node to node (A in the diagram) in the myelinated nerve rather than continuously along the entire membrane as is the case in the unmyelinated nerve fiber. In saltatory conduction the electrical current flows through the surrounding extracellular fluids (B) and the axoplasm (C) from node to node. Therefore, each node becomes excited in succession as the impulse jumps down the fiber. This mechanism increases the velocity of nerve transmission in myelinated fibers. It also conserves energy for the axon since only the nodes are depolarized and there is a subsequent loss of fewer ions than is the case in nonmyelinated fibers. Therefore, the axon requires less energy to retransport the lost ions across the membrane.

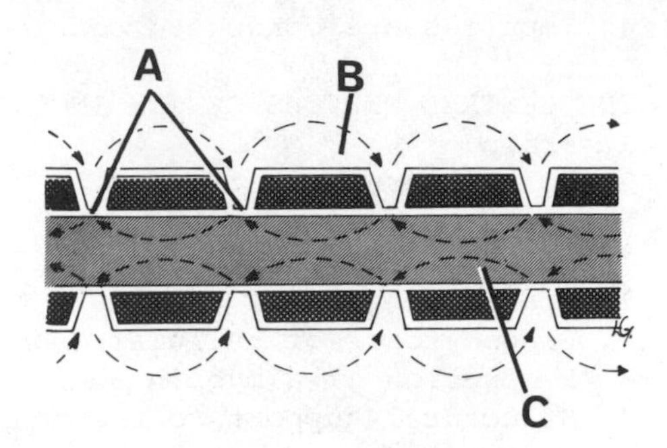

2. B, <u>1 and 3 are correct</u>. The central nervous system (CNS) consists of the brain and spinal cord. These two vital components of the nervous system are contained within protecting bone: the bones of the skull and the vertebrae of the vertebral column. The cranial and spinal nerves plus the autonomic nervous system (ANS) are considered to be components of the peripheral nervous system (PNS).

3. E, <u>All are correct</u>. The neuroglial cells of the CNS carry out a variety of functions. They are thought to play a role in providing some structural support to the neurons. However, the support provided by this group of cells is delicate since there are no typical connective tissue elements present (collagen or reticulin). Other possible functions include the production of a myelin sheath around certain neurons (by oligodendrocytes); the phagocytic activity of microglia in the presence of bacteria; a component of the blood-brain barrier (astrocytes) which protects the neurons from harmful substances in the vascular system; and the transporting of nutrients to the neurons may also be an additional role for glial cells.

4. A, <u>1, 2, and 3 are correct</u>. The basic functions of the nervous system are varied and complex. One of the most important functions of the nervous system is to control the homeostasis of the body (or the internal environment). The nervous system also receives impulses from the body and by the use of memory makes decisions on them. The brain is also capable of reasoning and uses all sources of incoming information and memory. It is also capable of disregarding any unimportant information and also provides a mechanism to act upon important bits of information. This mechanism for action is an effector system (muscle tissue and glands) which carries out appropriate responses as directed by the CNS.

5. A, 1, 2, and 3 are correct. The neurons of the nervous system are either excitatory or inhibitory, depending on the type of neurotransmitter substances they secrete at their terminals. It is thought that a neuron cannot be both an inhibitory and excitatory neuron. Well established excitatory transmitter substances are acetylcholine, dopamine, norepinephrine, and serotonin. Gamma aminobutyric acid (GABA) and glycine are inhibitory substances secreted by neurons. There are thought to be many other substances that may act as neurotransmitters.

6. E, All are correct. A receptor can be defined as the peripheral end of a sensory neuron. This neuron forms the peripheral portion of a reflex arc and conducts sensory impulses from the periphery to the CNS. Receptors are thought to be able to mediate more than one type of sensation, but each receptor usually responds best to one particular type of stimulus. Receptors therefore keep the body in touch with changes in its environment both external and internal. The sensitivity of a receptor is thought to depend either on the number (density) of receptors in a given area or a change in the threshold (level of sensitivity) of a given receptor.

7. C, 2 and 4 are correct. A ganglion of the nervous system consists of gray matter which is an accumulation of neuron cell bodies and dendrites. It is located outside of the central nervous system (i.e., brain and spinal cord). A similar accumulation of gray matter within the CNS is called a nucleus.

8. E, All are correct. The neuromuscular junction (motor end plate) is the area of contact between an efferent nerve and a muscle fiber. This junction forms the effector end of the reflex arc. Nerve impulses travel along the axon and cause the release of small vesicles from the axon terminals into the gap between the axon terminals and the skeletal muscle fiber or fibers. These vesicles contain acetylcholine which quickly diffuses across the gap, stimulates the sarcolemma of the muscle fiber and causes it to contract. A substance, called cholinesterase, located on the muscle side of the gap quickly destroys the acetylcholine, so that the muscle fiber may not be stimulated again before another impulse can be transmitted from the nerve.

9. A, 1, 2, and 3 are correct. White matter of the nervous system is located within the brain or the spinal cord. It consists of myelinated nerve fibers (axons) which appear white to the naked eye. These accumulations of myelinated axons form the various tracts of the CNS. Nerves are nothing more than tracts of the nervous system but they are located outside of the CNS.

10. D, Only 4 is correct. An accumulation of gray matter deep within the brain or spinal cord is called a nucleus. It consists of nerve cell bodies and dendrites. The best example of these masses of gray matter is the basal nuclei of each cerebral hemisphere. These nuclei are also commonly called the basal ganglia but the correct term is nuclei.

FIVE-CHOICE ASSOCIATION QUESTIONS

DIRECTIONS: Each group of questions below consists of a numbered list of descriptive words or phrases accompanied by a digram with certain parts indicated by letters, or by a list of lettered headings. For each numbered word or phrase, SELECT THE LETTERED PART OR HEADING that matches it correctly. Then insert the letter in the space to the right of the appropriate number. Sometimes more than one numbered word may be correctly matched to the same lettered part or heading.

ASSOCIATION QUESTIONS

1. _____ A process carrying an impulse towards the soma

2. _____ The axon hillock

3. _____ A node of Ranvier

4. _____ The telodendrion

5. _____ The neurilemma

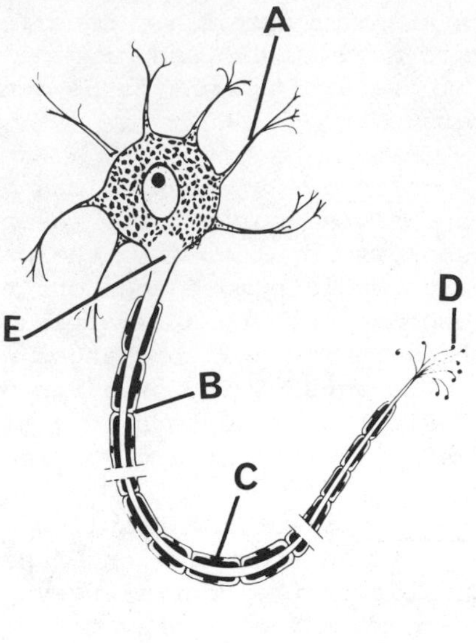

A
B
C
D
E

6. _____ A neuron carrying an impulse towards the CNS

7. _____ A cell sensitive to a change in the environment

8. _____ Implies a neuron carrying an impulse away from the CNS

9. _____ A neuron connecting two neurons in the spinal cord

10. _____ A cell which carries out a response

A. Motor

B. Effector

C. Sensory

D. Receptor

E. None of the above

------------------------ ANSWERS, NOTES AND EXPLANATIONS ------------------------

1. A. The dendrite is a process of a nerve cell (neuron), carrying an impulse towards (afferent) the cell body (soma). A typical motoneuron has many dendrites but only one axon. An axon is a similar process carrying an impulse away from the cell body and may be referred to as an efferent (motor) process.

2. E. The axon hillock is seen in large multipolar neurons as a region from which the axon arises. This region is either devoid of Nissl substance (RNA) or may contain small amounts. In other neurons the distinction between the axon hillock and the rest of the cell body is usually not apparent.

3. B. The node of Ranvier is a short gap between the successive neurilemmal (Schwann) cells and their myelin which surrounds the axons of peripheral nerves. At the node, the myelin is absent and it serves to increase the speed of conduction of nerve impulses by a mechanism known as saltatory conduction.

4. D. The telodendria are the terminal branches of the axon. They exhibit specializations that are related either to synaptic transmission or to neurosecretion.

5. C. In this diagram the upper figure represents a myelinated nerve fiber and the lower figure represents four unmyelinated nerve fibers. The neurilemma or Schwann cell (A in the diagram) of a peripheral myelinated nerve wraps itself around the axon (B). In myelinated nerves these cells are responsible for producing the myelin (C). The neurilemma refers to the outer layer of the Schwann cell plus the covering of glycoprotein surrounding it. These components of the neurilemma appear as one when viewed by the light microscope. In unmyelinated fibers the Schwann cell (D) produces little myelin and does not wrap itself around the axons.

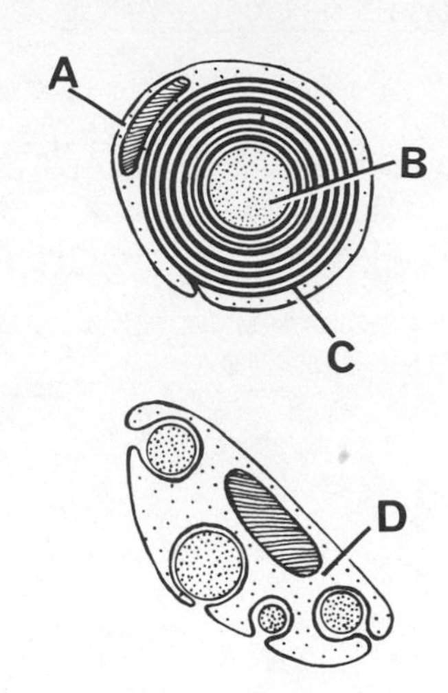

6. C. A neuron carrying an action potential towards the CNS is called an afferent or sensory neuron. It relays information from a receptor cell of the body to the spinal cord or brain. This neuron informs the CNS about the internal environment of the body or about the external environment.

7. D. A cell sensitive to the environment is called a receptor. Receptors serve two important general functions in that they perceive stimuli involved in all sensations and reflexes. These cells detect changes in our external and internal environments and initiate the responses necessary for adjusting the body to these changes so as to restore and maintain homeostasis.

8. A. The motor neuron is an efferent neuron that carries an action potential away from the CNS. This neuron stimulates the effector cell which may be a skeletal muscle or a gland cell.

9. E. The neuron connecting other neurons in the spinal cord is called an internuncial neuron (A in the diagram). Such a neuron may also be called an intercalated neuron or an interneuron. The internuncial neuron forms a reflex arc and connects an afferent neuron (B) bringing sensory information into the cord with a motor neuron (C) stimulating an effector. Internuncial neurons are also found in the brain.

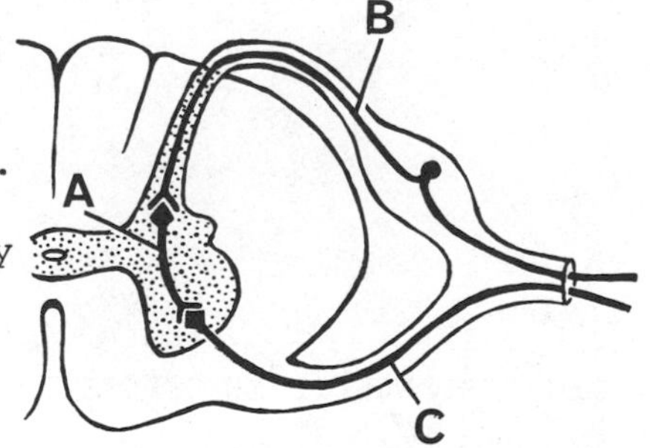

10. B. The effector is a cell or group of cells (muscle or glandular) carrying out a response which has been initiated by a receptor. An example of a neuroeffector junction would be the neuromuscular junction. Such a junction allows the muscle fibers to contract when stimulated by the motor neuron.

FIVE-CHOICE COMPLETION QUESTIONS.

DIRECTIONS: Each of the following questions or incomplete statements is followed by five suggested answers or completions. SELECT THE SINGLE BEST ANSWER in each case and then circle the appropriate letter at the lower right of each question.

1. THE VENTRAL ROOT OF A SPINAL NERVE IS INDICATED BY ________.

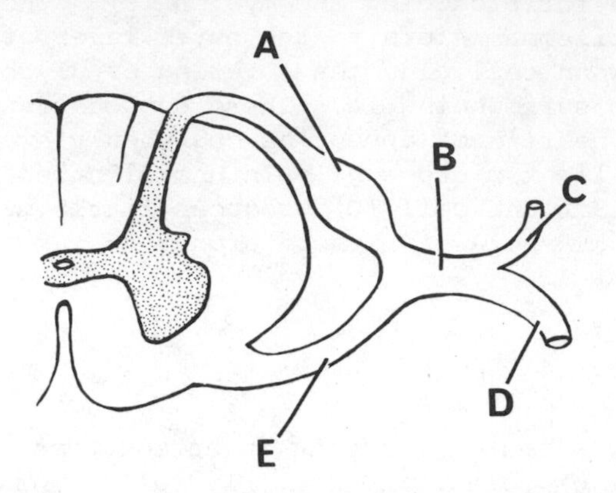

A B C D E

2. THE STRUCTURE FORMED FROM THE WALLS OF THE VESICLE INDICATED BY THE ARROW IS THE ________.
   A. Midbrain
   B. Cerebellum
   C. Pons
   D. Cerebral cortex
   E. None of the above

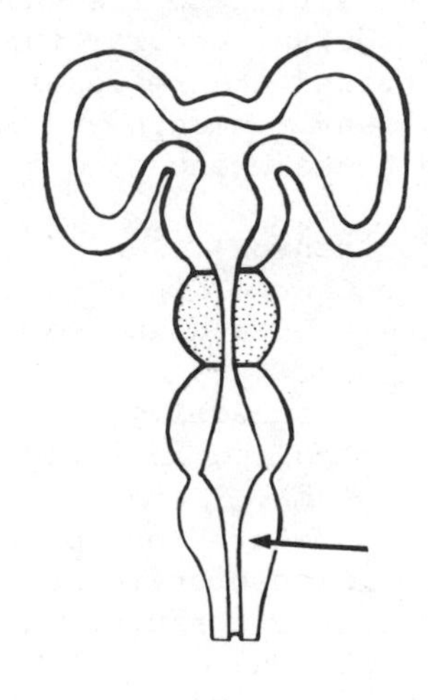

A B C D E

3. THE THALAMUS IS INDICATED BY ______.

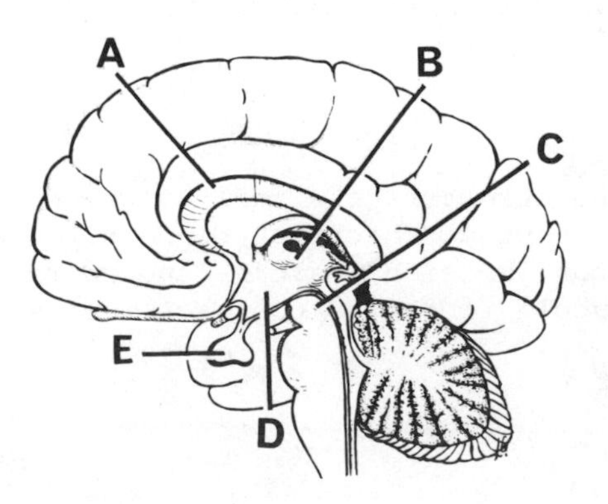

A B C D E

SELECT THE SINGLE BEST ANSWER

4. THE SHADED AREA IS THE _______ LOBE OF THE CEREBRAL HEMISPHERE.
   A. Frontal
   B. Insula
   C. Occipital
   D. Parietal
   E. Temporal

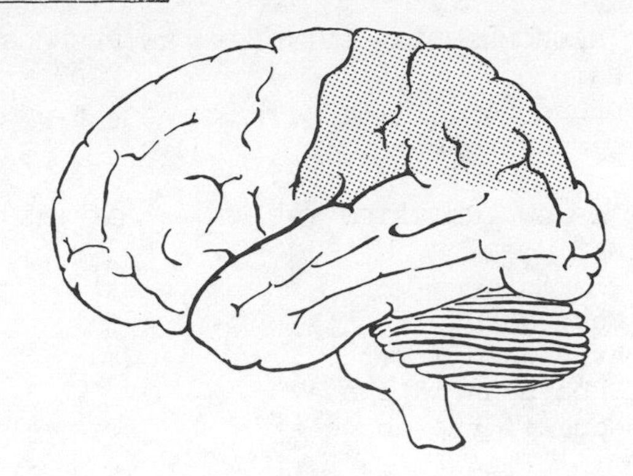

A B C D E

5. A TRACT OF THE CEREBRUM PASSING BETWEEN DIFFERENT PORTIONS OF THE SAME HEMISPHERE IS CALLED A(AN) _______ TRACT.
   A. Association
   B. Commissural
   C. Ascending projection
   D. Descending projection
   E. None of the above

A B C D E

6. THE ARROW INDICATES THE ______.
   A. Pons
   B. Thalamus
   C. Midbrain
   D. Cerebellum
   E. Medulla oblongata

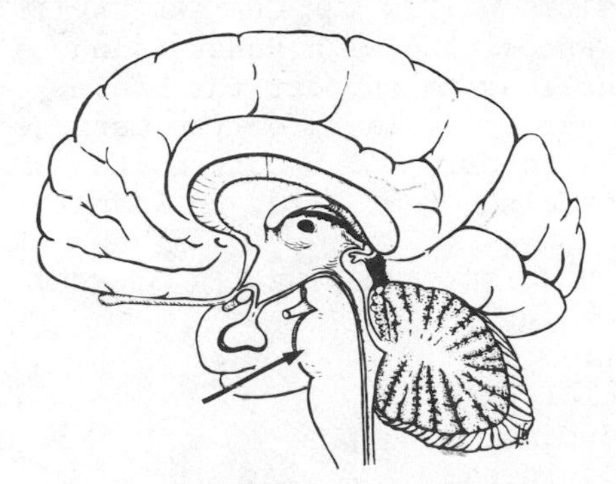

A B C D E

7. THE INTERVENTRICULAR FORAMEN IS INDICATED BY __________.

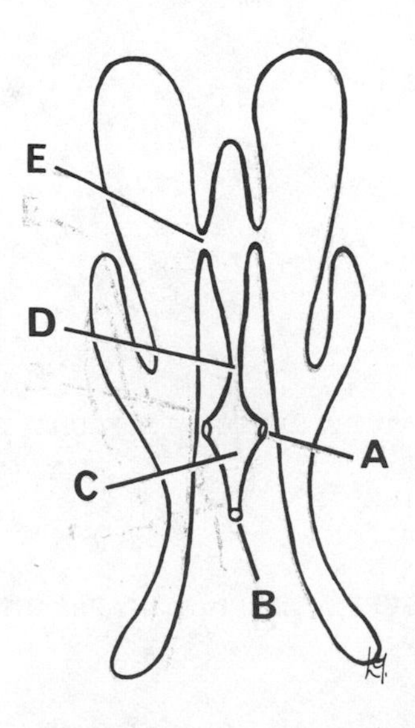

A B C D E

SELECT THE SINGLE BEST ANSWER

8. THE NEUROMUSCULAR SPINDLES ARE RESPONSIBLE FOR DETECTING ____.
   A. Pain
   B. Pressure
   C. Temperature
   D. Proprioception
   E. None of the above

A B C D E

9. THE ARROW INDICATES THE ______ OF THE SPINAL CORD.
   A. Pia mater
   B. Dura mater
   C. Arachnoid layer
   D. Subarachnoid space
   E. None of the above

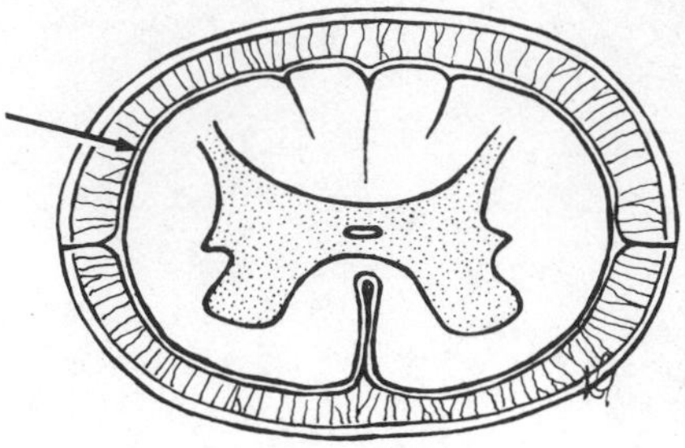

A B C D E

10. WHICH OF THE FOLLOWING IS INCORRECT FOR THE DURA OR ITS EXTENSIONS WITHIN THE CRANIAL CAVITY?
   A. Is associated with venous sinuses
   B. Functions to support the brain
   C. Is the outer layer of the meninges
   D. The tentorium lies between the cerebellar hemispheres
   E. The falx cerebri lies between the cerebral hemispheres

A B C D E

11. THE ARROW INDICATES THE ______ NERVE.
   A. Vagus
   B. Optic
   C. Facial
   D. Abducens
   E. Hypoglossal

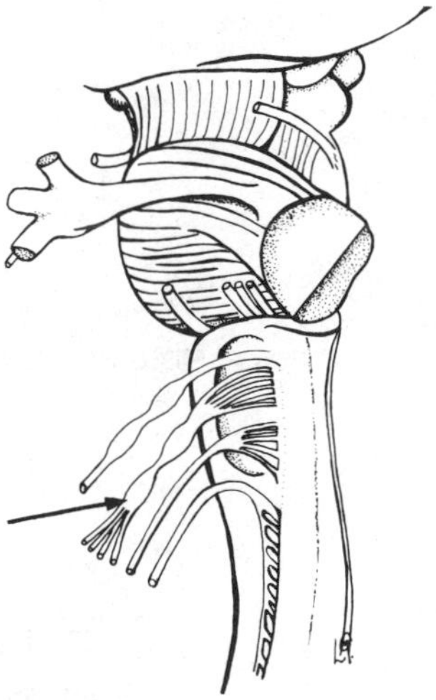

A B C D E

12. THE RESPIRATORY CENTER IS FOUND IN THE:
   A. Midbrain
   B. Cerebellum
   C. Diencephalon
   D. Cerebrum
   E. Medulla oblongata

A B C D E

13. THE NERVE SUPPLYING THE DIAPHRAGM IS THE _______.
   A. Axillary
   B. Femoral
   C. Median
   D. Phrenic
   E. Ulnar

A B C D E

SELECT THE SINGLE BEST ANSWER

14. THE LUMBOSACRAL PLEXUS ARISES FROM THE __________ SPINAL NERVE.
A. L_1-S_1
B. L_1-S_2
C. L_2-S_4
D. L_3-S_5
E. None of the above

A B C D E

15. THE CONUS MEDULLARIS IS INDICATED BY _______.

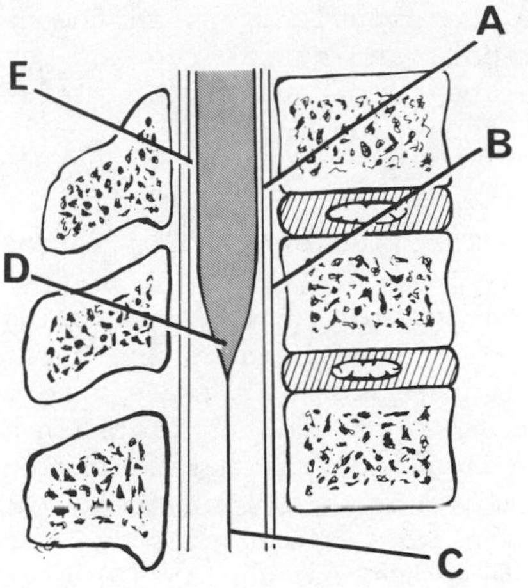

A
E
B
D
C

A B C D E

16. THERE ARE _______ PAIRS OF CERVICAL SPINAL NERVES.
A. Six
B. Seven
C. Eight
D. Nine
E. Ten

A B C D E

17. THE INTERMEDIOLATERAL GRAY COLUMN IS INDICATED BY ________.

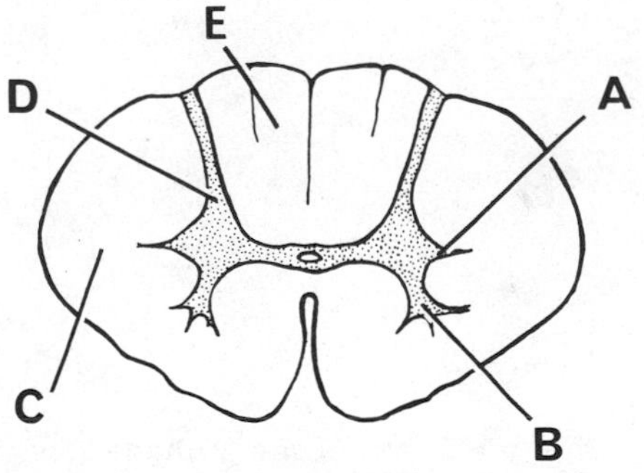

E
D
A
C
B

A B C D E

------------------------ ANSWERS, NOTES AND EXPLANATIONS ------------------------

1. E. The typical spinal cord segment has a ventral root (E) and dorsal (unlabelled) root which come together laterally to form a spinal nerve (B). The spinal nerve contains both sensory (afferent) fibers, whose cell bodies

are found in the dorsal root ganglion (A), and motor (efferent) fibers which exit the cord via the ventral root. The spinal nerve splits into two rami, the anterior ramus (D) that supplies the anterolateral thoracic wall, in the thoracic region, and the posterior ramus (C) that supplies the skin and muscles of the back. The anterior rami of the spinal nerves in the cervical, lumbar, and sacral regions contribute to the formation of the plexuses in each of these regions.

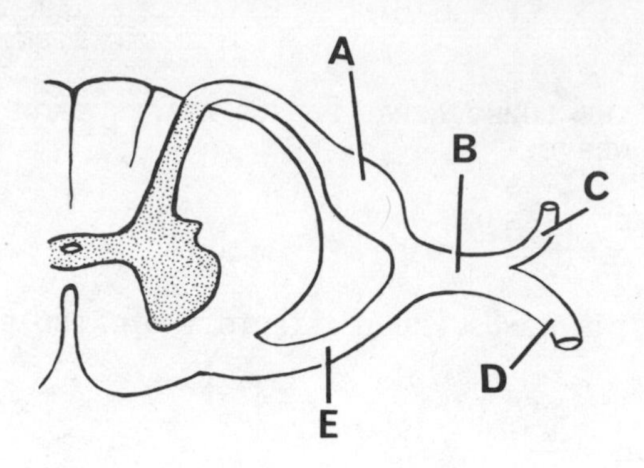

2. E. The diencephalon is formed from the vesicle indicated by B in the diagram. The diencephalon of the developing brain gives rise to the thalamus and hypothalamus. The telencephalon (A) gives rise to the cerebral hemispheres and the midbrain structures develop from the mesencephalon (C). The metencephalon (D) of the developing brain gives rise to the pons and cerebellum, whereas the myelencephalon (E) gives rise to the medulla oblongata.

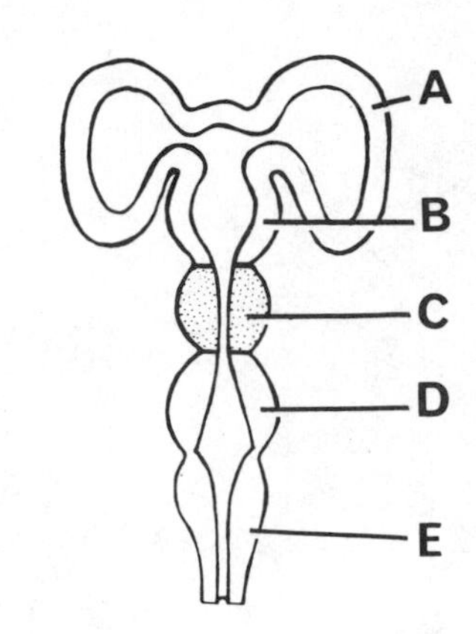

3. B. The thalamus of the diencephalon is indicated by B in the diagram. Other structures labelled are the corpus callosum (A), the midbrain (C), hypothalamus (D), and the pituitary gland (E).

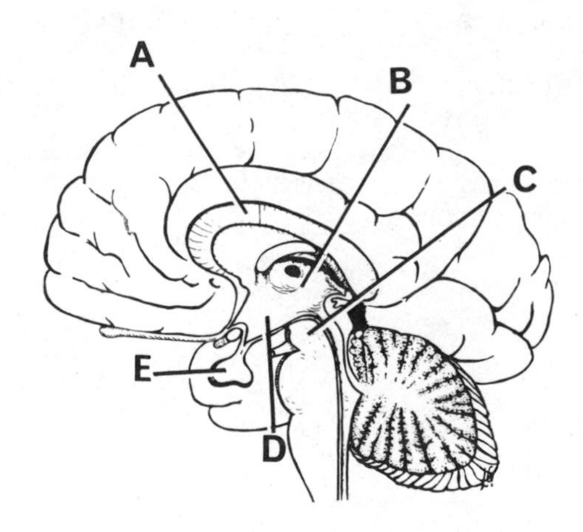

4. D. The parietal lobe (shaded in the diagram) is the area lying posterior to the central sulcus (C) and anterior to the parietooccipital fissure (D). The occipital lobe (E) forms the posterior aspect of the cerebral hemisphere. The temporal (A) and frontal (B) lobes form the remaining lobes of the hemisphere. The insula (not seen in this diagram) lies deep to the lateral fissure (F) and can be seen when the temporal and frontal lobes are separated.

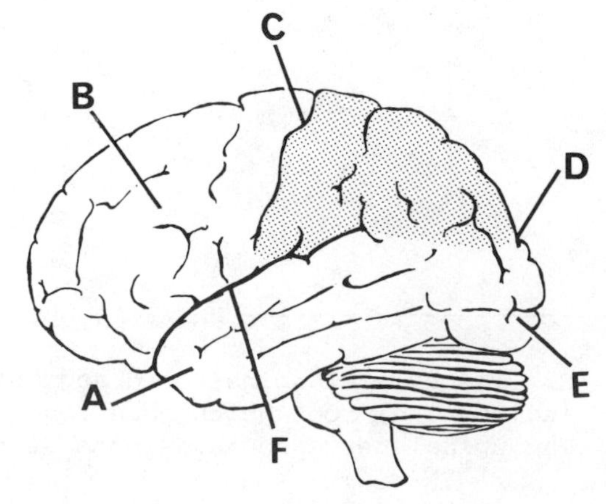

5. A. The myelinated nerve tracts of the cerebrum are of three basic types: a) commissural fibers which transmit impulses from one hemisphere to the other, primarily via the corpus callosum; b) association fibers which transmit impulses from one part of the cerebral cortex to the other; and c) projection fibers which pass through the internal capsule of the cerebrum either as ascending or descending tracts. Projection fibers carry messages from the cerebral cortex to lower levels of the CNS or from the lower level components of the CNS to the cerebral cortex.

6. A. The pons (labelled A in the diagram) forms part of the brain stem and lies between the midbrain (C) and the medulla oblongata (E). The cerebellum (D) lies superior and posterior to the brain stem, which traditionally consists of the midbrain, pons, and medulla oblongata. The thalamus (B) is part of the diencephalon.

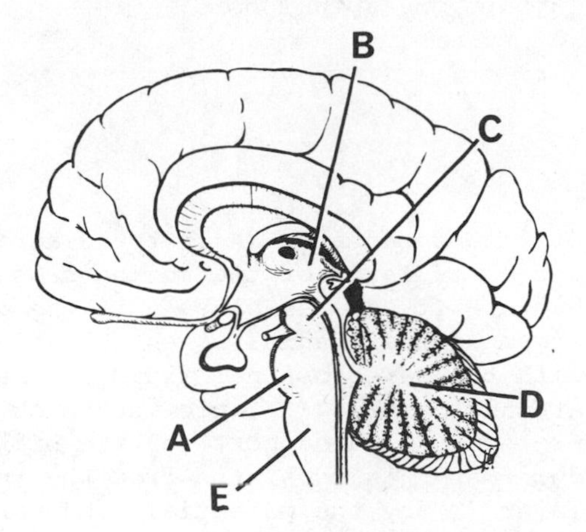

7. E. In this diagram of the ventricles of the brain, the interventricular foramen is indicated by E. The two large lateral ventricles and the third ventricle of the brain are not labelled but are connected by the interventricular foramen. The third ventricle is connected to the fourth ventricle (C) by the aqueduct of Sylvius (D). The cerebrospinal fluid, produced in ventricles of the brain, leaves the brain and passes into the subarachnoid space of the brain and spinal cord by the foramen of Magendie (B) and the foramina of Luschka (A).

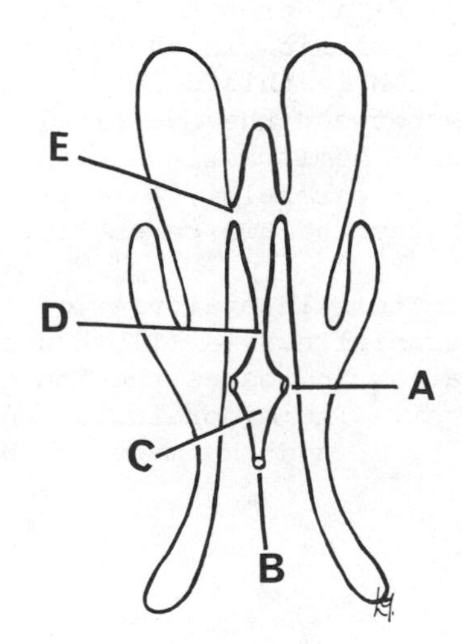

8. D. The neuromuscular and neurotendinous spindles and the Ruffini endings located in muscles, tendons, and joints all provide data for reflex adjustments of muscle action and for the awareness of position and movement. Nerve impulses from these receptors are conducted centrally by sensory neurons of either spinal or cranial nerves. These sensory neurons either distribute their impulses to the brain as data information or are involved in local cord reflex responses.

9. C. The arachnoid layer of the spinal cord meninges is indicated by E in the diagram. The dura mater (A) forms the outer layer of the meninges. The subarachnoid space (B) lies beneath the arachnoid layer which is attached to the pia mater (C) by trabeculae. The denticulate ligament (D) is a lateral extension of the pia mater and serves to support the spinal cord.

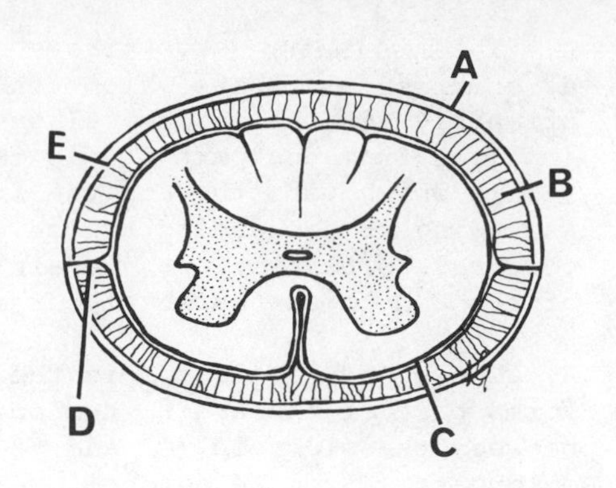

10. D. The dura mater (A in the diagram) is the outer layer of the meninges of the brain. It adheres closely to the periosteum of the cranial bones (B) and along with the cerebrospinal fluid of the subarachnoid space (C), provides much of the protection and support for the brain. The dura mater is separated from the arachnoid layer (D) by the potential subdural space. The falx cerebri (E) is an extension of the dura between the cerebral hemispheres. The inner layer of the meninges or pia mater (F) adheres to the brain. Another dural reflexion is the tentorium cerebelli (not seen in this diagram) which intervenes between the occipital lobes of the cerebral hemispheres and the cerebellum. The falx cerebelli, also a dural reflexion, separates the two cerebellar hemispheres.

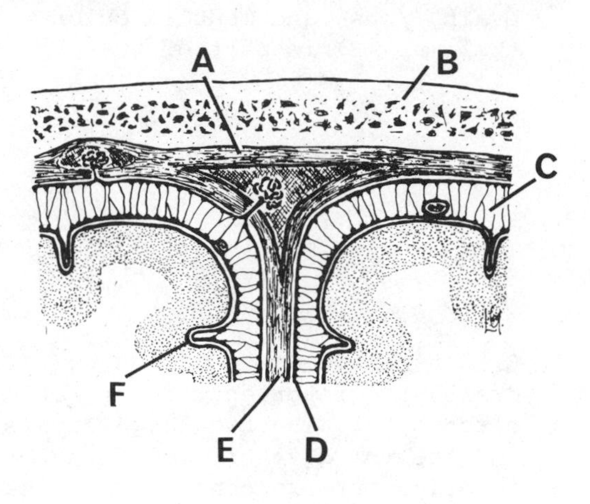

11. A. In this lateral view of the brain stem, the cranial nerves III through XI are labelled. The arrow indicates the Xth cranial nerve (vagus). The midbrain is indicated by A, the pons by B, and the medulla oblongata by C.

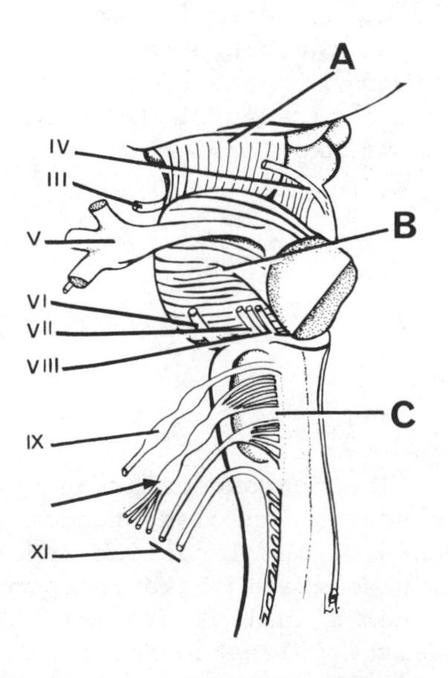

12. E. The respiratory center, along with the vasomotor and cardiac centers, is located in the medulla oblongata of the brain stem. However, a component of the respiratory center is also located in the pons. These three important centers of the medulla are called the vital centers because of the nature of

the functions they control. Trauma or disease affecting the lower end of the brain stem, particularly the medulla, will often lead to death. Other reflex centers found in the medulla are those of coughing, sneezing, vomiting, and swallowing.

13. D. The nerve responsible for innervating the diaphragm is the phrenic. It is a nerve that arises from the union of cervical spinal nerves four and five (may also have other contributions). The phrenic is the only motor nerve to the diaphragm; however, besides carrying motor fibers it also contains sensory and sympathetic fibers. Some of the intercostal nerves also contain sensory fibers from the diaphragm.

14. E. The lumbosacral plexus of nerves is formed by the anterior rami of the first three lumbar, most of the fourth lumbar, the first three sacral, and part of the fourth sacral nerves. The nerve branches from this plexus (lumbar and sacral) supply the lower limb and pelvic structures.

15. D. The lower end of the spinal cord is called the conus medullaris (D in the diagram). The conus medullaris reaches the lower level of the second lumbar vertebra. The dura mater (B) and the subarachnoid layer (A) of the spinal cord are indicated. The pia mater (not labelled) is closely adhered to the cord, and its extension, the filum terminale (C), projects down into the subarachnoid space (E) and fuses with the dura mater of the dural sac.

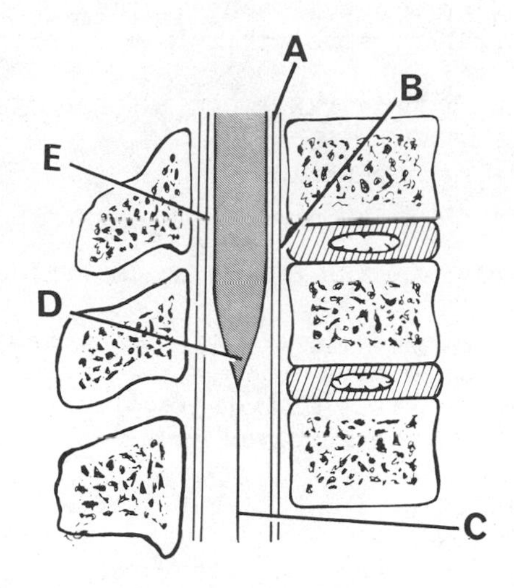

16. C. There are eight pairs of cervical spinal nerves. There are 12 pairs of thoracic, five pairs of lumbar, five pairs of sacral, and a single pair of coccygeal spinal nerves. There are, therefore, a total of 31 pairs of spinal nerves exiting from the spinal cord.

17. A. The intermediolateral gray column of the spinal cord is indicated by A. The posterior (D) and anterior (B) columns of gray matter consist, as does all gray matter of the central nervous system, mainly of neuron cell bodies, some of their processes, and neuroglial cells. The white matter of the cord consists of three columns or funiculi: posterior (E), lateral (C), and anterior (which is not labelled). White matter consists primarily of myelinated nerve fibers and neuroglial cells. Both the white and gray matter also contain many blood vessels.

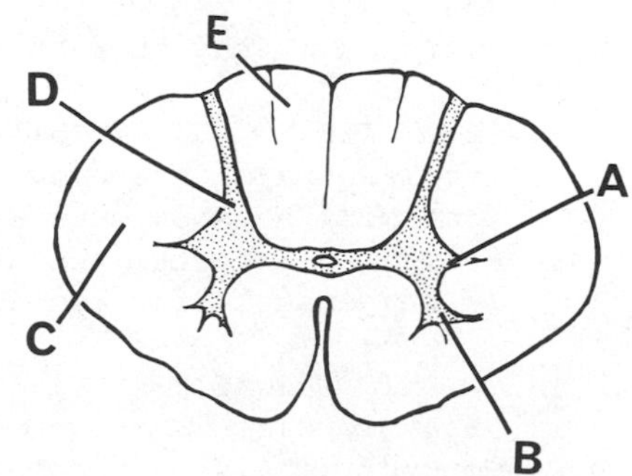

MULTI-COMPLETION QUESTIONS

DIRECTIONS: In each of the following questions or incomplete statements, ONE OR MORE of the completions given is correct. At the lower right of each question, circle A if 1, 2, and 3 are correct; B if 1 and 3 are correct; C if 2 and 4 are correct; D if only 4 is correct; and E if all are correct.

1. THE CEREBRAL CORTEX:
   1. Is the thin surface layer of the cerebrum
   2. Projects about a cm into each hemisphere
   3. Consists of gray matter
   4. Contains myelinated nerve tracts and bundles

   A B C D E

2. WHICH OF THE FOLLOWING STATEMENTS IS(ARE) TRUE FOR A RECEPTOR?
   1. May mediate more than one type of sensation
   2. Detect changes in the environment
   3. Form a component of the reflex arc
   4. Their sensitivity may depend on their density

   A B C D E

3. THE FOURTH VENTRICLE IS PRODUCED FROM WHICH OF THE FOLLOWING LUMINA?
   1. Metencephalon
   2. Telecephalon
   3. Myelencephalon
   4. Diencephalon

   A B C D E

4. WHICH OF THE FOLLOWING IS(ARE) CORRECT FOR THE RETICULOSPINAL TRACTS?
   1. Are a component of the extrapyramidal system
   2. Originate within the brain stem
   3. Consist of both facilitatory and inhibitory fibers
   4. May be involved in many repetitious movements

   A B C D E

5. WHICH OF THE FOLLOWING STATEMENTS IS(ARE) TRUE FOR THE MEDULLA OBLONGATA?
   1. Contains the respiratory center
   2. Houses the nuclei of the fifth to eighth cranial nerves
   3. Composed mainly of white matter
   4. Is continuous with the midbrain

   A B C D E

6. THE CEREBROSPINAL FLUID (CSF) IS:
   1. Produced only in the lateral ventricles
   2. Absorbed by the arachnoid villi
   3. Found within the ventricles exclusively
   4. Secreted by the choroid plexuses

   A B C D E

7. WHICH OF THE FOLLOWING IS(ARE) NOT A FUNCTION OF THE HYPOTHALAMUS?
   1. Involved in maintaining the wakeful state
   2. Controls the ANS centers in the cord
   3. Helps regulate the release of some anterior pituitary hormones
   4. Is responsible for emotions

   A B C D E

8. THE CORTICOSPINAL TRACTS:
   1. Relay impulses to the anterior horn motoneurons by interneurons
   2. Conduct impulses to somatic effectors
   3. Partially cross in the medulla oblongata
   4. Originate in the postcentral gyrus of the cerebral cortex

   A B C D E

9. THE CRANIAL NERVES INVOLVED IN MOVEMENTS OF THE EYEBALL INCLUDE THE:
   1. Facial
   2. Optic
   3. Trigeminal
   4. Oculomotor

   A B C D E

A	B	C	D	E
1,2,3	1,3	2,4	Only 4	All correct

10. WHICH OF THE FOLLOWING STATEMENTS IS(ARE) TRUE FOR THE PONS?
   1. Lies between the midbrain and the medulla oblongata
   2. Contains nuclei of cranial nerves V through VIII
   3. Is involved in the control of respiration
   4. Most descending fibers cross to the opposite side in this region

   A B C D E

11. THE BRACHIAL PLEXUS:
   1. Gives rise to the median nerve
   2. Provides nerves to joints
   3. Has contributions from spinal nerve T_1
   4. Contains motor fibers

   A B C D E

12. THE THALAMUS OF THE DIENCEPHALON:
   1. Has an active role as an endocrine organ
   2. Relays sensory impulses to the cerebrum
   3. Plays an important role in temperature regulation
   4. Is involved in the arousal or alerting mechanism

   A B C D E

13. THE SOMATIC SENSORY AREA OF THE BRAIN:
   1. Interprets movement and the position of parts of the body
   2. Is located in the precentral gyrus of the cerebral cortex
   3. Receives sensory input from the opposite side of the body
   4. Is connected by neurons to the hypothalamus

   A B C D E

14. WHICH OF THE FOLLOWING STATEMENTS ABOUT THE SPINAL CORD IS(ARE) CORRECT?
   1. Begins at the level of the foramen magnum
   2. Extends down to the lower end of the second sacral vertebra
   3. Has enlargements in the cervical and lumbar regions
   4. Consists of 32 segments

   A B C D E

15. THE DORSAL ROOT GANGLION:
   1. Contains the cell bodies of afferent neurons
   2. Has many synapses formed within it
   3. Is continuous with the dorsal root
   4. Usually has some efferent neurons

   A B C D E

16. THE FUNCTIONS OF THE CEREBELLUM INCLUDE THE:
   1. Control of the heart rate
   2. Synergistic control of muscle action
   3. Relaying of all kinds of sensory impulses
   4. Maintenance of equilibrium

   A B C D E

17. THE ANTERIOR RAMI:
   1. Are larger than the posterior rami
   2. Contain both motor and sensory fibers
   3. Of the cervical, lumbar, and sacral nerves form plexuses
   4. Give rise to white rami communicans in the thoracic and upper lumbar regions

   A B C D E

18. WHICH OF THE FOLLOWING STATEMENTS IS(ARE) CORRECT FOR A SENSORY PATHWAY?
   1. Terminates in the postcentral gyrus of the parietal lobe
   2. Includes a peripheral neuron that terminates in the CNS
   3. Has a neuron ascending within the cord or brainstem to the thalamus
   4. The cell body of the peripheral neuron lies in the dorsal root or the cranial nerve ganglion

   A B C D E

1. B, 1 and 3 are correct. The cerebral hemispheres are covered by a thin layer of gray matter called the cerebral cortex (A in the diagram). This layer of gray matter, less than 5 mm thick, is responsible for sensory, motor, and intellectual functions, only to list a few. Beneath the thin cortex lies the white matter (B) which consists of millions of myelinated nerve fibers forming tracts and bundles. These fibers connect all areas of the brain, spinal cord, and peripheral nerves into a functional unit known as the nervous system. Some of the basal (ganglia) nuclei (C) are seen in the center of the cerebral hemispheres.

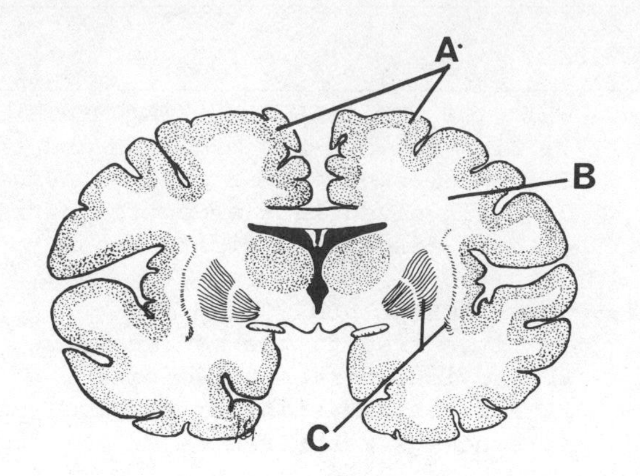

2. E. All are correct. A receptor is the distal end of a sensory neuron and is involved either in detecting sensations or initiating a reflexive response. The receptors are capable of detecting changes in both our external or internal environments. They also initiate the response (reflex arc) necessary to adjust to these changes so as to maintain or restore homeostasis. The receptors of the body may mediate more than one type of sensation and their sensitivity may in fact depend on their density in the skin or other tissues.

3. B, 1 and 3 are correct. The fourth ventricle develops from the lumen of the myelencephalon (E in the diagram) and the metencephalon (D) of the developing brain. The aqueduct of Sylvius arises from the lumen of the mesencephalon (C), whereas the third ventricle is derived from the lumen of the diencephalon (B). The lateral ventricles of the cerebral hemispheres develop from the lumen of the telencephalon (A).

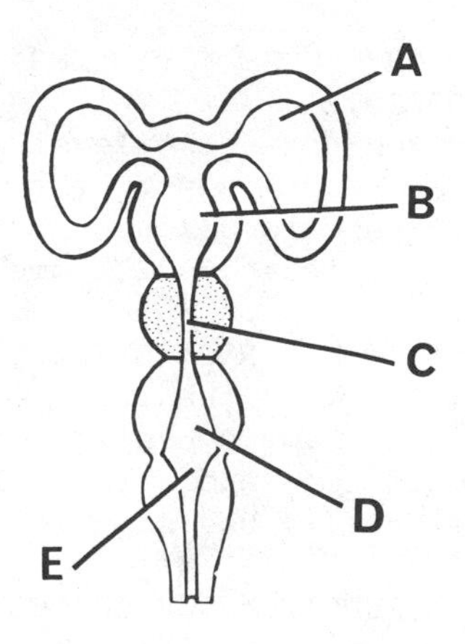

4. E, All are correct. The reticulospinal tracts are components of the ill-defined extrapyramidal motor system. These tracts originate from cell bodies located in the reticular formation of the brain stem and terminate by the synapsing of interneurons to the anterior horn cells of the spinal cord. These reticulospinal neurons function as facilitatory or inhibitory neurons. The summation of these opposing influences determines the response of the lower motoneuron. These tracts play a role in producing our larger, more automatic movements because of sequential or simultaneous contractions of muscles such as those used in walking or normal voluntary movements.

5. B, 1 and 3 are correct. The medulla oblongata is a component of the brain stem and lies between the pons above and the cervical portion of the spinal cord below. It contains the vital centers (breathing, cardiovascular) and reflex centers (coughing, sneezing, swallowing, and vomiting) of the brain. The medulla also consists of large numbers of nerve fibers (white matter) that transmit impulses to or from the higher centers of the brain. Many of these

fibers cross (decussate) to the opposite side of the brain or cord in the medulla. The nuclei of the last four cranial nerves (IX, X, XI, and XII) are found in the medulla oblongata.

6. C, 2 and 4 are correct. The cerebrospinal fluid (CSF) is produced in the choroid plexuses of the lateral ventricles of the brain. Smaller plexuses of the third and fourth ventricles also contribute to the total volume of the CSF. The main site of absorption of the CSF into the venous system is through the arachnoid villi which project into the dural venous sinuses. The CSF is found within the ventricles of the brain, the central canal of the spinal cord, the subarachnoid spaces, and the cisternae of the brain and the spinal cord.

7. D, Only 4 is correct. The hypothalamus of the diencephalon links the psyche (mind) with the body and the nervous system with the endocrine system of the body. This component of the brain is responsible for control over the autonomic centers located both in the brainstem and the spinal cord. It serves as a regulator and coordinator of autonomic activities and helps to control and integrate the responses made by the body's visceral effectors. Neurons of the hypothalamic nuclei synthesize the hormones secreted by the posterior pituitary (neurohypophysis) gland. Other neurons of the hypothalamus release hormones that control the release of anterior pituitary hormones which are responsible for controlling the secretion of hormones from the adrenal cortex, thyroid gland, and sex glands. The hypothalamus also plays a role in maintaining the waking state (alerting mechanism), regulating appetite (food intake), and provides a mechanism for controlling the normal body temperature. The thalamus, and not the hypothalamus, plays an important role in the mechanisms responsible for interpreting emotions.

8. A, 1, 2, and 3 are correct. The corticospinal tracts are well-defined somatic motor pathways which consist of upper and lower motoneurons that conduct impulses from the brain to the skeletal muscles (somatic effectors) of the body. Most of the upper motor neurons begin in the gray matter of the precentral gyrus of the cerebral cortex and descend through the cerebrum and the brain stem. In the medulla, about three-quarters of the fibers cross to the opposite side of the white matter of the cord, whereas the remaining fibers continue to descend uncrossed in the white matter. Neurons of both tracts (crossed and uncrossed) synapse with interneurons of the gray matter, which in turn synapse with the anterior horn cells. Stimulation of the anterior horn motoneurons by impulses of the corticospinal tract results in the stimulation of individual groups of muscles.

9. D, Only 4 is correct. There are three cranial nerves involved in coordinating the movements of the eyeball. The oculomotor nerve innervates all the extraocular eye muscles except the superior oblique and the lateral rectus muscles which are innervated by the trochlear and abducens nerves, respectively. The nuclei for the oculomotor and trochlear cranial nerves are located in the midbrain and the nucleus of the abducens cranial nerve is located in the pons.

10. A, 1, 2, and 3 are correct. The pons is the middle component of the brain stem, lying between the midbrain, above, and the medulla oblongata, below. The pons consists primarily of white matter and serves as part of the synaptic or relay station of the brain stem, providing connections between the cerebral cortex and the cerebellum. The pons contains the nuclei of the cranial nerves V through VIII. It also contains a nucleus which is responsible for controlling the rate of respiration and helps control the medullary respiratory nucleus which is responsible for the basic rhythm of respiration. However, the corticospinal fibers cross to the opposite side of the cord in the medulla, not the pons.

11. E, All are correct. The brachial plexus arises, most commonly, from the anterior primary rami of the lower four cervical ($C_{5,6,7}$, and $_8$) and the first

thoracic spinal nerves. It primarily supplies the muscles and skin of the upper limb and shoulder. The plexus has five major nerve branches: the axillary, median, ulnar, radial, and musculocutaneous nerves. These nerves contain motor and sensory fibers and each also receives sensory fibers from the various joints of the upper limb.

12. C, 2 and 4 are correct. The thalamus, due to its strategic position, is involved in a number of important pathways. It is responsible for the recognition of sensation such as pain, temperature, and touch, and relays most sensory impulses on to the cerebrum. The thalamus plays a role in mechanisms that produce complex reflex movements and those that maintain the body in an alert or an aroused state. The thalamus also plays a role in emotions by associating sensory impulses with feelings of pleasantness and unpleasantness.

13. B, 1 and 3 are correct. The somatic sensory area of the brain is shaded in the diagram and lies in the gray matter of the postcentral gyrus which is located behind the central sulcus (A). This area of the cortex produces general sensation which is involved in pain, temperature, touch (both crude and fine discrimination), pressure, stereognosis (form and identification of objects by touch), kinesthesia (changes in the position of joints), vibratory sense, and two-point and weight discrimination. The general sensations of the right side of the body are predominately perceived by the left cerebral cortex because the ascending fibers cross (decussate) from the left side to right side in the brain stem. The opposite is true for sensations coming from the left side of the body. The last neurons of the sensory chain transmit impulses from the thalamus of the diencephalon up through the cerebral white matter (internal capsule) to the cortex of the postcentral gyrus.

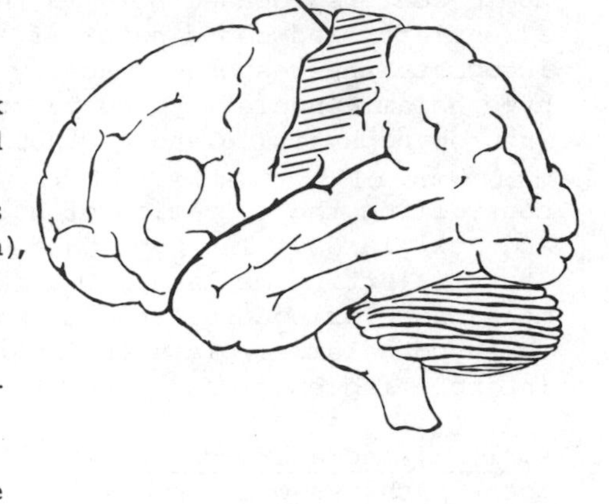

14. B, 1 and 3 are correct. The spinal cord is about the size of the little finger and extends from the level of the foramen magnum to the lower edge of the body of the second lumbar vertebra. It is continuous with the medulla and has two enlargements, one each in the cervical and lumbar regions, where the brachial and lumbar plexuses to the respective limbs find their origin. The cord gives off 31 pairs of spinal nerves and the exit of each nerve defines a segment of the spinal cord.

15. B, 1 and 3 are correct. The dorsal root ganglion (A in the diagram) contains the cell bodies of the afferent (sensory) neurons (B). The processes of these neurons are located in the dorsal root (C) of the spinal nerve and carry sensory impulses into the spinal cord. Within the cord the afferent neuron synapses with an efferent (motor) neuron (D) that carries impulses away from the spinal cord by the anterior root (E). The dorsal and anterior roots come together to form a mixed (containing both efferent and afferent fibers) spinal nerve (F).

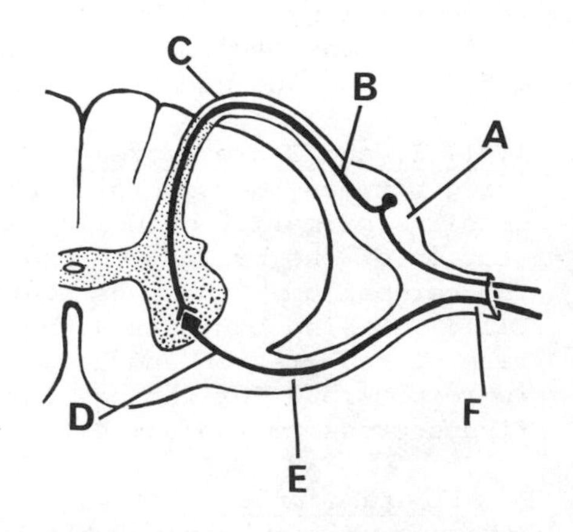

16. C, 2 and 4 are correct. The cerebellum is thought to function by controlling groups of skeletal muscle. In conjunction with the cerebral cortex the cerebellum is important in coordinating the skilled movements of muscle groups. It also controls skeletal muscle so as to maintain both equilibrium and posture. The cerebellum functions below the level of consciousness to produce movements that are smooth, steady, and efficient.

17. E, All are correct. The anterior primary ramus (A in the diagram) of a spinal nerve is larger than the posterior primary ramus (E). The anterior primary rami of the cervical, lumbar, and sacral nerves form plexuses. The posterior (C) and anterior (B) roots form a spinal (mixed) nerve (D), which then splits into a posterior and anterior primary ramus. These rami therefore contain both motor (efferent) and sensory (afferent) nerve fibers. From anterior rami, of the second thoracic to the second lumbar spinal nerves, exit the white rami communicans of the sympathetic division of the autonomic nervous system.

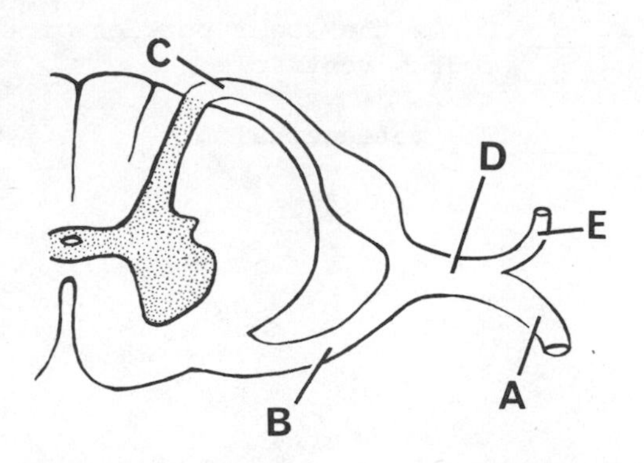

18. E, All are correct. Sensory pathways consist of a chain of three neurons that transmit an impulse from the periphery of the body to the CNS and through the thalamus to the cerebral cortex. The first (peripheral) neuron of the chain has its dendrites in the spinal or cranial nerves, its cell body in the dorsal root or cranial nerve ganglion and its axon terminates in the gray matter of the cord or the brain stem. The second neuron ascends from either a segment of the cord or the brain stem, crosses to the opposite side of the brain and terminates in the thalamus. The third neuron carries the impulse from the thalamus to the postcentral gyrus of the parietal lobe of the cerebral cortex.

FIVE-CHOICE ASSOCIATION QUESTIONS

DIRECTIONS: Each group of questions below consists of a numbered list of descriptive words or phrases accompanied by a diagram with certain parts indicated by letters, or by a list of lettered headings. For each numbered word or phrase, SELECT THE LETTERED PART OR HEADING that matches it correctly. Then insert the letter in the space to the right of the appropriate number. Sometimes more than one numbered word may be correctly matched to the same lettered part or heading.

1. _____ A layer forming the inner boundary of the subdural space
2. _____ The falx cerebri
3. _____ The dura mater
4. _____ Site for reabsorption of CSF
5. _____ The arachnoid layer

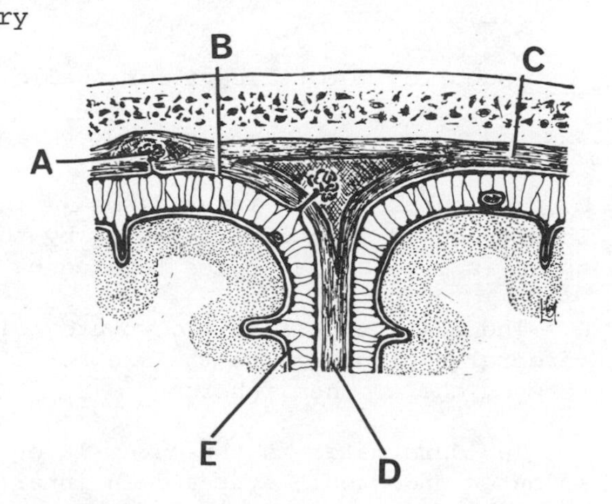

ASSOCIATION QUESTIONS

6. _____ Develops into the midbrain

7. _____ The telencephalon

8. _____ Contains the third ventricle

9. _____ Forms the lower portion of the fourth ventricle

10. _____ The metencephalon

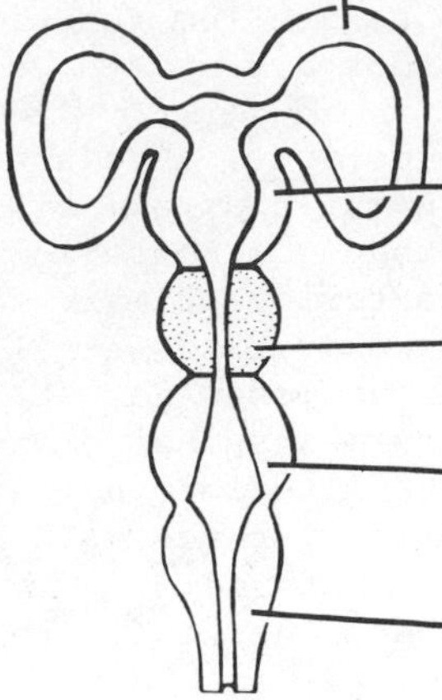

A
B
C
D
E

11. _____ A purely sensory nerve

12. _____ Involved in the sensation of balance

13. _____ A nerve involved with eye movments

14. _____ A motor nerve to the tongue

15. _____ A nerve whose cell bodies are located in the midbrain

16. _____ Contains afferent fibers for taste

17. _____ A nerve whose nucleus is located in the medulla

18. _____ Supplies motor fibers for the lateral rectus muscle

19. _____ A nerve whose motor nucleus is located in the pons

20. _____ Supplies motor fibers for swallowing

A. Facial
B. Optic
C. Hypoglossal
D. Trochlear
E. None of the above

------------------------ ANSWERS, NOTES AND EXPLANATIONS ------------------------

1. B. The subarachnoid layer forms the inner boundary of the subdural space. This space is limited externally by the inner layer of the dura. The subdural space is a potential space located between the above two meninges.

2. D. The falx cerebri is a downward fold of the dura matter between the two cerebral hemispheres of the brain. It lies within the longitudinal fissure between the two hemispheres.

3. C. The dura mater is the outer layer of the meninges and may be separated into an outer endosteal layer and an inner meningeal layer. The endosteal (endocranium) adheres to the inner surface of the cranial bones. The meningeal layer has four inward extensions: the falx cerebri, the tentorium cerebelli, the falx

cerebelli, and the diaphragma sellae. The dura of the brain is also continuous with the dura surrounding the spinal cord.

4. A. In many dural venous sinuses there are small projections of the subarachnoid layer called the arachnoid granulations. These villi project into the venous sinuses and are involved in returning the cerebrospinal fluid (CSF) to the blood stream.

5. B. The arachnoid layer is the middle layer of the meninges. It is attached to the pia mater, lying on the brain, by small projections called trabeculae. The space between the arachnoid layer and the pia mater forms the subarachnoid space and contains the CSF.

6. C. The midbrain of the mature brain develops from the mesencephalon. The lumen of the midbrain forms the aqueduct that joins the third and fourth ventricles of the brain.

7. A. The telencephalon forms from the forebrain or prosencephalon (A in the diagram) a primary brain vesicle. This primary brain vesicle forms two secondary brain vesicles: the diencephalon (B) which forms the thalamus and hypothalamus; and the telencephalon (C) which forms the cerebral hemispheres. The lateral ventricles (D) form from the lumen of the telencephalon and the third ventricle forms from the lumen of the diencephalon (E).

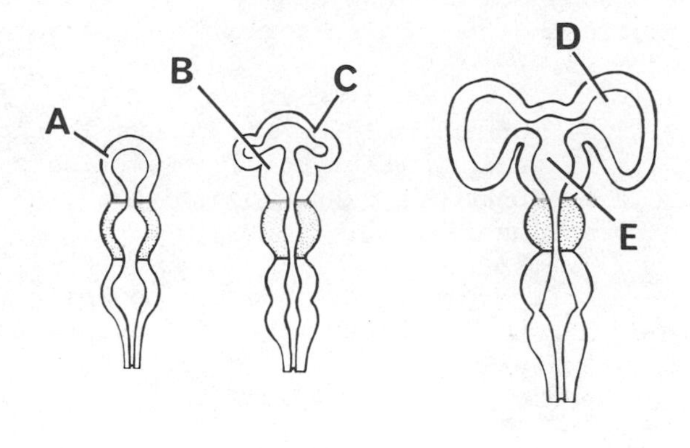

8. B. The lumen of the diencephalon forms the third ventricle. This lumen is therefore surrounded by the thalamus and the hypothalamus.

9. E. The rhombencephalon (A in the diagram), a primary brain vesicle, divides to form two secondary brain vesicles: metencephalon (B) and the myelencephalon (C). The lumen of the myelencephalon forms the lower portion of the fourth ventricle (D) and the walls develop into medulla. The lumen of the metencephalon forms the upper portion of the fourth ventricle and the walls form the pons and the cerebellum.

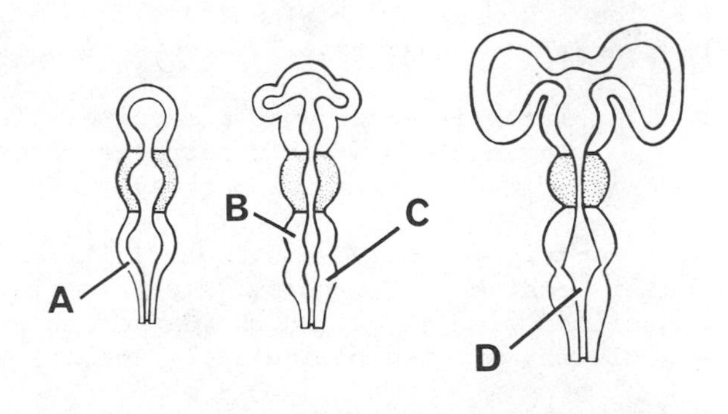

10. D. The metencephalon is derived from the rhombencephalon. Its lumen forms the upper portion of the fourth ventricle and its walls form the pons and cerebellum.

11. B. The optic or the second cranial nerve is considered to be purely sensory. The receptors of this nerve are located in the retina of the eyeball and are involved with the perception of images (vision). Nerve fibers from the retina

(optic tract) terminate in the thalamus and from there, relay fibers continue on to the occipital lobe of the cortex.

12. E. The vestibular nerve, involved in the perception of balance or imbalance, is a division of the eighth cranial nerve (vestibulocochlear). Fibers from the semicircular canals of the inner ear course their way to the vestibular nuclei of the pons and medulla. These fibers, like the cochlear division of this nerve, are purely sensory. The cochlear division is involved in relaying sensations of sound from the organ of Corti, located in the cochlea of the inner ear, to the cochlear nuclei located midway between the pons and the medulla of the brain stem.

13. D. The trochlear (fourth) cranial nerve, along with the oculomotor and abducens cranial nerves, are involved in movements of the eyeball. These three nerves are considered to be pure motor but may have some proprioceptive (muscle sense) fibers. The trochlear nerve is responsible for innervating the superior oblique muscle, the abducens nerve innervates the lateral rectus muscle and the oculomotor nerve innervates the remaining four external eye muscles.

14. C. The motor muscle of the tongue is the hypoglossal or twelfth cranial nerve. This nerve innervates both the extrinsic and intrinsic muscles of the tongue. Some evidence has been proposed that this pure motor nerve may carry some proprioception fibers.

15. D. The trochlear (IV) and the oculomotor (III) nerves have their nuclei in the midbrain (mesencephalon). These two nerves carry motor fibers to all of the external eye muscles, except the lateral rectus muscle which is innervated by the abducens (VI) nerve.

16. A. The facial (VII) nerve contains taste fibers originating from the taste buds of the anterior two-thirds of the tongue. The other nerve relaying the sensation of taste to the brain is the glossopharyngeal (IX). This nerve receives impulses from the posterior one-third of the tongue. The facial nerve also contains fibers which are distributed to the muscles of facial expression and to the salivary glands (secretion).

17. C. The hypoglossal (XII) nerve plus the accessory (XI), vagus (X), and glossopharyngeal (IX) nerves all have their nuclei located in the medulla.

18. E. The lateral rectus muscle receives efferent fibers via the abducens (VI) nerve. The nucleus of this nerve is situated in the pons. This nerve is considered to be purely motor but may carry some proprioceptive fibers.

19. A. The trigeminal (V), abducens (VI), and facial (VII) nerves all have motor nuclei located in the pons. The trigeminal nerve also has a sensory ganglion associated with it called the Gasserian nucleus; whereas the facial nerve is associated with the geniculate (sensory) ganglion.

20. E. The nerve supplying motor (efferent) fibers which are involved in swallowing is the glossopharyngeal (IX). This nerve is also involved in controlling the secretion of saliva and the reflex control of blood pressure and respiration. This nerve also has sensory (afferent) fibers which are involved with the sensation of taste.

FIVE-CHOICE COMPLETION QUESTIONS

DIRECTIONS: Each of the following questions or incomplete statements is followed by five suggested answers or completions. SELECT THE SINGLE BEST ANSWER in each case and then circle the appropriate letter at the lower right of each question.

1. WHICH OF THE FOLLOWING NEURONS IS ADRENERGIC?
   A. Sympathetic preganglionic
   B. Parasympathetic postganglionic
   C. Sympathetic postganglionic
   D. Parasympathetic preganglionic
   E. None of the above

   A B C D E

2. THE WHITE RAMUS COMMUNICANS IS INDICATED BY ________.

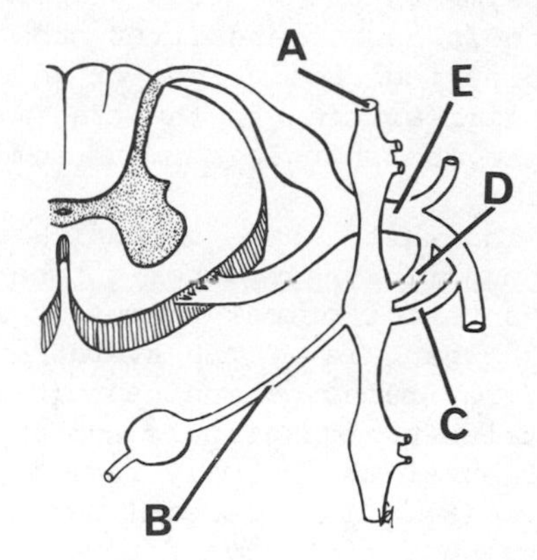

   A B C D E

3. WHICH OF THE FOLLOWING IS NOT A PARASYMPATHETIC GANGLION?
   A. Otic
   B. Celiac
   C. Ciliary
   D. Sphenopalatine
   E. Submandibular

   A B C D E

4. WHICH OF THE FOLLOWING CRANIAL NERVES CARRIES PREGANGLIONIC PARASYMPATHETIC FIBERS:
   A. Facial
   B. Abducens
   C. Hypoglossal
   D. Trochlear
   E. None of the above

   A B C D E

5. WHICH OF THE FOLLOWING IS NOT TRUE FOR THE SYMPATHETIC DIVISION OF THE AUTONOMIC NERVOUS SYSTEM (ANS)?
   A. Inhibits peristalsis
   B. Accelerates the heart rate
   C. Causes constriction of the pupil
   D. Produces relaxation of the bronchioles
   E. Constricts vessels in the cerebrum

   A B C D E

6. WHICH OF THE FOLLOWING IS NOT CORRECT FOR THE AUTONOMIC NERVOUS SYSTEM?
   A. All preganglionic neurons are cholinergic
   B. The parasympathetic system is dominant in stressful situations
   C. Most autonomic fibers are tonically active
   D. The two divisions of the ANS may be considered to be antagonistic
   E. Functions in the maintenance of homeostasis

   A B C D E

SELECT THE SINGLE BEST ANSWER

7. WHICH OF THE FOLLOWING IS NOT A FUNCTIONAL COMPONENT OF THE SYMPATHETIC DIVISION OF THE ANS?
   A. Superior cervical ganglion
   B. Celiac ganglion
   C. Splanchnic nerve
   D. Hypogastric plexus
   E. Otic ganglion

A B C D E

------------------------- ANSWERS, NOTES AND EXPLANATIONS -------------------------

1. C. The sympathetic postganglionic neurons are for the most part considered to be adrenergic. This means that most axon-effector synapses use norepinephrine as the neurotransmitter substance. Two exceptions to this rule are the postganglionic sympathetic vasodilator and postganglionic neurons to the sweat glands. All preganglionic sympathetic and preganglionic and postganglionic parasympathetic neurons are considered to be cholinergic (secrete acetylcholine as their neurotransmitter substance). Many autonomic drugs act by mimicking or blocking cholinergic or adrenergic discharges. However, despite the apparent similarities in the transmitter chemistry of preganglionic and postganglionic cholinergic neurons some drugs may act at these sites differently.

2. C. The white ramus communicans (C) contains preganglionic sympathetic fibers and joins the anterior primary ramus of a spinal nerve to the ganglia of the sympathetic chain. The gray ramus communicans (D in the diagram) contains postganglionic sympathetic neurons that enter the primary rami of the spinal nerve (E). The visceral nerve (B) contains postganglionic nerve fibers that innervate the viscera of the body. The sympathetic chain (A) contains preganglionic neurons that transmit impulses up or down the sympathetic chain to higher or lower ganglia.

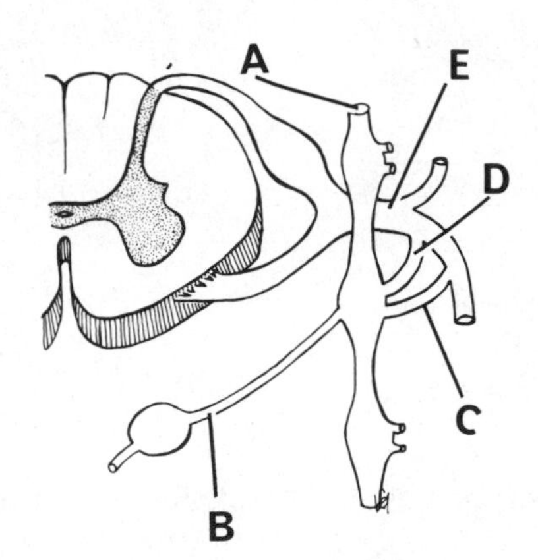

3. B. The celiac ganglion is a collateral ganglion of the sympathetic division of the ANS. The other four are parasympathetic ganglia. The ciliary ganglion provides postganglionic fibers to the ciliary muscle and pupillary sphincter of the eye. The pterygopalatine (sphenopalatine) ganglion innervates the lacrimal and palatine glands, whereas the submandibular (submaxillary) ganglion innervates the submandibular and sublingual salivary glands. The otic ganglion supplies the parotid salivary gland with postganglionic parasympathetic fibers.

4. A. The cranial nerves carrying parasympathetic outflow (preganglionic neurons) are the oculomotor, facial, glossopharyngeal, and vagus nerves. The vagus nerve distributes its autonomic fibers to the thoracic and abdominal viscera by way of the prevertebral plexuses. The other three cranial nerves distribute their parasympathetic fibers to the head.

5. C. Both cardiac and pulmonary functions are inhibited by parasympathetic vagal fibers, in that they decrease the force and rate of the heart beat and produce relaxation of the smooth muscle in the bronchioles of the lung. In general the parasympathetic division of the ANS is essentially anabolic since it is directed towards the preservation, accumulation, and storage of energies in the body. The sympathetic division on the other hand is considered to be catabolic in nature. It uses up body energy and inhibits the intake and assimilation of nutrient material. So the sympathetic division among other functions produces

acceleration of the heart, constricts blood vessels in the cerebrum, inhibits peristalsis, and produces dilation of the pupil.

6. B. A general function of the ANS is to maintain the general homeostasis of the body. The autonomic nervous system (ANS) is formed by the sympathetic and parasympathetic divisions. The sympathetic division is considered to be involved in catabolic activities, whereas the parasympathetic division is generally involved in anabolic activities. The sympathetic division is dominant in stressful situations and the parasympathetic division is primarily responsible for returning the body to its normal unstressed position. Therefore, the two divisions of the ANS are considered to be functionally antagonistic towards each other. All preganglionic fibers along with the parasympathetic postganglionic fibers are considered to be cholinergic. The postganglionic sympathetic fibers are adrenergic. Most autonomic fibers are continually active and this tonicity allows a single neuron system to increase or decrease the activity of a stimulated organ.

7. E. The otic ganglion is not a functional component of the sympathetic division but rather of the parasympathetic division of the ANS. The superior sympathetic ganglion is the uppermost ganglion of the sympathetic chain and provides the site for synapses of the preganglionic neuron with that of the postganglionic neuron. The celiac ganglion receives preganglionic fibers from the spinal nerves T_5-T_{12}. The postganglionic fibers arising from this ganglion innervate the gastrointestinal tract. The splanchnic nerves for the most part carry preganglionic fibers to the celiac and the superior and inferior mesenteric ganglia. The hypogastric plexus is located in the lower abdominal region and contains postganglionic sympathetic and preganglionic parasympathetic fibers.

MULTI-COMPLETION QUESTIONS

DIRECTIONS: In each of the following questions or incomplete statements, ONE OR MORE of the completions given is correct. At the lower right of each question, circle A if 1, 2, and 3 are correct; B if 1 and 3 are correct; C if 2 and 4 are correct; D if only 4 is correct; and E if all are correct.

1. THE SYMPATHETIC DIVISION OF THE ANS HAS:
   1. Its origin in spinal cord segments T_1-L_2
   2. 22 paired ganglia forming a chain
   3. An involvement with the white rami communicans
   4. More white rami communicans than gray

   A B C D E

2. POSTGANGLIONIC PARASYMPATHETIC NEURONS:
   1. Have long axons
   2. Originate in the intramediolateral gray column of the sacral cord
   3. Use norepinephrine at the synapse with the effector
   4. Usually involve a response in a single organ (effector)

   A B C D E

3. THE AUTONOMIC NERVOUS SYSTEM (ANS) IS:
   1. Independent of an afferent system
   2. A visceral effector system
   3. A single neuron system
   4. Not under control of the will

   A B C D E

4. THE PARASYMPATHETIC DIVISION OF THE ANS STIMULATES THE:
   1. Secretion of saliva
   2. Peristaltic movements of the intestine
   3. Contraction of the urinary bladder
   4. Accommodation of the lens for near vision

   A B C D E

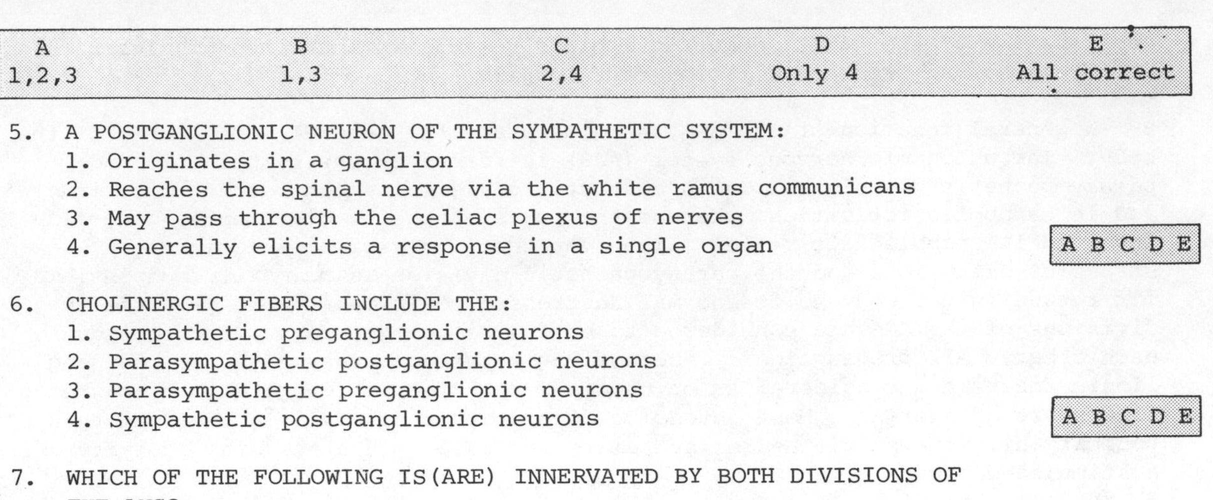

A	B	C	D	E
1,2,3	1,3	2,4	Only 4	All correct

5. A POSTGANGLIONIC NEURON OF THE SYMPATHETIC SYSTEM:
   1. Originates in a ganglion
   2. Reaches the spinal nerve via the white ramus communicans
   3. May pass through the celiac plexus of nerves
   4. Generally elicits a response in a single organ

   A B C D E

6. CHOLINERGIC FIBERS INCLUDE THE:
   1. Sympathetic preganglionic neurons
   2. Parasympathetic postganglionic neurons
   3. Parasympathetic preganglionic neurons
   4. Sympathetic postganglionic neurons

   A B C D E

7. WHICH OF THE FOLLOWING IS(ARE) INNERVATED BY BOTH DIVISIONS OF THE ANS?
   1. All blood vessels
   2. Bronchioles
   3. Sweat glands
   4. Eyes

   A B C D E

8. WHICH OF THE FOLLOWING STATEMENTS OF THE AUTONOMIC SYSTEM IS(ARE) CORRECT?
   1. Parasympathetic neurons are dominant in the digestive glands
   2. Tonicity is lacking in the fibers of this system
   3. Stress mediates an increase in sympathetic impulses
   4. Is not controlled by the higher centers of the somatic system

   A B C D E

------------------------ ANSWERS, NOTES AND EXPLANATIONS ------------------------

1. A, <u>1, 2, and 3 are correct</u>. The sympathetic division of the ANS originates in the interomediolateral gray column of the spinal cord segments T_1-L_2 and so may also be called the thoracolumbar division. This division has 22 pairs of ganglia (paravertebral) forming the sympathetic chains, located on either side of the vertebral column. The preganglionic neurons originating in the spinal cord reach the paravertebral ganglia by the white rami communicans. There is usually one white ramus communicans for each spinal nerve (T_1-L_2); however there are many more gray rami communicans since these connections leave the 22 sympathetic ganglia at all levels of the spinal cord. The gray rami communicans carry postganglionic sympathetic fibers back to the spinal nerves for distribution.

2. D, <u>Only 4 is correct</u>. Postganglionic parasympathetic neurons as a rule are unmyelinated fibers that evoke a local response to a single organ. This is opposite to the sympathetic postganglionic neurons which have a broad response. The parasympathetic postganglionic neurons are shorter, originate in ganglia near or in the organ they innervate and use acetylcholine as their neurotransmitter substance.

3. C, <u>2 and 4 are correct</u>. The autonomic nervous system (ANS) is a two-neuron system (preganglionic and postganglionic) which functions as a visceral efferent (motor) system. This system is typically described as an autonomous system, that is, it is not considered to be under the control of the will. However, there are some exceptions. The ANS is usually described as a visceral efferent system and little attention is given to a visceral afferent component. There is no doubt that such visceral and vascular afferent systems exist, but their precise anatomical distribution and physiological significance are by no means fully understood.

4. E, All are correct. The parasympathetic division is basically an anabolic system because its energies are directed towards the preservation, accumulation, and the storage of energies in the body. This system inhibits and causes the vessels in the cerebrum and abdominal viscera to dilate; inhibits the anal sphincter so that feces may escape; and inhibits the urinary sphincters so they will open and permit the emptying of the bladder. The parasympathetic division also stimulates the bronchi to constrict, the gastrointestinal tract to increase peristalsis, the bladder to contract, the pupil to constrict, the bulging of the lens for accommodation, and the secretion of saliva, gastric juice, pancreatic juice and insulin.

5. B, 1 and 3 are correct. The postganglionic neuron of the sympathetic division of the ANS originates in a ganglion (either a paravertebral or a collateral ganglion). From the paravertebral ganglion the postganglionic fibers join the spinal nerve by passing through the gray ramus communicans. They innervate the blood vessels, sweat glands, and erector pilorum muscles of the skin, and the blood vessels of skeletal muscle and bone. Postganglionic nerve fibers from collateral ganglia go to form either distinct nerves or perivascular (celiac) plexuses and innervate the glands and smooth muscle fibers found in the walls of blood vessels of the intestines. The postganglionic sympathetic fibers elicit responses over a larger number of glands, whereas the postganglionic parasympathetic fibers tend to elicit a response to a single organ.

6. A, 1, 2, and 3 are correct. The autonomic nervous system is thought to have two major neurotransmitter substances: acetylcholine and norepinephrine. All preganglionic neurons and the postganglionic neurons of the parasympathetic system secrete acetylcholine at the synapses and are called cholinergic neurons. The postganglionic sympathetic neurons secrete norepinephrine and are called adrenergic neurons. However, the postganglionic sympathetic neurons to the sweat glands and the blood vessels of skeletal muscle are thought to be cholinergic rather than adrenergic.

7. C, 2 and 4 are correct. Both divisions of the ANS are involved in the functioning of the smooth muscle of cerebral, abdominal viscera, and genital blood vessels; of hollow organs such as bronchi, digestive tract, anal and urinary sphincters, bladder, and eye; and glands such as found in the digestive tract and pancreas. Parasympathetic fibers are not involved in the smooth muscle functions of skin and skeletal blood vessels, arrector pili muscles of hair, or such glands as the sweat, liver, and adrenal medulla.

8. B, 1 and 3 are correct. Parasympathetic impulses to the digestive glands and the muscles of the digestive tract are dominant over sympathetic impulses under normal conditions. Under stressful conditions impulses from the sympathetic system dominate. The ANS is for the most part not under the control of the will but this does not mean it fails to come under control of higher centers in the brain such as the hypothalamus or the cerebral cortex. These higher brain centers along with centers in the spinal cord help control or coordinate the ANS. Most autonomic fibers are continually active. Such tonicity allows a single neuron system to increase or decrease the activity of a stimulated organ.

FIVE-CHOICE ASSOCIATION QUESTIONS

DIRECTIONS: Each group of questions below consists of a numbered list of descriptive words or phrases accompanied by a diagram with certain parts indicated by letters, or by a list of lettered headings. For each numbered word or phrase, SELECT THE LETTERED PART OR HEADING that matches it correctly. Then insert the letter in the space to the right of the appropriate number. Sometimes more than one numbered word may be correctly matched to the same lettered part or heading.

ASSOCIATION QUESTIONS

1. _____ An adrenergic neuron
2. _____ Neurons representing the sympathetic division
3. _____ A cholinergic postganglionic neuron
4. _____ A neuron originating in the sacral region of the spinal cord
5. _____ A neuron originating in a ganglion within the organ innervated

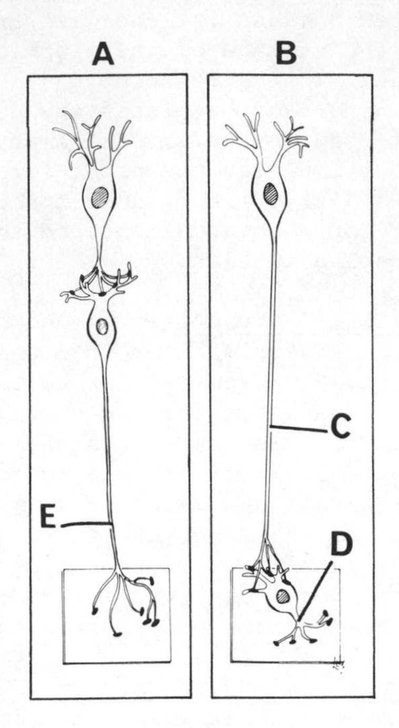

------------------------ ANSWERS, NOTES AND EXPLANTIONS ------------------------

1. E. The postganglionic neurons of the sympathetic division of the ANS are considered to be adrenergic. This means that its axon synapses with the effector organ by the use of norepinephrine as the neurotransmitter substance. These neurons have long axons when compared to the short axons of sympathetic preganglionic fibers.

2. A. The two neurons representing the sympathetic division of the ANS show a short preganglionic neuron with many collaterals and a long postganglionic neuron which leads to the effector. The parasympathetic neuron system as indicated by B (in the diagram) reveals a long preganglionic neuron with few collaterals and a short postganglionic neuron which synapses with the effector. In general, the sympathetic division of the ANS has a wider, more general distribution than the parasympathetic division which tends to be more local in its distribution.

3. D. The postganglionic parasympathetic neuron is cholinergic. Other cholinergic neurons include the preganglionic neurons of the sympathetic and parasympathetic divisions, and certain anatomical sympathetic postganglionic neurons that innervate the smooth muscle cells of blood vessels supplying sweat glands and skeletal muscle. All other anatomical sympathetic postganglionic neurons are classified as adrenergic.

4. C. All preganglionic neurons of the sacral portion of the parasympathetic division of the ANS arise from the interomediiolateral gray column of spinal cord segments S_2-S_4.

5. D. The postganglionic parasympathetic neuron originates from a ganglion located near or within the walls of the organ innervated. Consequently this neuron is extremely short as compared to its preganglionic neuron (C in the diagram).

9. SPECIAL SENSES

LEARNING OBJECTIVES

Be Able To

- Describe and identify on simple diagrams the three coats (tunica) of the eyeball with special reference to the following features: sclera, cornea, canal of Schlemm, ciliary body, ciliary muscle, suspensory ligament, iris, fovea centralis, macula lutea, and optic disc (papilla).
- Describe the chambers of the eyeball, giving their contents and function.
- Describe the functions of and nerves controlling the external and internal muscles of the eye.
- Describe the structure of the eyelids and the lacrimal apparatus, and give their functions.
- Discuss the functional layers of the retina, its stimulation, and describe how refraction, accommodation, and convergence are involved in the formation of a retinal image.
- Describe the components of the visual pathway and how the image is conducted to the visual cortex.
- Define the following terms as they relate to vision: myopia, hypermetropia, astigmatism, presbyopia, diplopia, heterophoria, exophoria, and strabismus.
- Describe the anatomy of the external and middle ear with special reference to the following: auricle, external acoustic meatus, tympanic membrane, auditory ossicles, fenestra (vestibuli) ovalis, fenestra (cochlea) rotunda, and eustachian (auditory) tube.
- Describe the anatomy, location, and function of the following components of the inner ear: bony labyrinth (vestibule, cochlea, and semicircular canals), and membranous labyrinth (utricle, saccule, cochlear duct, organ of Corti, tectorial membrane, semicircular ducts, perilymph, endolymph, ampulla, and crista ampullaris).
- Describe the components of the auditory pathway and the apparatus involved in the maintenance of equilibrium.

- Describe the anatomy, location, and function of the olfactory and gustatory (taste) sense organs.

RELEVANT READINGS

Anthony & Kolthoff - Chapter 10.
Chaffee & Greisheimer - Chapter 10.
Crouch & McClintic - Chapters 17 and 18.
Jacob, Francone & Lossow - Chapter 9.
Landau - Chapter 12.
Tortora - Chapter 16.

FIVE-CHOICE COMPLETION QUESTIONS

DIRECTIONS: Each of the following questions or incomplete statements is followed by five suggested answers or completions. SELECT THE SINGLE BEST ANSWER in each case and then circle the appropriate letter at the lower right of each question.

1. THE AREA OF THE RETINA WHERE THE CONES ARE THE MOST DENSE IS THE:
   A. Fovea centralis
   B. Ora serrata
   C. Macula lutea
   D. Optic disc
   E. None of the above

   A B C D E

2. THE LACRIMAL SAC IS INDICATED BY ________.

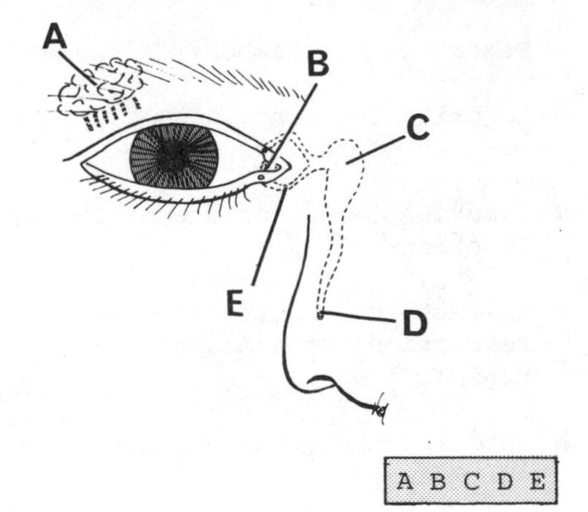

   A B C D E

3. WHICH OF THE FOLLOWING IS NOT A COMPONENT OF THE INNER EAR?
   A. Utricle
   B. Saccule
   C. Membranous ampulla
   D. Endolymphatic duct
   E. Mastoid air cells

   A B C D E

4. THE LAYER OF THE EYE WHOSE ANTERIOR PORTION IS MODIFIED TO FORM THE CILIARY BODY AND THE IRIS IS THE _______.
   A. Lens
   B. Retina
   C. Choroid
   D. Sclera
   E. None of the above

   A B C D E

5. THE STRUCTURE OVERLYING THE OVAL WINDOW IS THE:
   A. Oval membrane
   B. Stapes
   C. Malleus
   D. Tympanic membrane
   E. None of the above

   A B C D E

SELECT THE SINGLE BEST ANSWER

6. THE SACCULE IS INDICATED BY _________.

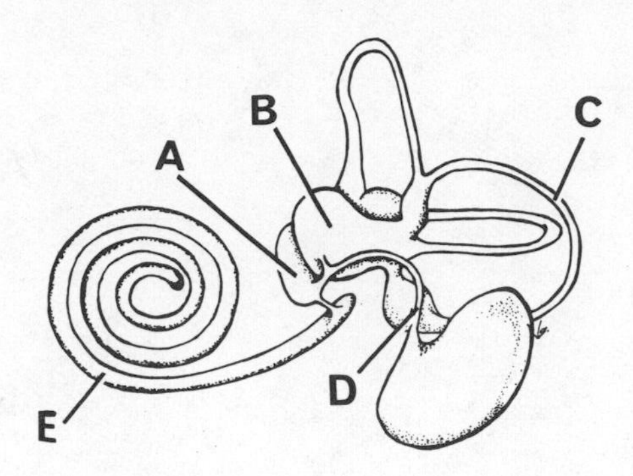

A B C D E

7. THE CRANIAL NERVE RESPONSIBLE FOR OLFACTION IS THE ________.
   A. Second
   B. Third
   C. Fourth
   D. Fifth
   E. None of the above

A B C D E

8. IN THIS DIAGRAM OF THE RIGHT EYE, THE INFERIOR OBLIQUE MUSCLE IS INDICATED BY __________.

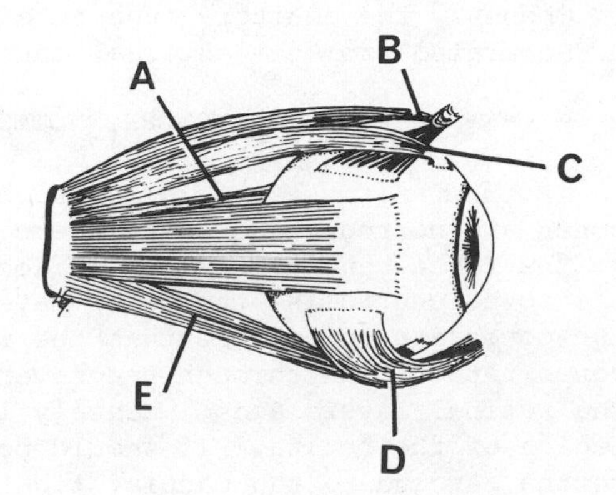

A B C D E

9. THE AUDITORY PORTION OF THE CEREBRAL CORTEX IS INDICATED BY __________.

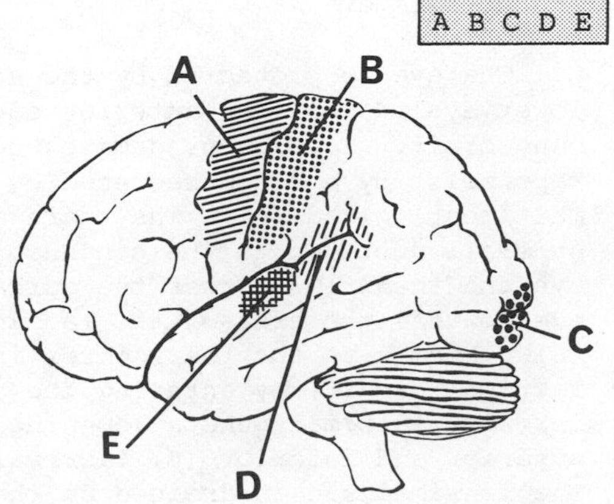

A B C D E

SELECT THE SINGLE BEST ANSWER

10. THE SCLERA IS INDICATED BY ______.

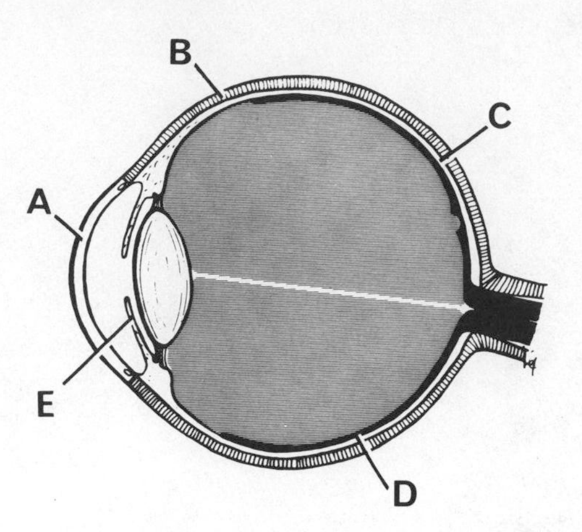

A B C D E

11. WHICH OF THE FOLLOWING IS NOT TRUE FOR THE MIDDLE EAR?
    A. Contains three ossicles
    B. Located in the sphenoid bone
    C. Is an epithelial lined cavity
    D. Opens by the auditory tube into the nasopharynx
    E. Separated from the external ear by the tympanic membrane

A B C D E

------------------------- ANSWERS, NOTES AND EXPLANATIONS -------------------------

1. A. Both the fovea centralis and the macula lutea of the retina have only cones and no rods. However, the cones are narrower and more densely packed at the fovea centralis, which is located at the center of the macula lutea. At the fovea centralis, the usual layers of the retina which normally lie above the cones are thinned so that the incoming light passes more directly to the cones rather than through the several layers of the retina. This thinning of the retinal layers aids immensely in the acuity of visual perception by this region of the retina. It should be realized that in other regions of the retina, excluding the macula, light must pass through the various layers of the retina before it hits or stimulates the rods and cones. The passage of light through such a nonhomogeneous tissue reduces the visual acuity for most of the retina.

2. C. The eye is protected by the upper and lower eyelids and the anterior edges of the bony orbit. To prevent damage the eye is kept moist by a fluid secreted by the lacrimal (tear) gland (A). The tears are secreted by small ducts on to the conjunctiva of the eye and then flow across the surface of the eye towards the area of the lacrimal lake (B). From here the tears enter into the lacrimal sac (C) by entering the superior and inferior lacrimal puncta (openings) into the superior and inferior (E) lacrimal canaliculi. The lacrimal sac is drained by the nasolacrimal duct (D) into the inferior meatus of the nose.

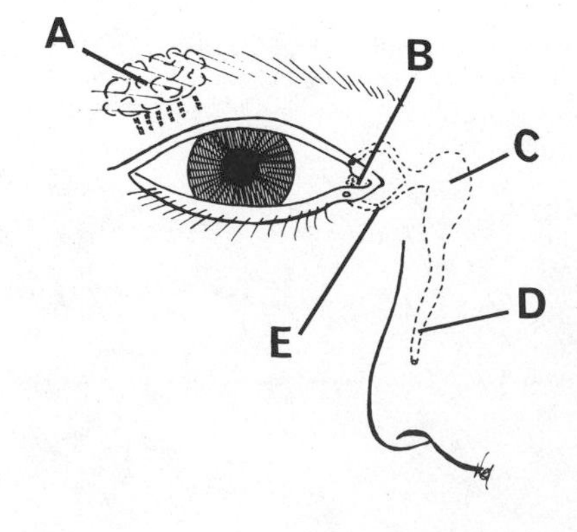

3. E. The inner ear consists of a bony labyrinth (chamber) which is lined by a membranous labyrinth. The bony labyrinth is made up of the vestibule, cochlea, and semicircular canals. The bony vestibule contains the utricle and saccule of the membranous labyrinth, the bony cochlea contains the cochlear duct portion of the membranous labyrinth, and the bony semicircular canals contain the membranous semicircular ducts. The mastoid air cells are not a component of the inner ear but are continuous with the middle ear.

4. C. The uvea, the middle vascular layer of the eyeball, has three parts to it: the choroid, the ciliary body, and the iris. The ciliary body and its extension, the iris, are both made up mainly of smooth muscle cells. The ciliary body moves forward when its smooth muscle contracts, decreasing the tension on the lens which then becomes more curved in the center. This allows the eye to focus on near objects, a process called accommodation. When the smooth muscle cells of the iris contract they constrict the pupil (miosis). The iris contracts reflexly when light strikes the retina (light reflex) and during the process of accommodation (focusing on a near object).

5. B. The foot plate of the stapes lies over the oval window (fenestra vestibuli). The foot plate of the stapes conducts vibrations which began at the tympanic membrane, passed through the ossicles to the perilymph of the bony cochlea. The movement of the perilymph is transmitted by the vestibular (Reissner's) membrane to the endolymph of the cochlea duct. The endolymph sets the basilar membrane moving and this in turn stimulates the hair cells of the organ of Corti. This is briefly a description of the physical apparatus that transmits the sound waves of the air to the end organ of hearing (organ of Corti) in the inner ear.

6. A. All the structures in this diagram form part of the membranous labyrinth of the inner ear. These membranous structures contain endolymph and are surrounded by perilymph which is contained within the bony labyrinth of the temporal bone. The saccule (A in the diagram) and the utricle (B) lie within the vestibule of the bony labyrinth. The membranous semicircular ducts (C, only the lateral one is labelled) are located within the bony semicircular canals and the cochlear duct (E) is contained within the bony cochlea. The endolymphatic duct (D) and sac are part of the endolymphatic system. The sac is associated with the meninges of the brain in the posterior portion of the temporal bone and it is here that the endolymph is absorbed into the subarachnoid space surrounding the brain. The endolymph is produced by the vascular-like cells of the cochlear duct.

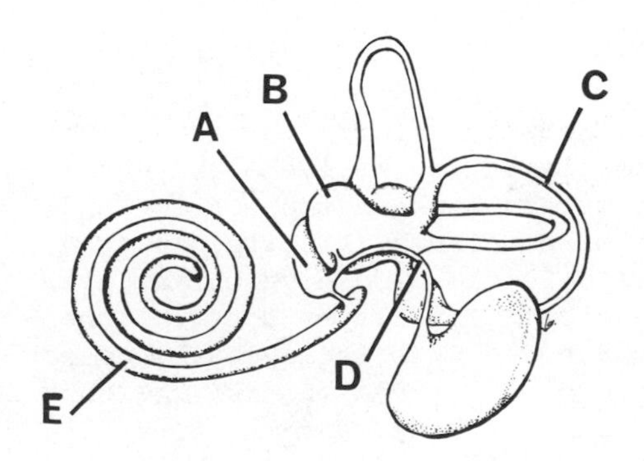

7. E. The olfactory or first cranial nerve is responsible for the sense of smell or the detection of odors. The olfactory membrane lies in the upper reaches of the nasal cavities. The bipolar cells responsible for the sensation of smell are located within this membrane and by cellular extensions (olfactory hairs) projecting to the surface of the mucous membrane, sample the air brought into the nose. The substances to be detected are carried to these sensory hairs by blasts of inspired air.

8. D. There are six muscles that are responsible for moving the eyeball. The superior rectus (C in the diagram), the inferior rectus (E), lateral rectus (not labelled) and medial rectus (A) muscles are responsible for turning the eyeball up, down, laterally and medially, respectively. The superior oblique (B) and the inferior oblique (D) muscles individually rotate the eyeball. The movements of the eyeball do not depend solely on an individual muscle but rather the complicated combinations of the six muscles. Their main task is to move the eyeballs so they both look exactly on the same object.

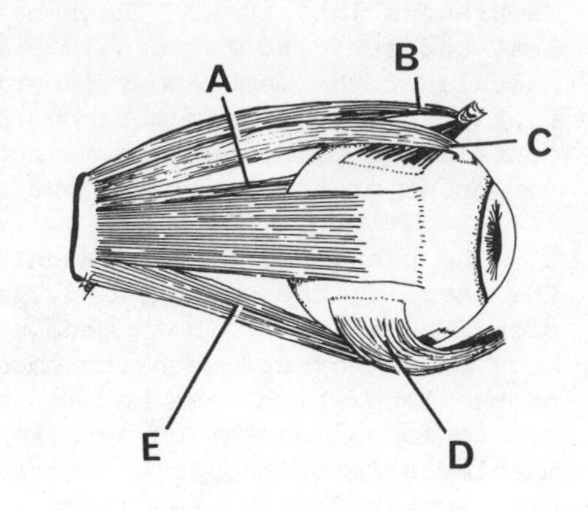

9. E. The auditory area of the cerebral cortex is the superior temporal gyrus and is indicated by E in the diagram. The motor area (A) is found anterior to the central sulcus and somatic sensory area (B) is located behind this sulcus. The speech area (D) of the cortex is located also in the temporal lobe but beneath the auditory area. The visual (area) cortex (C) is located in the occipital lobe.

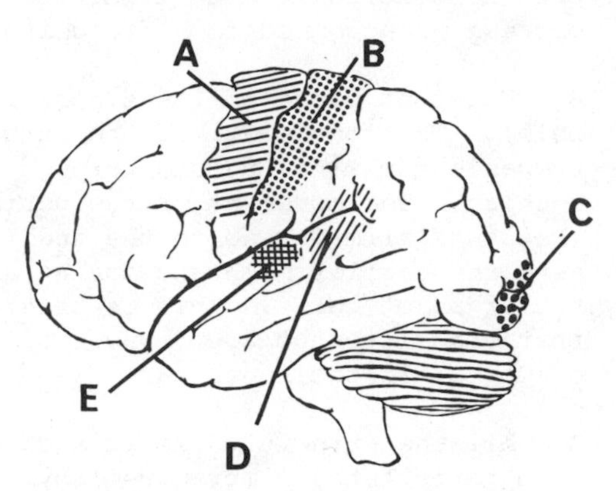

10. B. The sclera forms the tough opaque outer layer of the eyeball. Anteriorly the sclera becomes transparent and forms the cornea (A). Deep to the sclera is the middle vascular layer or uvea (C). This layer contains the blood vessels to the eyeball and consists of three components: the choroid, the ciliary body, and the iris (E). The inner layer of the eyeball is the retina (D) and contains the photoreceptors of the eye.

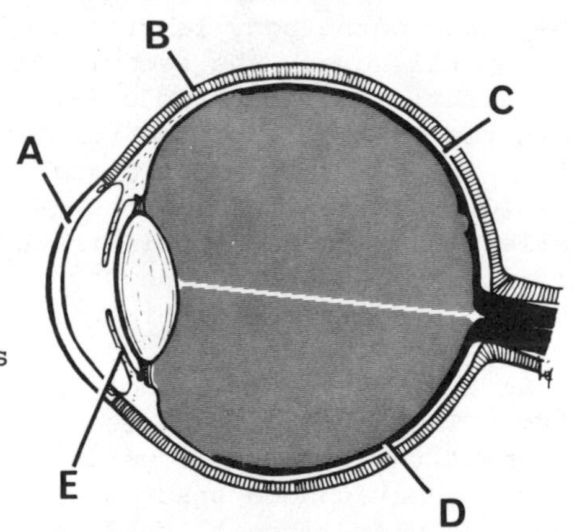

11. B. The middle ear is an epithelial lined bony cavity found within the temporal bone of the skull. It is separated from the external ear by the tympanic membrane, to which is attached a chain of three ossicles. The malleus, one of the ossicles, is attached to the inner surface of the tympanic membrane and at its other end articulates with the incus. The incus in turn articulates with the stapes which covers the oval window. The middle ear communicates with the inner ear by the fenestra vestibuli (oval window) and the fenestra

cochlea (round window) which is covered by a membrane. The middle ear communicates with the nasopharynx by the auditory (eustachian) tube. The middle ear is also continuous with the many mastoid air cells of the temporal bone.

MULTI-COMPLETION QUESTIONS

DIRECTIONS: In each of the following questions or incomplete statements, ONE OR MORE of the completions given is correct. At the lower right of each question, circle A if 1, 2, and 3 are correct; B if 1 and 3 are correct; C if 2 and 4 are correct; D if only 4 is correct; and E if all are correct.

1. WHICH OF THE FOLLOWING STATEMENTS IS(ARE) TRUE FOR THE ANTERIOR CHAMBER OF THE EYE?
   1. Its fluid drains into the venous system
   2. Fails to communicate with the posterior chamber
   3. Located between the iris and the cornea
   4. Contains vitreous humor

   A B C D E

2. THE SPECIAL SENSES INCLUDE:
   1. Equilibrium
   2. Smell
   3. Hearing
   4. Taste

   A B C D E

3. THE NERVE(S) INVOLVED WITH THE EXTRINSIC MUSCLES OF THE EYE INCLUDE THE:
   1. Glossopharyngeal
   2. Optic
   3. Trigeminal
   4. Abducens

   A B C D E

4. WHICH OF THE FOLLOWING STATEMENTS IS(ARE) CORRECT FOR THE RODS OF THE RETINA?
   1. Uses rhodopsin as its light-sensitive chemical
   2. Requires vitamin D for the synthesis of photochemicals
   3. Are thought to be functional in dark conditions
   4. Involved in color vision

   A B C D E

5. WHICH OF THE FOLLOWING STATEMENTS IS(ARE) TRUE FOR THE AQUEOUS HUMOR?
   1. Found in the space between the cornea and the iris
   2. Located in the posterior chamber of the eye
   3. Produced by the ciliary processes
   4. Contains a mesh of collagen fibrils

   A B C D E

6. THE AUDITORY TUBE:
   1. Is continuous with the inner ear
   2. Equalizes the pressure on both sides of the tympanic membrane
   3. Is a total bony structure
   4. Is normally collapsed

   A B C D E

7. THE TRANSMISSION OF ODORS INCLUDES WHICH OF THE FOLLOWING COMPONENTS?
   1. An area anterior to the hypothalamus
   2. The mitral cell
   3. The uncus
   4. The olfactory tract

   A B C D E

8. TASTE IMPULSES REACH THE BRAIN STEM BY WHICH OF THE FOLLOWING NERVES?
   1. Trigeminal
   2. Facial
   3. Accessory
   4. Vagus

   A B C D E

A	B	C	D	E
1,2,3	1,3	2,4	Only 4	All correct

9. WHICH OF THE FOLLOWING IS(ARE) NOT CORRECT FOR THE INTRINSIC MUSCLES OF THE EYE?
   1. Includes the ciliary muscle
   2. Regulates the size of the pupil
   3. Consists of smooth muscle fibers
   4. Are innervated in part by the trigeminal nerve

   A B C D E

10. THE TASTE BUDS ARE:
   1. Associated with the papillae of the tongue
   2. Thought to respond to all taste stimuli
   3. Modified epithelial cells
   4. More sensitive in individuals over 45 years

   A B C D E

11. THE UTRICLE AND THE SACCULE:
   1. Are filled with perilymph
   2. Both contain a macula
   3. Have continuous tectorial membranes
   4. Are involved in the righting reflexes

   A B C D E

12. WHICH OF THE FOLLOWING IS(ARE) TRUE FOR THE AUDITORY PATHWAY?
   1. The organ of Corti acts as an auditory receptor
   2. Impulses from the right ear pass up both sides of the brain
   3. Auditory neurons synapse in the inferior colliculus
   4. Its primary neurons synapse in the pons

   A B C D E

ANSWERS, NOTES AND EXPLANATIONS

1. B, 1 and 3 are correct. The anterior chamber of the eye contains aqueous humor and is located between the posterior surface of the cornea and the anterior surface of the iris. It communicates posteriorly with the posterior chamber, lying between the posterior surface of the iris and the anterior surface of the lens and the ciliary body. The fluid contained in these two compartments drains by the sinus venosus into the ciliary veins.

2. E, All are correct. The senses of the body can be divided into two types: general senses and special senses. The receptors for the general senses are usually simple end organs, found throughout the body, which perceive sensations such as cold, touch, and pain, to name only a few. Their afferent fibers are contained within both spinal and cranial nerves. On the other hand, the special senses such as sight, hearing, equilibrium, smell, and taste are more complicated in that they are highly developed and some are extremely complicated. Each has its own neuronal pathway and most are thought to have their own cortical receiving areas. The special senses are all found in the head and their afferent fibers travel in certain of the cranial nerves.

3. D, Only 4 is correct. The external muscles of the eye are innervated by the oculomotor (III), trochlear (IV), and abducens (VI) cranial nerves. The superior oblique muscle is innervated by the trochlear nerve, the lateral rectus muscle by the abducens and the remaining four (inferior oblique, medial rectus, superior rectus, and inferior rectus) muscles by the oculomotor nerve.

4. B, 1 and 3 are correct. The rods of the retina contain a photochemical substance called rhodopsin which is a complex derived from a protein (opsin) and a derivative of vitamin A. Rhodopsin is highly sensitive to light and breaks down rapidly when it is exposed and in doing so transmits an impulse. During dark periods rhodopsin is resynthesized and becomes functional again. Idopsin

and possibly other photochemicals are found in the cones. These photochemicals of the cones require a greater quantity of light to 'stimulate' them and so they function effectively during the daylight hours. Therefore the cones are responsible for daylight and color vision, whereas the rods are much more effective during the night or when the light is not so bright. When we go from a bright area to a dark area it takes a few moments for rhodopsin to reform and allow the rods to function. This time lag we call adaptation. In certain patients where there is a marked decrease in the amount of vitamin A contained in the pigmented layer of the retina night blindness is common.

5. A, 1, 2, and 3 are correct. The aqueous humor fills the anterior and posterior chambers of the eye. Its composition is approximately that of protein-free plasma and is produced by the ciliary processes. The aqueous humor produces an intraocular pressure that may help to maintain the shape of the eyeball. It is removed at the sinus venosus which is located at the iridocorneal angle. Any failure in resorption of the aqueous humor produces an increased intraocular pressure and eventually will produce a condition called glaucoma. The vitreous body fills the posterior four-fifths of the eyeball and is similar in composition to the aqueous humor but contains large quantities of collagen fibrils and mucopolysaccharides and because of this forms a more gelatinous mass.

6. C, 2 and 4 are correct. The auditory tube, consisting of a medial hollow bony segment and a lateral hollow cartilaginous segment, joins the middle ear cavity to the nasopharynx. This tube is normally collapsed but becomes open when one yawns or swallows. This opening of the auditory tube allows the pressure of the middle ear to equalize with the external air pressure. In other words it equalizes the pressure on both sides of the tympanic membrane. Without this equalization the membrane is not free to vibrate with the sound waves and hearing is impaired. Ironically, the auditory tube also provides a potential path for the spread of an infection from the throat region to the middle ear (as may be the case in tonsillitis).

7. E, All are correct. A detailed understanding of the pathways and specific regions of the brain involved in smell is still vague. The sensation of smell is detected by the olfactory cells of the upper reaches of the nasal cavity, which in turn synapse with mitral cells of the olfactory bulb. From here the stimuli pass up the olfactory tract to a number of centers in the brain. The primary brain centers for the sensation of smell are thought to be in an area anterior to the hypothalamus and in the area of the uncus. Secondary olfactory tracts are thought to pass to the hypothalamus, thalamus, and some of the brain stem nuclei. These secondary areas are thought to be involved in the automatic responses of the body to olfactory stimuli which deal with feeding, fear, excitement, pleasure, and sexual drives.

8. C, 2 and 4 are correct. The neuronal pathways for the transmission of taste sensations from the tongue and the oropharyngeal region into the CNS are by way of the facial (VII), glossopharyngeal (IX), and vagus (X) nerves. The facial nerve carries the impulses from the taste buds of the anterior two-thirds of the tongue and the glossopharyngeal nerve carries these sensations from the posterior one-third of the tongue. The vagus nerve transmits sensations from the taste buds found in the epiglottic region. Once in the brain stem the taste fibers pass up to the thalamus and then on to the insular area (deep to the temporal lobe) of the cerebral cortex.

9. D, Only 4 is correct. The intrinsic muscles of the eye include: the ciliary muscles that contract and relax the tension on the lens, the sphincter pupillae muscles which contract and constrict the pupil, and the dilator pupillae muscles which contract and dilate (open) the pupil. All three consist of

smooth muscle fibers and are actively involved in accommodation. The ciliary muscles are under control of both divisions of the ANS, whereas the sphincter pupillae is controlled by the parasympathetic division, and the dilator pupillae is under control of the sympathetic division.

10. A, 1, 2, and 3 are correct. The taste buds are modified epithelial cells and are located on many papillae situated on the dorsum of the tongue, the palate, the epiglottis, and the tonsillar pillars. The number of functional taste buds decreases with age and the remaining taste sensations are less critical. The individual taste buds are thought to respond to all four of the primary taste stimuli: sour, salty, bitter, and sweet.

11. C, 2 and 4 are correct. The membranous utricle and saccule are located within the bony vestibule of the inner ear. Each of these membranous sacs contain endolymph. At the base of these sacs is a group of hair receptors (macula) which contain tiny calcium carbonate particles called otoliths. These structures are involved in the vestibular (balance) functions of the inner ear. With changes in position the otoliths move and stimulate the receptor hairs, that in turn stimulate the receptors of the vestibular nerve. Its fibers conduct the impulses to the brain and produce a sense of head position and a sensation of a change in the pull of gravity. Stimulation of the macula also evokes the righting reflexes, that is the muscular responses to restore the body and its parts to their normal position when they have been displaced.

12. A, 1, 2, and 3 are correct. The organ of Corti, of the cochlear duct, is the receptor of hearing. The nerve fibers (primary neurons) from the spiral ganglion of the organ of Corti enter the cochlear nuclei of the medulla where they synapse. From the medulla, the secondary neurons then either cross to the opposite side of the brain or remain on the same side, passing upward with successive neurons synapsing in the superior olivary nucleus (cerebellum), the inferior colliculus (midbrain), the medial geniculate nucleus (telencephalon), and then to the auditory cortex of the superior temporal gyrus of the cerebrum.

FIVE-CHOICE ASSOCIATION QUESTIONS

DIRECTIONS: Each group of questions below consists of a numbered list of descriptive words or phrases accompanied by a diagram with certain parts indicated by letters, or by a list of lettered headings. For each numbered word or phrase, SELECT THE LETTERED PART OR HEADING that matches it correctly. Then insert the letter in the space to the right of the appropriate number. Sometimes more than one numbered word may be correctly matched to the same lettered part or heading.

1. _____ A clinical term for short-sightedness
2. _____ A condition where the lens loses its elasticity
3. _____ Results most frequently from an irregular cornea
4. _____ An exaggerated esophoria that cannot be overcome by neuromuscular effort
5. _____ Far-sightedness due to a shortened eyeball is called

A. Astigmatism

B. Presbyopia

C. Myopia

D. Strabismus

E. None of the above

ASSOCIATION QUESTIONS

6. _____ The organ of Corti

7. _____ Contains perilymph

8. _____ Contains endolymph

9. _____ The scala tympani

10. _____ Tectorial membrane

11. _____ Cochlear duct

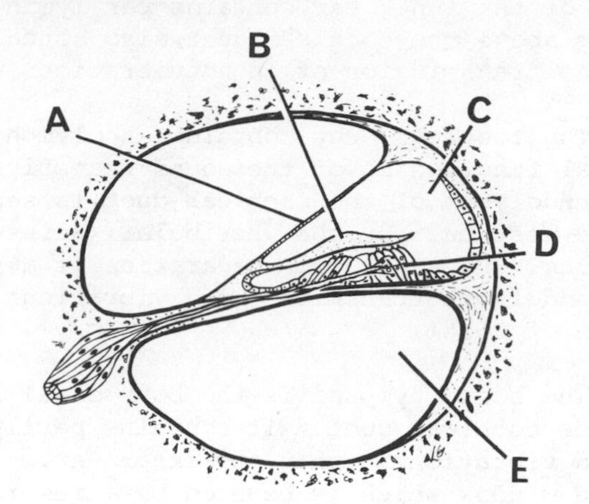

--------------------------- ANSWERS, NOTES AND EXPLANATIONS ---------------------------

1. C. In a normal eye, light rays from an object are refracted (deflection of light) by the cornea, aqueous humor, lens, and vitreous humor and then converge on the retina where an inverted image is formed. Myopia (near-sighted) is a clinical term referring to an error of refraction, in which the object appears as a blurred image. This is produced when the image falls short of the retina because of an elongated eyeball. A biconcave lens (glasses) placed in front of the eyeball will lessen the refraction and cause the image to fall properly on the retina.

2. B. Presbyopia is a condition produced when the lens loses some of its elasticity and is more difficult to accommodate for both near and far vision. Therefore the person is far-sighted and this condition is common in older people.

3. A. An astigmatism results most frequently from an irregular cornea, causing horizontal and vertical rays to be focused at two different points on the retina. Often the lens can compensate for the irregularities of the cornea. An irregularly-shaped lens can also produce an astigmatism.

4. D. Strabismus (cross-eye or squint) is an exaggerated esophoria that cannot be overcome by neuromuscular effort. Esophoria occurs when the internal rectus muscle of the eyeball is stronger than the external rectus, causing the eye to be pulled nasalward (exophoria is the opposite condition). Surgery is often required to correct this condition.

5. E. Far-sightedness, commonly due to a shortened eyeball, is called hyperopia. In this condition the image fails to fall on the retina but rather 'overshoots' it. Biconvex lens will rectify this error in refraction.

6. D. The organ of Corti is the receptor for hearing and is located in the cochlear duct. It lies on the basilar membrane (unlabelled) and consists of hair cells which are joined to the spiral ganglion by nerve fibers. From this ganglion axons extend outward to form the cochlear portion of the eighth cranial nerve.

7. E. The portion of the bony canal (scala tympani) located below the cochlear duct of the inner ear contains perilymph. The scala vestibuli, a bony canal lying above the cochlear duct, also contains perilymph. Perilymph is important in the transmission of sound vibrations within the cochlea of the inner ear.

8. C. The cochlear duct contains endolymph, a special fluid required for the normal functioning of the sound receptive hair cells of the organ of Corti. The endolymph of the cochlear duct is separated from the perilymph of the scala vestibuli by the vestibular (Reissner's) membrane, labelled A in the diagram. However, this separation is a physical one not a functional one. The endolymph transmits sound vibrations originating from the perilymph to the organ of Corti.

9. E. The scala tympani is the bony canal located beneath the basilar membrane of the cochlear duct. It contains perilymph and functions in transmitting sound vibrations within the inner ear. The scala tympani terminates at the round window which is covered by a membrane.

10. B. The tectorial membrane of the cochlear duct (scala media) can be thought of functionally as a stationary structure. The hairs or cilia of the hair cells of the organ of Corti are either embedded or touch the tectorial membrane. When the basilar membrane vibrates in response to the movements of the fluids in the cochlea it pushes the hairs of the hair cells into or against the tectorial membrane. The bending of the hairs excites the hair cells, and this in turn excites the nerve fibers entering their bases. The bending of the hairs is thought to cause a change in the electrical potential across the hair cell membrane. This alternating potential is the receptor potential of the hair cell, and in turn stimulates the cochlear nerve endings.

11. C. The cochlear duct (scala media) contains the organ of Corti and the endolymph which transmits the sound vibrations from the perilymph of the scalae tympani and vestibuli to the hair cells of the organ of Corti.

10. THE ENDOCRINE SYSTEM

LEARNING OBJECTIVES

Be Able To:

- Define the general characteristics of an endocrine gland and the principles involved in the control of homeostasis by the endocrine system.
- For the pituitary (hypophysis cerebri) describe the location and target organs of its two components (adenohypophysis and neurohypophysis).
- Describe the hormones secreted from the adenohypophysis and neurohypophysis, their target organs, functions, and how their secretions are controlled.
- For the hormones of the adenohypophysis and neurohypophysis, describe the result of hypersecretion and hyposecretion upon the target organs and give the name of the disease states.
- Describe the location and structures of the parathyroid and thyroid glands, giving the hormones secreted by each, and their normal and abnormal functions.
- Give the location and structure of the suprarenal (adrenal) glands, describing the hormones secreted from each zone and their functions.
- Describe the location, and the function and control of the hormones secreted by the following organs: ovaries, pancreas (islets of Langerhans), pineal gland, placenta, and testes.
- Describe the effects of hypersecretion and hyposecretion of hormones secreted by the suprarenal, ovaries, pancreas, placenta, and testes.
- Describe the possible functions of the hormones secreted by the kidneys, and the pineal and thymus glands.

RELEVANT READINGS

Anthony & Kolthoff - Chapter 11.
Chaffee & Greisheimer - Chapter 20.
Crouch & McClintic - Chapter 31.
Jacob, Francone & Lossow - Chapter 15.
Landau - Chapter 15.
Tortora - Chapter 17.

FIVE-CHOICE COMPLETION QUESTIONS

DIRECTIONS: Each of the following questions or incomplete statements is followed by five suggested answers or completions. SELECT THE SINGLE BEST ANSWER in each case and then circle the appropriate letter at the lower right of each question.

1. THE POSTERIOR PITUITARY (NEUROHYPOPHYSIS) SECRETES:
   A. Thyrotropin
   B. Thyroxine
   C. Prolactin
   D. Insulin
   E. None of the above

   A B C D E

2. WHICH OF THE FOLLOWING IS NOT CONSIDERED TO BE AN ENDOCRINE GLAND?
   A. Pineal
   B. Pancreas
   C. Parathyroid
   D. Spleen
   E. Ovary

   A B C D E

3. WHICH OF THE FOLLOWING IS NOT A HORMONE SECRETED BY THE ANTERIOR PITUITARY (ADENOHYPOPHYSIS)?
   A. Oxytocin
   B. Luteinizing hormone
   C. Prolactin
   D. Thyroid-stimulating hormone
   E. Growth hormone

   A B C D E

4. THE ENDOCRINE PANCREAS SECRETES WHICH OF THE FOLLOWING HORMONES?
   A. Calcitonin
   B. Glucagon
   C. Glucocorticoids
   D. Aldosterone
   E. None of the above

   A B C D E

5. THE _______ HORMONE IS RESPONSIBLE FOR THE STIMULATION OF TESTOSTERONE SECRETION BY THE INTERSTITIAL CELLS OF THE TESTIS.
   A. Renin
   B. Oxytocin
   C. Luteinizing
   D. Follicle-stimulating
   E. None of the above

   A B C D E

6. THE THYROID GLAND IS NORMALLY LOCATED _______.
   A. Deep to the sternum
   B. Anterior to the trachea
   C. Superior to the hyoid bone
   D. At the base of the tongue
   E. Adjacent to the thyroid cartilage

   A B C D E

7. WHICH OF THE FOLLOWING IS NOT A HORMONE SECRETED BY THE PLACENTA?
   A. Estrogen
   B. Lactogen
   C. Progesterone
   D. Luteinizing
   E. Chorionic gonadotropin

   A B C D E

SELECT THE SINGLE BEST ANSWER

8. THE ENDOCRINE GLAND NORMALLY LOCATED IN THE PELVIS IS THE __________.

A. Thyroid
B. Pancreas
C. Pineal gland
D. Adrenal gland
E. None of the above

A B C D E

9. THE HORMONE ACTIVELY INVOLVED IN DECREASING THE CALCIUM LEVELS OF THE BLOOD IS __________.

A. Thyroxine
B. Epinephrine
C. Parathyroid
D. Oxytocin
E. None of the above

A B C D E

10. WHICH OF THE FOLLOWING FUNCTIONS IS NOT ATTRIBUTABLE TO TESTOSTERONE?

A. Decreases the thickness of the skin
B. Causes the normal descent of the testis
C. Increases the muscle mass of the body
D. Increases the total quantity of bone matrix
E. Decreases the growth of hair on top of the head

A B C D E

11. THE __________ HORMONE IS SECRETED BY THE JUXTAGLOMERULAR CELLS OF THE KIDNEY.

A. Insulin
B. Renin
C. Calcitonin
D. Oxytocin
E. None of the above

A B C D E

------------------------ ANSWERS, NOTES AND EXPLANATIONS ------------------------

1. E. The posterior pituitary (neurohypophysis) secretes two hormones: the antidiuretic hormone (ADH, vasopressin) and oxytocin. The posterior pituitary consists of supporting cells (pituicytes), large numbers of terminal nerve fibers (A), and terminal nerve endings (B). These nerve fibers originate from the supraoptic (D) and paraventricular (C) nuclei of the hypothalamus. ADH is formed primarily in the supraoptic nuclei, whereas oxytocin is formed primarily in the paraventricular nuclei. These hormones move down the nerve tracts (E) of the pituitary stalk and accumulate in the nerve endings. When nerve impulses are transmitted down these nerve tracts the hormones are immediately released from the nerve endings and are absorbed into adjacent capillaries. ADH produces antidiuresis or decreased excretion of water by the kidney tubules. Oxytocin functions by stimulating the pregnant uterus to contract and expel its contents at the end of gestation. It also causes milk to be expressed from the alveoli into the ducts of the mammary gland.

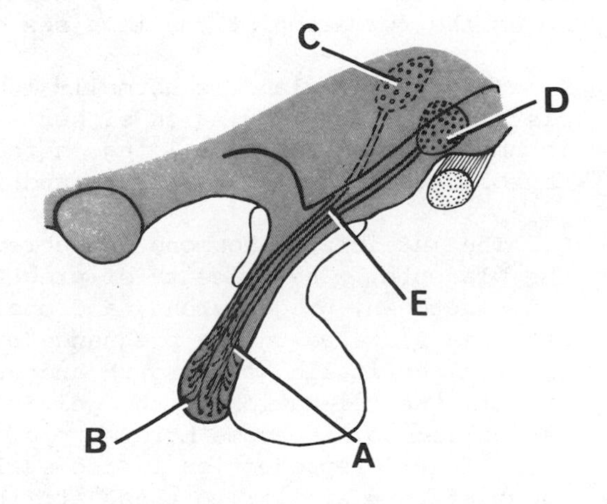

2. D. The endocrine glands of the body include the pituitary, pineal, thyroid, parathyroid, thymus, adrenal (suprarenal), pancreas (islet of Langerhans), kidney, ovary (female), and testis (male). The spleen does not have an endocrine function but is rather the largest organ of the lymphatic system in the body.

3. A. Oxytocin, a hormone secreted by the posterior pituitary (neurohypophysis), is involved in stimulating the pregnant uterus to contact and expel its contents at the end of gestation. It is also involved in delivering milk from the glands of the breast to the nipples during suckling. The main hormones secreted by the anterior pituitary are: growth hormone, corticotropin, thyroid-stimulating hormone, prolactin, and two separate gonadotropic hormones (follicle-stimulating and luteinizing hormones).

4. B. The pancreas is both an exocrine and an endocrine organ. The endocrine portion of the gland is formed by small masses of cells called the islets of Langerhans. In this mass two cell types predominate: the alpha or A cell which secretes the hormone glucagon, and the beta or B cell which secretes the hormone insulin. Glucagon has several functions which are diametrically opposed to those of insulin. It produces a hyperglycemic effect in that it can quickly elevate the blood glucose level by about 20 per cent in 20 minutes. Glucagon produces two important effects on glucose metabolism in that it causes the break down of glycogen to glucose (glycogenolysis) and stimulates the formation of glucose from other sources (gluconeogenesis) such as amino acids and lipids. Insulin, on the other hand, enhances the rate of glucose metabolism, decreases the blood glucose concentration, and increases the glycogen stores in the tissues of the body.

5. C. The interstitial cells of Leydig, in the testis, begin to secrete testosterone when they are stimulated by the luteinizing hormone (LH) from the anterior pituitary (adenohypophysis). The quantity of testosterone secreted is almost proportional to the amount of LH secreted, in fact an excess of testosterone shuts down the production of LH by the anterior pituitary. A hormone called the chorionic gonadotropin, produced by the placenta, stimulates the development of the interstitial cells of the testis in the fetus and causes a temporary secretion of testosterone. This fetal testosterone promotes the formation of the male sex organs.

6. B. The thyroid gland is normally located anterior to the trachea. It consists of two lobes lying on either side of the trachea and joined by an isthmus. Embedded in or lying on the posterior surface of each lobe of the thyroid gland are the superior and inferior parathyroid glands.

7. D. The luteinizing hormone is secreted by the anterior pituitary and not by the placenta. The placenta of pregnancy produces large quantities of estrogen, lactogen, progesterone, and chorionic gonadotropin. The large amount of estrogen secreted during pregnancy produces an enlargement of the uterus and external genitalia, the growth and enlargement of the glandular tissue of the breast, the relaxation of the pelvic ligaments so the symphysis pubis and the sacroiliac joint become more supple, and it is also thought to enhance the rate of cell reproduction in the early embryo. The placental lactogen produces effects similar to those attributed to the growth hormone, in that it causes the deposition of protein for the growth of tissues. This lactogen also promotes additional growth of the breast tissue and can stimulate milk to be secreted. Progesterone, secreted by the placenta, is essential for the normal progression of the pregnancy. It is involved in preparing the endometrium of the uterus for implantation, decreases contractions by the gravid (pregnant) uterus, and increases secretion in the mucous membranes of the uterus and uterine tubes to provide nutrients for the developing embryo. The chorionic gonadotropin, secreted by the early beginnings of the placenta, prevents the normal involution of the corpus luteum at the end of the female sex cycle. This corpus luteum produces increased quantities of estrogen and progesterone which helps maintain the embryo during its early development. Failure of the chorionic gonadotropin to be secreted will allow the involution of the corpus luteum which will likely produce a spontaneous abortion of the developing embryo.

8. E. The ovary of the female is located in the pelvis. The testes of the male are not located within the pelvis but are located within the scrotum, a specialized pouch of skin.

9. E. A hormone secreted by the parafollicular cells of the thyroid gland works against the effects of the parathyroid hormone to increase the blood calcium levels. This hormone is called calcitonin. Calcitonin reduces plasma calcium levels in the blood in three ways: 1) it decreases the activity of osteoclasts which normally breaks down bone and releases calcium into the blood, 2) it stimulates the osteoblasts of bone to produce new bone and thus removes calcium from the blood stream, and 3) it inhibits the formation of new osteoclasts from mesenchymal stem cells.

10. A. Testosterone, the hormone secreted by the interstitial cells of the testis, is secreted during the fetal life (last two months of gestation) of a male child and is responsible for the normal descent of the testis. A lack of testosterone production during this period of fetal life leaves the testis undescended. After puberty the testis again secretes testosterone, which is responsible for the development of the secondary male characteristics such as: the increased size of the external genitalia (penis, scrotum, and testis), the distribution of body hair characteristic to males, the decrease in the growth of hair on top of the head (complete baldness is also dependent on inheriting a genetic factor for the development of baldness), the increased muscle mass of a male, the increased thickness and roughness of the skin, the increased mass of bone, and the hypertrophy of the laryngeal mucosa and enlargement of the larynx producing a bass voice. Lack of testosterone secretion produces a decreased or complete lack of development of the above male characteristics.

11. B. The kidney secretes two hormones: renin by the juxtaglomerular cells and erythropoietin thought to be by cells of the glomeruli or possibly the juxtaglomerular cells. Renin is released when the arterial blood pressure falls too low and it then reacts with the renin substrate to release the substance angiotensin I. Angiotensin I is quickly converted to angiotensin II which raises the arterial blood pressure by constricting the arterioles. Angiotensin is also involved in regulating the body fluid volumes by working through the kidneys. Erythropoietin stimulates the production of new red blood cells.

MULTI-COMPLETION QUESTIONS

DIRECTIONS: In each of the following questions or incomplete statements, ONE OR MORE of the completions given is correct. At the lower right of each question, circle A if 1, 2, and 3 are correct; B if 1 and 3 are correct; C if 2 and 4 are correct; D if only 4 is correct; and E if all are correct.

1. THE ENDOCRINE SYSTEM:
   1. Regulates body functions
   2. Secretes chemical substances called hormones
   3. Is interrelated functionally with the nervous system
   4. Consists of a large number of glands with ducts

   A B C D E

2. HORMONES SECRETED BY THE SUPRARENAL MEDULLA INCLUDE:
   1. Somatotropin
   2. Thyroxin
   3. Oxytocin
   4. Norepinephrine

   A B C D E

3. WHICH OF THE FOLLOWING IS(ARE) GENERAL MECHANISMS BY WHICH MANY HORMONES FUNCTION?
   1. Stimulation of DNA synthesis
   2. Activation of cyclic AMP
   3. Induction of an action potential
   4. Activation of protein synthesis

   A B C D E

A	B	C	D	E
1,2,3	1,3	2,4	Only 4	All correct

4. THE BASOPHILIC EPITHELIAL CELLS OF THE ANTERIOR PITUITARY SECRETE WHICH OF THE FOLLOWING HORMONES?
   1. Thyrotropin
   2. Growth hormone
   3. Melanocyte-stimulating hormone
   4. Prolactin

   A B C D E

5. WHICH OF THE FOLLOWING IS(ARE) CONSIDERED TO BE METABOLIC EFFECTS OF THE GROWTH HORMONE?
   1. Increased mobilization of fats
   2. Stimulates protein synthesis
   3. Decreased rate of carbohydrate utilization
   4. Greater use of fats as a source of energy

   A B C D E

6. WHICH OF THE FOLLOWING STATEMENTS IS(ARE) TRUE FOR THE FACTORS CONTROLLING THE ANTERIOR PITUITARY?
   1. Originates in the hypothalamus
   2. Initiated by hormones
   3. Utilizes a portal system
   4. Mediated by nerve impulses

   A B C D E

7. THE THYROID GLAND:
   1. Secretes thyroxine continuously
   2. Is stimulated by an anterior pituitary hormone
   3. Stores its hormones in a globulin form
   4. Requires iodine to produce thyroxine

   A B C D E

8. WHICH OF THE FOLLOWING STATEMENTS IS(ARE) CORRECT FOR INSULIN?
   1. Causes glycogen to be metabolized to glucose
   2. Stimulates glucose transport by facilitated diffusion
   3. Promotes the release of free fatty acids from fat cells
   4. Is secreted by the beta cells of the islet

   A B C D E

9. THE FEMALE HORMONE(S) RESPONSIBLE FOR THE SECRETORY CHANGES IN THE ENDOMETRIUM INCLUDE(S):
   1. Prolactin
   2. Estrogen
   3. Follicle-stimulating
   4. Progesterone

   A B C D E

10. THE PARATHYROID GLANDS:
    1. Become stimulated when the blood calcium level is low
    2. Are found beneath the capsule of the thyroid gland
    3. Contain hormone-secreting cells called chief cells
    4. Secrete a hormone that stimulates bone absorption

    A B C D E

11. WHICH OF THE FOLLOWING STATEMENTS IS(ARE) TRUE FOR THE CHORIONIC GONADOTROPIN?
    1. Stimulates the interstitial cells of the male fetus
    2. Secreted into the fluids of the mother
    3. Prevents normal involution of the corpus luteum
    4. Stimulates secretion of milk from the breast

    A B C D E

12. THE SUPRARENAL CORTEX:
    1. Secretes the hormone epinephrine
    2. Produces a hormone involved with sodium resorption
    3. Normally consists of two different layers of cells
    4. Is involved with increasing blood glucose levels

    A B C D E

A	B	C	D	E
1,2,3	1,3	2,4	Only 4	All correct

13. GLANDS KNOWN TO SECRETE THEIR HORMONES IN RESPONSE TO THE APPROPRIATE NERVE STIMULI INCLUDE THE:
    1. Suprarenal medulla
    2. Testis
    3. Posterior pituitary
    4. Thyroid

    A B C D E

------------------------ ANSWERS, NOTES AND EXPLANATIONS ------------------------

1. A, 1, 2, and 3 are correct. The endocrine system consists of a large number of ductless glands, whose cells secrete a chemical substance or hormone into the blood stream. These hormones exert a physiological control effect on most cells of the body and along with some nerve stimuli from the nervous system represent the two major controlling factors of bodily functions.

2. D, Only 4 is correct. Sympathetic stimulation of the suprarenal (adrenal) medulla produces large quantities of norepinephrine and epinephrine which are secreted into the blood stream. These two hormones produce effects that last ten times as long as their local neuronal secretion because they are slowly catabolized from the blood stream. They have the same effect as direct sympathetic stimulation, that is, constriction of blood vessels throughout the body, increased activity of the heart, inhibition of the gastrointestinal tract, plus many other effects. However, these hormones have a much more generalized effect since they have been secreted into the blood stream rather than within the substance of a target organ.

3. C, 2 and 4 are correct. Many hormones function by controlling the activity levels of specific cells within the target tissues. The different hormones achieve these effects in many different ways. Two important mechanisms by which hormones function are: 1) the activation of cellular cyclic AMP (cyclic 3',5'-adenosine monophosphate), and 2) the activation of specific genes which in turn activate protein synthesis. In the first instance the stimulating hormone (first messenger) excites the target cell through an appropriate receptor which in turn causes cyclic AMP to be formed in the cell. Then the cyclic AMP (second messenger) causes the hormonal effects inside the cell. In the case of the steroid hormones (from the adrenal cortex, ovaries, and testes), they activate specific genes within the target cells to form intracellular proteins. These proteins, in the form of enzymes, activate the specific functions of the cell.

4. B, 1 and 3 are correct. The anterior pituitary (adenohypophysis) contains two types of epithelial cells: basophils and acidophils. Acidophils, having an affinity for acidic stains, are responsible for secreting growth hormone (GH) and prolactin. Basophils, having an affinity for basic stains, secrete the following hormones: thyrotropin (TH), melanocyte-stimulating hormone (MSH), follicle-stimulating hormone (FSH), luteinizing hormone (LH), and adrenocorticotropic hormone (corticotropin or ACTH).

5. E, All are correct. Growth hormone, in contrast to other hormones, does not function through a specific target gland but instead exerts effects on all or almost all of the tissues of the body. Therefore, it is often called a somatotropic hormone (Gr. soma, body; trope, turning). Growth hormone causes the release of fatty acids from adipose tissue into the blood stream and enhances the conversion of these fatty acids to acetyl-CoA (a source of cellular energy). In fact, under the influence of GH, fatty acids are a preferred source of energy in the body over both carbohydrates and proteins. GH effects protein metabolism by: enhancing amino acid transport through cell membranes, enhances ribosomal protein synthesis, increases RNA synthesis, and at the same

time reduces the break down of proteins. This hormone also effects the cellular metabolism of glucose by: decreasing the utilization of glucose for energy, increasing the glycogen deposition in cells, and decreasing the uptake of glucose by the cells.

6. A, 1, 2, and 3 are correct. Most secretion from the pituitary (hypophysis) is controlled either by nerve impulse or hormones originating in the hypothalamus of the brain (diencephalon). The secretion by the posterior pituitary is controlled by nerve fibers originating from the supraoptic or paraventricular nuclei of the hypothalamus. In contrast, secretion by the anterior pituitary is controlled by hypothalamic releasing and inhibitory factors which are secreted by the hypothalamus. These factors are carried to the anterior pituitary by the hypothalamic-hypophyseal portal vessels where they reach the glandular cells and control their secretion.

7. E, All are correct. The thyroid gland secretes two important iodinate hormones: thyroxine and triiodothyronine. Basically these two hormones function in the same way but differ in the rapidity and intensity of their action. Thyroxine is continuously being produced by the follicular epithelial cells of the gland and requires about 1 mg of iodine to be ingested per week for its normal synthesis. The thyroxine is coupled to globulin molecules to form thyroglobulin which is either stored in this form or broken down to release the free thyroxine into the blood stream. Once in the blood stream the free thyroxine recombines to plasma protein and is carried to the tissues of the body. Since thyroxine is stored in combination with globulin, a decrease in thyroxine production is not functionally noticed until all the stores of thyroglobulin are used up. The thyroid gland is stimulated to produce thyroxine by the thyrotropin hormone secreted by the anterior pituitary. This hormone increases thyroglobulin production, the rate of 'iodide trapping', the release of thyroxine into the blood stream, the size and secretory activity of the thyroid cells, and the actual number (hypertrophy) of thyroid-secreting cells. To maintain a normal basal metabolic rate the precise amount of thyroid hormone must be continuously secreted. To provide the exact amount of thyroxine a specific feedback mechanism operates both through the hypothalamus and the anterior pituitary to control the rate of thyroid secretion in proportion to the metabolic needs of the body.

8. C, 2 and 4 are correct. Insulin, a small protein secreted by the beta cell of the endocrine pancreas, stimulates glucose transport into the cells of most tissues of the body. This hormone stimulates some type of carrier molecule to transport glucose (facilitated diffusion) through the cell membrane in either direction. Insulin, however, does not enhance the transport of glucose into the brain or red blood cells, the intestinal mucosa, or through the tubular epithelium of the kidney. When there is excessive amounts of insulin or glucose or both, the liver cells take up large quantities of glucose from the blood by simple diffusion rather than by facilitated diffusion. Insulin promotes the uptake of glucose by the liver cells by producing a special enzyme that phosphorylates the glucose so it will not pass back out of the liver cell by facilitated diffusion as seen in other tissues. In the absence of insulin or when the blood glucose falls below a certain level the liver releases glucose back into the blood. So the liver functions as an important blood glucose buffer mechanism. The other pancreatic hormone, glucagon, stimulates the liver cell to break down the stored glycogen to glucose (glycogenolysis) which is then released into the blood. Insulin strongly enhances the uptake of glucose by both the liver and the fat cells. The presence of excess glucose in these cells promotes the storage of fat. A lack of insulin causes the reverse effect in that free fatty acids are released from the fat cells. Insulin also promotes an active transport of amino acids into cells and an increase in protein synthesis, both of which tend to promote the growth of the body.

9. D, Only 4 is correct. Progesterone is one of two hormones secreted by the ovary. The other is estrogen which is responsible for cellular proliferation and growth of tissues of the female sexual organs and other tissues related to reproduction. The most important function of progesterone is to promote secretory changes in the endometrial lining of the uterus in preparation for the implantation of a fertilized ovum. Progesterone also promotes secretory changes in the mucosal lining of the uterine (fallopian) tubes which nurture the dividing zygote as it traverses the tube prior to its implantation in the uterus. Progesterone is also important in developing the lobules and alveoli of the breast. It further causes the alveolar cells to proliferate, enlarge, and to become secretory in nature; but not to secrete milk.

10. E, All are correct. The parathyroid glands are formed by four small masses of cells, often embedded within the posterior surface of the two lobes of the thyroid gland and beneath its capsule. The parathyroid hormone (parathormone) is a major regulator of the calcium levels found within the blood. This hormone is secreted by the chief cells, which are stimulated to produce more hormone when the calcium level of the blood falls. The parathyroid hormone is essential to life because it controls the level of calcium in the body fluids, which in turn is involved in the normal excitability of muscle and nerve tissue. A low level of calcium in the blood stimulates the secretion of parathyroid hormone by the parathyroids. This hormone stimulates the resorption of bone which frees up calcium and releases it into the blood stream. The parathyroid hormone also increases the absorption of calcium through the intestinal wall into the blood stream, providing the diet contains food with sufficiently high levels of calcium. Another site of action for the parathyroid hormone is the kidney tubules where it increases the reabsorption of calcium from the glomerular filtrate back into the blood stream.

11. A, 1, 2, and 3 are correct. The chorionic gonadotropin is secreted by the trophoblastic cells (of the early zygote or fertilized ovum) into the fluids of the mother. The rate of secretion peaks at about seven weeks after ovulation and then declines. Its most important function is to prevent the normal involution of the corpus luteum at the end of the female sex cycle. If the corpus luteum is allowed to involute the fetus would be spontaneously aborted, but instead the corpus luteum produces large quantities of estrogen and progesterone. These excessive amounts of estrogen and progesterone cause the endometrium of the uterus to grow and store large quantities of nutrients. During the latter part of pregnancy the placenta takes over from the corpus luteum and produces the required amounts of estrogen and progesterone. Chorionic gonadotropin also exerts an interstitial cell stimulating effect on the testes of a male fetus, which results in the production of testosterone.

12. C, 2 and 4 are correct. The encapsulated suprarenal gland is formed by an outer cortical portion, consisting of three layers or zones of variable distinctness, and an inner medullary portion. The cortical layers are responsible for elaborating three different types of hormones: glucocorticoids, mineralocorticoids, and sex hormones. The medullary portion of the gland secretes two catecholamines (epinephrine and norepinephrine). The glucocorticoids accelerate the break down of cellular proteins to amino acids which are then utilized to produce blood glucose and they also shift the cellular dependence for a source of energy from glucose to lipids. Glucocorticoids also help to maintain normal blood pressure, decrease the number of eosinophils in the blood, and aid in repairing tissue damage produced by the inflammatory response. This hormone is also known to increase its concentration in the blood during stressful situations, but exactly how this increase helps the body is unknown. The mineralocorticoids, especially aldosterone, are involved in regulating the electrolyte metabolism. Its main concern is the homeostasis of blood sodium levels. Aldosterone is controlled by the renin-angiotensin system of the

kidneys, whereas the glucocorticoid secretion is controlled by ACTH (adreno-corticotropic hormone) secreted by the anterior pituitary.

13. B, 1 and 3 are correct. The hormones of the suprarenal (adrenal) medulla and the posterior pituitary are known to secrete their hormones only in response to the appropriate nerve stimuli. Stimulation of the sympathetic nerves of the suprarenal medulla causes large quantities of epinephrine and norephinephrine to be released into the circulating blood. The two hormones (antidiuretic hormone and oxytocin) of the posterior pituitary are released into the surrounding capillaries from nerve endings which have been stimulated by impulses transmitted down from the hypothalamic nuclei to the nerve endings. Similarly, most hormones of the anterior pituitary are not secreted to any significant extent unless they are stimulated by nervous activity from the hypothalamus, although minimal quantities of hormone secretion are independent of any nervous stimulation.

FIVE-CHOICE ASSOCIATION QUESTIONS

DIRECTIONS: Each group of questions below consists of a numbered list of descriptive words or phrases accompanied by a diagram with certain parts indicated by letters, or by a list of lettered headings. For each numbered word or phrase, SELECT THE LETTERED PART OR HEADING that matches it correctly. Then insert the letter in the space to the right of the appropriate number. Sometimes more than one numbered word may be correctly matched to the same lettered part or heading.

1. _____ Secretes progesterone
2. _____ Secretes a hormone that stimulates the pregnant uterus to contract
3. _____ The thymus gland
4. _____ In part is functionally related to the ANS
5. _____ The pineal gland

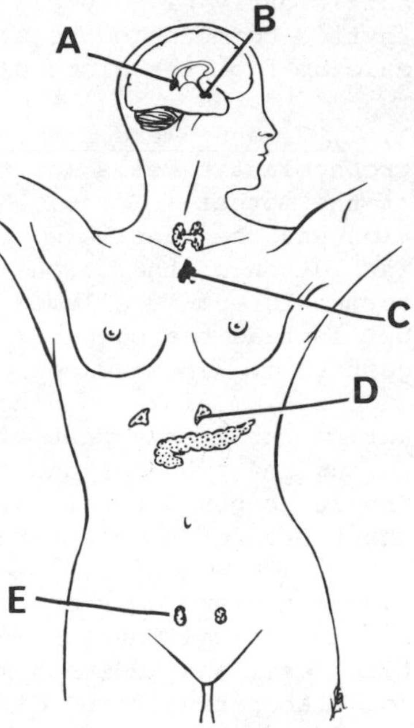

A
B
C
D
E

6. _____ An oversecretion of growth hormone
7. _____ A chronic excess of glucocorticoids
8. _____ A hypersecretion of thyroxine
9. _____ A deficiency of parathyroid hormone
10. _____ A hyposecretion of thyroxine

A. Cretinism
B. Tetany
C. Acromegaly
D. Cushing's syndrome
E. None of the above

1. E. The ovaries located in the pelvis secrete two hormones: estrogen and progesterone. The ovary is stimulated to secrete these hormones by the follicle-stimulating (FSH) and luteinizing hormones (LH) secreted by the anterior pituitary. The comparable organ to the ovary in the male is the testis, which is contained not in the pelvis but within the scrotum.

2. B. The posterior pituitary secretes oxytocin which aids in stimulating the smooth muscle of the pregnant uterus to contract and expel the fetus at the end of gestation. This hormone is also involved in expressing milk from the alveoli into the ducts of the mammary gland.

3. C. The thymus gland is often a bilobed structure lying beneath the sternum in the anterior mediastinum. The thymus is primarily a lymphoid organ but also secretes at least one hormone called thymosin. This hormone is thought to induce stem cells, in other lymphoid organs, to become immunologically competent lymphocytes capable of forming antibodies against foreign protein. Some abnormality of the thymic mechanism is thought to be involved in the 'autoimmune diseases', in which there appear to be rejection-like reactions against some of the body's own cells.

4. D. The medulla of the suprarenal (adrenal) gland is functionally related to and controlled by the sympathetic division of the ANS. This portion of the suprarenal gland secretes epinephrine which in general has the same functions as the sympathetic division of the ANS but it also serves to enhance these activities and prolong them. Epinephrine also stimulates ACTH production, which in turn stimulates the production of the hormones of the adrenal cortex. Norepinephrine, also produced by the medulla of the suprarenal gland, has a widespread vasoconstrictor effect in such tissues as skeletal muscle, skin, and viscera.

5. A. The pineal gland, which arises from the roof of the third ventricle, has been given numerous functions none of which are supported by any real strong evidence. It has been suggested that it is a source of growth-inhibiting factor, a source of a compound that 'cures' schizophrenia, and a possible source of a hormone that antagonizes the secretion or effects of ACTH. Other functions listed for the pineal hormone also include: inhibition of the onset of puberty, regulation of aldosterone secretion by the suprarenal gland, and a source of melatonin, a substance that causes a lightening of skin color.

6. C. Acromegaly develops in individuals after they are full grown. This condition is produced by an oversecretion of growth hormone after epiphyseal plates of the long bones are fused. Acromegaly is characterized by an enlargement of the bones of the feet, hands, jaws, and cheeks. If, in contrast, the anterior pituitary secretes excessive amounts of growth hormone before the epiphyseal plates are fused the individual will grow to a giant size. If, on the other hand, there is a hyposecretion of the hormone the individual will be a dwarf.

7. D. The glucocorticoids accelerate the mobilization of lipids from adipose tissue and they also accelerate the catabolism (breakdown) of lipids by most cells of the body. Hypersecretion of glucocorticoids over a long period of time (chronic) results in a characteristic redistribution of body fat called Cushing's syndrome. This syndrome is characterizied by a redistribution of body fat from the limbs to the face, shoulders, trunk, and abdomen. The face becomes moon-shaped and the shoulders appear humped as in a buffalo.

8. E. Hypersecretion of thyroxine produces a disease called Graves' disease or exophthalmic goiter. This disease is characterized by an increase in the protein bound iodine (PBI), an increased appetite, loss of weight, an increased

metabolism, an increased nervous irritability, and possibly the most noticed: exophthalmos. This exophthalmos is produced by a marked edema in the extraocular fat pad found behind the eyeball.

9. B. The major function of the parathyroid hormone is to help maintain the homeostasis of blood calcium concentration by promoting calcium absorption by the blood. Normal calcium concentrations are necessary for normal neuromuscular irritability, cell membrane permeability, blood clotting, and the proper functioning of many enzyme-induced reactions. Hypocalcemia (decreased calcium levels) due to low parathyroid hormone levels increases the neuromuscular irritability so much that muscle spasms and convulsions occur (this condition is called tetany).

10. A. Hyposecretion of thyroxine during the developing years leads to a condition called dwarfism and cretinism. This condition goes hand in hand with a low metabolic rate, retarded growth and sexual development, and often mental retardation. If the decrease in the amount of thyroxine occurs later in adult life the individual will develop a low metabolic rate, a slow mental and physical vigor, often a gain in weight, loss of hair, and a thickening of the skin because of an accumulation of subcutaneous fluids. This condition is called myxedema.

11. THE CARDIOVASCULAR SYSTEM

LEARNING OBJECTIVES

BLOOD

Be Able To:

- Describe the functions and the following normal characteristics of blood: color, specific gravity, sedimentation rate, pH, and volume.
- Describe the red blood cell (erythrocyte) as to its size, chemical composition, formation, numbers, life span, and functions.
- Define the following terms as they pertain to red blood cells (RBC): anisocytosis, poikilocytosis, cyanosis, hematocrit, pernicious anemia, and anemia.
- Describe the types, formation, functions, numbers, and life span of the granular and nongranular leukocytes found in the blood.
- Give the following characteristics of the blood platelets: origin, functions, number, and life span.
- Define the following conditions as they pertain to leukocytes and blood platelets: septicemia, leukocytosis, leukopenia, leukemia, infectious mononucleosis, agranulocytosis, thrombocytopenia, and thrombocytosis.
- Discuss the composition and functions of normal blood plasma.
- Describe the mechanism of blood coagulation and the following abnormal states: thrombosis, thrombus, embolus, thrombophlebitis, hemorrhage, and hemophilia.
- Describe the ABO blood grouping and the Rh factor and give their significance.

THE HEART AND ITS PHYSIOLOGY

Be Able To:

- Describe the heart as to its location, its fibrous covering, and the layers making up its wall.
- Using simple drawings of the heart label its four chambers, valves, superior and inferior venae cavae, azygos vein, aorta, brachiocephalic and left common carotid and subclavian arteries, pulmonary arteries and veins, papillary muscles, chordae tendineae, and the interatrial and interventricular septa.

- Using a simple diagram of the heart indicate the flow of blood through the heart to the pulmonary and systemic systems.

- Describe the extrinsic control and intrinsic excitatory and conducting systems of the heart.

- Identify on suitable diagrams the following: right and left coronary arteries, anterior and posterior descending arteries, marginal artery, circumflex artery, coronary sinus, and the major coronary veins.

- Define the following terms as they pertain to the heart or its functions: stenosis, pericarditis, myocarditis, endocarditis, angina pectoris, coronary thrombosis, and infarct.

- Describe the physiological properties of cardiac muscle and describe the meaning of the various waves observed on a normal electrocardiogram (ECG).

- Define the terms: cardiac cycle, systole, diastole, cardiac output, and stroke volume.

- Discuss the factors involved in varying the heart rate.

- Describe the pressure changes found within the heart during the cardiac cycle and describe the heart sounds.

THE BLOOD VESSELS AND THEIR PHYSIOLOGY

Be Able To:

- Define the terms artery and vein.

- Describe and compare the three tunica (layers) forming the wall of a medium-sized artery and vein.

- Describe the nerve supply to blood vessels and the following terms: vasa vasorum, end-artery, anastomoses, capillary bed, pulmonary and systemic circulations, and portal system.

- Identify on simple diagrams and give the regions of the body supplied by each of the following branches of the aorta: coronary, brachiocephalic, left common carotid, left subclavian, intercostal, celiac trunk, suprarenal, lumbar, superior mesenteric, renal, gonadal, inferior mesenteric, and common iliac arteries.

- Identify on simple diagrams and give the area supplied by the following arteries: common carotid, external carotid, facial, maxillary, occipital, posterior auricular, superficial temporal, internal carotid, ophthalmic, and middle cerebral.

- Identify the following arteries of the neck and upper limb: subclavian, thyrocervical trunk, internal thoracic, axillary, radial recurrent, brachial, radial, ulnar, and common interosseous arteries.

- Identify the branches of the celiac trunk, superior and inferior mesenteric arteries and the organs they supply.

- Identify the following arteries of the pelvis and the lower limb: external iliac, internal iliac, femoral, profunda femoris, popliteal, anterior tibial, posterior tibial, and dorsalis pedis arteries.

- Define the terms: aneurysm, arteriolosclerosis, and atherosclerosis.

- Identify on appropriate diagrams the following veins and sinuses of the head or neck: superior and inferior sagittal, straight, transverse, sigmoid, cavernous, and superior and inferior petrosal sinuses; pterygoid plexus; superior temporal, posterior auricular, occipital, retromandibular, vertebral, external jugular, internal jugular, facial, and brachiocephalic veins.

- Identify the following venous branches of the upper limb: subclavian, axillary, brachial, cephalic, basilic, and median cubital veins.

- Identify on simple drawings the following veins: brachiocephalic, superior and inferior venae cavae, hepatic, azygos, and portal veins and describe the area drained by each.

- Identify the following venous branches of the lower limb or pelvis: external, internal, and common iliac, great saphenous, femoral, popliteal, anterior tibial, and posterior tibial.

- Discuss the factors involved in producing blood pressure and blood flow.

- Define the following and give their normal ranges: systolic and diastolic pressure of the systemic and pulmonary arteries, and the pulse pressure.

- Describe the factors involved in the venous pressure and venous return of blood to the heart.

- Discuss the factors involved in maintaining blood flow and producing the arterial pulse.

- Discuss the regulation of circulation with reference to the following: cardiac and vasomotor reflex centers, vasoconstrictor fibers, vasodilator fibers, pressoreceptors, chemoreceptors, and the control exerted by the higher centers of the brain.

RELEVANT READINGS

Anthony & Kolthoff - Chapter 13.
Chaffee & Greisheimer - Chapters 11, 12, 13, and 14.
Crouch & McClintic - Chapters 22, 23, 24, and 25.
Jacob, Francone & Lossow - Chapter 10.
Landau - Chapters 14, 15, 16, 17, and 18.
Tortora - Chapter 11.

BLOOD

FIVE-CHOICE COMPLETION QUESTIONS

DIRECTIONS: Each of the following questions or incomplete statements is followed by five suggested answers or completions. SELECT THE SINGLE BEST ANSWER in each case and then circle the appropriate letter at the lower right of each question.

1. THE STEM CELL RESPONSIBLE FOR THE CELLULAR ELEMENTS OF BLOOD IS THE ______.
   A. Monoblast
   B. Fibroblast
   C. Osteoblast
   D. Hemocytoblast
   E. None of the above

   A B C D E

SELECT THE SINGLE BEST ANSWER

2. THE CONDITION OF RED BLOOD CELLS WHERE THERE IS A GREAT VARIATION IN SIZE IS CALLED ________.
   A. Anemia
   B. Cyanosis
   C. Anisocytosis
   D. Agranulocytosis
   E. Poikilocytosis

   A B C D E

3. WHICH OF THE FOLLOWING IS NOT A FUNCTION OF BLOOD?
   A. Transports glandular secretions
   B. Provides structural support for the body
   C. Carries waste products to the kidneys
   D. Contains buffers that help maintain the acid-base balance
   E. Protects the body against the invasion of foreign bacteria

   A B C D E

4. THE MOST NUMEROUS WHITE BLOOD CELL (LEUKOCYTE) IN THE CIRCULATION IS THE __________.
   A. Basophil
   B. Eosinophil
   C. Lymphocyte
   D. Monocyte
   E. Neutrophil

   A B C D E

5. THE NUMBER OF PLATELETS PER cu mm OF BLOOD IS NORMALLY ABOUT ________.
   A. 100,000
   B. 300,000
   C. 500,000
   D. 700,000
   E. None of the above

   A B C D E

6. WHICH OF THE FOLLOWING CONDITIONS IS CAPABLE OF PRODUCING INTRAVASCULAR CLOTTING (THROMBOSIS)?
   A. Calcium deficiency
   B. Platelet deficiency
   C. Low levels of vitamin K
   D. Injury to the blood vessel wall
   E. None of the above

   A B C D E

7. VITAMIN ______ IS REQUIRED FOR THE MATURATION OF RED BLOOD CELLS.
   A. A
   B. C
   C. D
   D. K
   E. None of the above

   A B C D E

8. A STATIONARY BLOOD CLOT IS CALLED A(AN)_________.
   A. Embolus
   B. Hemorrhage
   C. Thrombosis
   D. Thrombus
   E. None of the above

   A B C D E

9. LYMPHOCYTES NORMALLY REPRESENT ABOUT _______ PER CENT OF ALL WHITE BLOOD CELLS.
   A. One
   B. Five
   C. Fifteen
   D. Twenty-five
   E. Forty-five

   A B C D E

------------------------- ANSWERS, NOTES AND EXPLANATIONS -------------------------

1. D. The cellular components of blood are formed from a stem cell called a hemocytoblast. This cell, which is capable of forming red and white blood cells and blood platelets, is found in the red marrow of certain bones of the skeleton and in some of the organs of the lymphatic system. The monoblast is a derivative of a hemocytoblast and forms the monocyte cell line of white blood cells. The fibroblast and the osteoblast are collagen and bone forming cells, respectively.

2. C. Anisocytosis is the condition of red blood cells where there is a great variation in size. If there is a great variation in shape, it is called poikilocytosis. Cyanosis is a condition where the mucous membranes or the skin appear a blue color. This blue color is due to an abnormally low level of oxygen carried by the red blood cells. Anemia is the term used when the number of red blood cells falls below its normal level (4.5-5.5 million/cu mm of blood). Anemia may be produced by a loss of blood, an increased destruction of red blood cells, only to name a few. Agranulocytosis refers to a low granular leukocyte (neutrophils, eosinophils, and basophils) count and may be caused by persistent fever.

3. B. The major function of the blood is to maintain the homeostasis of the internal cellular environment. All cellular activities utilize food and oxygen and produce waste products. To this end the blood transports oxygen to the cells from the lungs and carbon dioxide from the cells to the lungs to be eliminated; distributes products of digestion from the small intestine to all parts of the body; carries waste products to the kidneys to be eliminated; transports glandular secretions to their target organs; has buffers that aid in the maintenance of the acid-base balance; regulates body heat by distributing the heat produced by muscle activity; functions to protect the body against foreign bacteria; and aids in the development and function of body immunity.

4. E. Neutrophils are the most numerous leukocyte and account for 65-75 per cent of all white blood cells. Eosinophils represent 2-4 per cent, basophils almost one per cent, lymphocytes about 20-25 per cent, and monocytes about 2-6 per cent of all white blood cells.

5. B. Platelets are cytoplasmic fragments of the large megakaryocytes found in bone marrow. There are almost 250,000-300,000 platelets per cubic millimeter of blood, which represent about 30-50 times the normal number of leukocytes. Platelets are instrumental in initiating the formation of a blood clot.

6. D. Injury to the endothelial layer of the blood vessel wall is thought to be a major cause of intravascular clotting. It is now known that tiny blood vessels often rupture under normal conditions and intravascular clots quickly form to plug them. These small clots are normally dissolved after a short time by an enzyme of the blood called fibrinolysin. The other choices in this question all would have the opposite effect, that is, they would make it difficult for the formation of a clot to occur. A decrease in the normal number of platelets would hinder the initiation of clot formation and a deficiency of calcium would greatly hinder the conversion of prothrombin to thrombin. Vitamin K is needed by the liver to produce prothrombin at its normal rate and with low levels of this vitamin in the blood the prothrombin concentration would soon fall below the normal levels. Vitamin K is both absorbed from foods taken in by the diet and produced by certain of the intestinal bacteria. The absorption of this important vitamin by the small intestine requires bile to act as the transport media.

7. E. Vitamin B_{12} (extrinsic factor) is required for the maturation of red blood cells. An intrinsic factor, secreted by cells lining the stomach, aids vitamin B_{12} to be absorbed across the wall of the small intestine. Absence of this vitamin produces a lack of nuclear maturation and division, which greatly inhibits the rate of red blood cell production. Once vitamin B_{12} is absorbed into the circulation it is stored in the liver in large quantities (about 1000 micrograms) until it is needed by the bone marrow. The daily requirement of vitamin B_{12} is about one microgram. Failure of the stomach to produce adequate amounts of the intrinsic factor produces a condition where vitamin B_{12} is not absorbed by the small intestine. In this condition, called pernicious anemia, the bone marrow fails to produce a sufficient number of mature red blood cells.

8. D. A thrombus is a blood clot which is attached to the inner wall of a blood vessel, whereas a clot that breaks free and moves with the circulation is called an embolus. Intravascular coagulation that leads to the formation of such a thrombus is called a thrombosis, while the escape of blood from the vessels into the intracellular spaces is called a hemorrhage.

9. D. Lymphocytes normally represent about 25 per cent of the white cell total. The lymphocytes are agranular cells and are very motile. These cells are not phagocytic but actively produce antibodies that play a central role in the mechanisms of immunity. The other white blood cells are present in the following percentages: neutrophils, 70; monocytes, 2-6; eosinophils, three; and basophils, one.

MULTI-COMPLETION QUESTIONS

DIRECTIONS: In each of the following questions or incomplete statements, ONE OR MORE of the completions given is correct. At the lower right of each question, circle A if 1, 2, and 3 are correct; B if 1 and 3 are correct; C if 2 and 4 are correct; D if only 4 is correct; and E if all are correct.

1. WHICH OF THE FOLLOWING IS(ARE) CONSIDERED TO BE NORMAL FOR BLOOD?
   1. Represents about seven per cent of the weight of the body
   2. A specific gravity of 1.060
   3. A hematocrit of 45 per cent
   4. A pH of 6.6

   A B C D E

2. PLATELETS ARE:
   1. Multinucleated cells
   2. Fragments of erythrocytes
   3. Formed in some lymph nodes
   4. Involved in blood clotting

   A B C D E

3. IN BLOOD CLOTTING:
   1. Ca^{++} plays an important role
   2. Prothrombin is converted to thrombin
   3. The clot formation is triggered by thromboplastin
   4. Fibrin is formed immediately after injury

   A B C D E

4. WHICH OF THE FOLLOWING STATEMENTS ABOUT MONOCYTES IS(ARE) CORRECT?
   1. Break up to form platelets
   2. Are the largest cell of the blood
   3. Contain small numbers of granules in their cytoplasm
   4. Constitute about five per cent of the white cell population

   A B C D E

5. BLOOD PLASMA CONTAINS:
   1. Water
   2. Albumin
   3. Glucose
   4. Urea

   A B C D E

6. WHICH OF THE FOLLOWING IS(ARE) CORRECT FOR THE REGULATION OF RED BLOOD CELL PRODUCTION?
   1. The rate, in part, depends on the physical activity
   2. Hypoxia stimulates the kidneys to release an erythropoietic factor
   3. The hemocytoblast is stimulated by erythropoietin
   4. Takes about two to three days before new cells are produced

   A B C D E

7. THE BLOOD TYPE O HAS ______ ON ITS RED BLOOD CELLS.
   1. Antigen A
   2. No antigen A
   3. Antigen B
   4. No antigen B

   A B C D E

A	B	C	D	E
1,2,3	1,3	2,4	Only 4	All correct

8. WHICH OF THE FOLLOWING IS(ARE) CLASSIFIED AS A GRANULAR LEUKOCYTE?
   1. Monocyte
   2. Neutrophil
   3. Lymphocyte
   4. Eosinophil

   A B C D E

9. THE Rh FACTOR IS:
   1. Attached to red blood cells
   2. Named for an antibody
   3. Found in about 90 per cent of the population
   4. Unimportant when considering a blood transfusion

   A B C D E

10. RED BLOOD CELLS:
    1. In their mature form do not have a nucleus
    2. Have an average diameter of 10 microns
    3. Are in the form of a biconcave disc
    4. Circulate within the blood stream for about 240 days

    A B C D E

ANSWERS, NOTES AND EXPLANATIONS

1. A, 1, 2, and 3 are correct. The total volume of blood in an individual is roughly seven per cent of the body weight. It has an average specific gravity of 1.060 (the specific gravity of water is 1.0) and a sedimentation rate (index) of 4-10 mm/hour. The sedimentation rate of blood is the speed at which the blood cells sink to the bottom of a tube (an increased rate may indicate an infection). The hematocrit is approximately 45 per cent and represents the total cell volume of a blood sample with the remaining liquid plasma having a volume of 55 per cent. The pH of blood is slightly alkaline and has a narrow range of 7.35-7.45. Oxygenated blood is a bright red color, whereas unoxygenated blood is dark red.

2. D, Only 4 is correct. Blood platelets are small cytoplasmic fragments of large megakaryocytes found in bone marrow. These fragments are not cells and have no nuclei, but are small disc-like bodies which are slightly smaller in diameter than the erythrocyte. They are dark staining bodies with few structural features. The platelets appear in blood smears as small groups which stick together. They are responsible for initiating the formation of blood clots.

3. A, 1, 2, and 3 are correct. The formation of a clot begins when the tissues and a blood vessel are damaged. The injured cells are thought to release thromboplastin, a substance that triggers the formation of a clot. Other factors are also involved at this early stage: a) an adequate Ca^{++} concentration, b) the release of a platelet factor which comes from the platelets that try to plug the hole in the damaged vessel, and c) other accessory factors. The platelet factor is thought to be involved with the formation of thromboplastin by combining with an antihemophilic factor present in the blood. Thromboplastin plus adequate amounts of Ca^{++}, vitamin K, and some accelerator factors are responsible for the formation of thrombin from prothrombin.

4. C, 2 and 4 are correct. Monocytes are the largest of the blood cells and account for two to six per cent of the total leukocyte population. The monocyte nucleus is poorly stained with a large amount of cytoplasm which is devoid of any granules. The monocytes do not arise directly from the primitive stem cell (hemocytoblast) but probably are derived from a cell in the lymphocytic line. Monocytes are mobile and markedly phagocytic cells, especially when they leave the circulating blood to enter the tissues. Many of the monocytes on entering the tissues, enlarge greatly and become fixed

where they perform phagocytic activity (tissue macrophages). These cells become fixed in blood and lymphatic channels and function to help protect the body against foreign bacteria. Such cells are located in the lymph nodes, the liver (Küpffer cells), the lungs (alveolar macrophages), and the spleen. These cells collectively form the reticuloendothelial system.

5. E, All are correct. Plasma is the liquid component of blood and consists of about 90 per cent water and 10 per cent solutes. About 60 to 80 per cent of the solute portion of plasma consists of three proteins: albumin, involved in the colloid osmotic pressure of the plasma; globulins, involved in the transport of substances and the immune response; and fibrinogen, involved in the coagulation of plasma. The other 20 to 40 per cent of the solute portion of plasma consists of: nutrients such as glucose (blood sugar), amino acids and lipids; gases in solution such as oxygen and carbon dioxide; inorganic salts such as calcium, phosphates, chlorides and others; trace elements such as iron, iodine, and others; waste products such as urea, uric acid, creatinine, and lactic acid; and other special substances such as hormones and antibodies. The normal constituents of plasma play an important role in the maintenance of homeostasis. The laboratory analysis of blood plasma in many diseased states will show either a decrease or an increase in a number of constituents of plasma and this will often help lead to a correct diagnosis.

6. E, All are correct. The total number of erythrocytes in the blood stream is controlled within rather narrow limits. The main concern of red blood cells is to provide enough oxygen to the cells so the body can function properly. The bone marrow begins to produce more new red blood cells when a person becomes anemic, moves to a higher altitude or becomes more physically active. In all these cases there is a need for more oxygen to be carried to the tissues, therefore more red blood cells need to be produced. Hypoxia (lack of oxygen) in the blood stimulates the kidneys to release a renal erythropoietic factor which produces erythropoietin from a plasma protein. Erythropoietin in turn stimulates the hemocytoblast of the bone marrow to produce an increased number of red blood cells. Erythropoietin is produced almost immediately upon a drop in the oxygen levels of the blood. However, it takes anywhere from two to three days before new red blood cells first appear in the circulation and it is only after five or more days that the maximum rate of new red blood cell production is reached.

7. C, 2 and 4 are correct. Blood type O has neither antigen A or B on its red blood cells and therefore has both anti-A and anti-B antibodies in its plasma. Blood type A has antigen A on its red cells and anti-B antibodies in its plasma, whereas blood type B has antigen B on its red cells and anti-A antibodies. Blood type AB, on the other hand, has both antigen A and B on its red cells and no anti-A nor anti-B antibodies in its plasma. O type blood has been called the universal donor, but this term is misleading because it assumes the absence of all antigens and antibodies and this is usually not the case. Before all transfusions, cross-matches should be made to check for agglutination.

8. C, 2 and 4 are correct. Leukocytes are divided into two groups, granulocytes and agranulocytes, depending on whether or not there are prominent granules in their cytoplasm. Granulocytes include neutrophils, basophils, and eosinophils, and are named for the staining characteristics of their cytoplasmic granules. The granules in neutrophils stain with neutral dyes, the granules in basophils stain with basic dyes, and the granules of eosinophils stain with acid dyes.

9. B, 1 and 3 are correct. The Rh factor is named for an antigen. An Rh-positive individual (about 90 per cent of the population) has Rh antigens on its red blood cells, whereas an Rh-negative individual (about 10 per cent of the population) has no Rh antigens. Normally no blood contains anti-Rh antibodies

but they may develop under certain conditions. Anti-Rh antibodies can appear in Rh-negative individuals who have received an Rh-positive blood transfusion. The Rh-negative individual will then form anti-Rh antigens in the blood. If at a later date the Rh-negative individual receives another blood transfusion from an Rh-positive donor, the anti-Rh antigens will destroy the donor's red blood cells containing the Rh antigens. The destroyed cells clump together (agglutinate) and may plug small blood vessels vital to survival. Anti-Rh antigens can also appear in Rh-negative women who are carrying an Rh-positive fetus. The fetal blood may leak through the placenta and cause the mother to produce the anti-Rh antigens in her blood. A subsequent Rh-positive fetus may be in jeopardy since the anti Rh-antigens of the mother may cross the placenta and destroy the red blood cells of the fetus (erythroblastosis fetalis).

10. B, 1 and 3 are correct. Red blood cells (erythrocytes), as do all the cellular elements of blood, arise from hemocytoblasts of the bone marrow. From this stem cell the immature erythrocyte progresses through a series of nuclear changes so that the mature erythrocyte is devoid of a nucleus. The mature erythrocyte is a biconcave disc, with an average diameter of 7.5 microns. Each erythrocyte contains about 34 per cent hemoglobin which combines with oxygen to form a rather loose bond. This bond is easily broken and replaced by carbon monoxide which forms a strong bond. This strong bond with carbon monoxide prevents oxygen from being carried to the cells of the body and anoxia occurs. Red blood cells circulate in the blood for approximately 120 days and then they are removed from the circulation by the liver and spleen.

FIVE-CHOICE ASSOCIATION QUESTIONS

DIRECTIONS: Each group of questions below consists of a numbered list of descriptive words or phrases accompanied by a diagram with certain parts indicated by letters, or by a list of lettered headings. For each numbered word or phrase, SELECT THE LETTERED PART OR HEADING that matches it correctly. Then insert the letter in the space to the right of the appropriate number. Sometimes more than one numbered word may be correctly matched to the same lettered part or heading.

1. _____ A component of blood that dissolves clots.
2. _____ The loss of hemoglobin from a cell.
3. _____ Large quantities of reduced hemoglobin produce.
4. _____ Produced by a low level of intrinsic factor.
5. _____ A disease characterized by a much prolonged coagulation time.

A. Pernicious anemia
B. Cyanosis
C. Hemophilia
D. Fibrinolysin
E. None of the above

6. _____ Secrete serotonin.
7. _____ A site where vitamin B_{12} is stored.
8. _____ The 'grave yard' of blood cells.
9. _____ The rarest blood cells.
10. _____ Contains macrophages called Küpffer cells.

A. Liver
B. Basophils
C. Spleen
D. Platelets
E. Eosinophils

1. D. Fibrinolysin is a normal component of blood that dissolves small clots formed on imperfections of the blood vessel walls. This process (fibrinolysis) involves the digestion of the fibrin threads in the clot by the proteolytic substance fibrinolysin, particularly in the small peripheral vessels where many intravascular clots are thought to occur.

2. E. The loss of hemoglobin from a red blood cell is called hemolysis (laking). This occurs when red blood cells are placed in a hypotonic solution and the cell takes up so much water that it bursts. Once the cell membrane ruptures the hemoglobin escapes.

3. B. Reduced hemoglobin is hemoglobin that has given up its oxygen and has a characteristic blue color as opposed to oxyhemoglobin which is red. Excessive amounts of reduced hemoglobin are found in the blood stream when there is a low availability of oxygen. When this occurs the capillaries of the skin and mucous membranes appear blue in color and this condition is called cyanosis.

4. A. The failure of the cells lining the stomach to produce enough intrinsic factor usually results in a condition called pernicious anemia. This condition is characterized by a marked reduction in the number of circulating red blood cells. Normally the intrinsic factor of the stomach allows the small intestine to absorb vitamin B_{12} (extrinsic factor) into the blood stream. This extrinsic factor is required in order for the red blood cells to mature. A lack of vitamin B_{12} absorption into the blood stream produces a decreased number of mature red blood cells in the circulation.

5. C. Hemophilia is a sex-linked hereditary disease where there is an absence of the antihemophiliac factor. Males are affected and the defect is transmitted by the female. This antihemophiliac factor is important in the initiating of thromboplastin in the early stages of clot formation.

6. D. Platelets contain two substances, one called the platelet factor and the other called serotonin. The platelet factor is an activator substance that initiates blood clotting. The other substance, serotonin, is released from disintegrating platelets and causes the constriction of the local blood vessels. In this way serotonin helps to control the bleeding of the affected vessels.

7. A. Vitamin B_{12} once absorbed by the small intestine, with the aid of the intrinsic factor secreted by the lining cells of the stomach, is stored in large quantities in the liver. This vitamin is slowly released by the liver into the blood stream where it is utilized by the bone marrow. Less than one microgram of vitamin B_{12} is utilized each day to maintain the normal red blood cell maturation, whereas the liver stores 1000 times this amount. Any depletion of vitamin B_{12} would take an extended period of time before anemia would appear.

8. C. The spleen has been called the 'grave yard' of the red blood cells. It is in the sinusoids of the red pulp that the fixed and free macrophages phagocytose cell debris, dying red blood cells, and foreign bacteria. The spleen is also responsible for the production of lymphocytes in the white pulp and for the storage of blood cells in the red pulp sinusoids.

9. B. The basophils are the rarest of all blood cells and form about one per cent of all leukocytes. These large cells are quite motile and have sparsely spaced blue granules which are thought to contain heparin. Heparin is an anticoagulent that prevents the formation of clots. Basophils are not actively phagocytic.

10. A. The liver is formed by hepatic cells which form cords that are interspersed by sinusoids. These sinusoids are lined by fixed macrophages called Küpffer cells. These cells destroy worn out erythrocytes and remove any bacteria from the circulation that reaches the liver from the intestines.

THE HEART AND ITS PHYSIOLOGY

FIVE-CHOICE COMPLETION QUESTIONS

DIRECTIONS: Each of the following questions or incomplete statements is followed by five suggested answers or completions. SELECT THE SINGLE BEST ANSWER in each case and then circle the appropriate letter at the lower right of each question.

1. THE SAC SURROUNDING THE HEART IS CALLED THE ________.
   A. Epicardium
   B. Myocardium
   C. Endocardium
   D. Pericardium
   E. None of the above

   A B C D E

2. THE ARROW INDICATES THE ______ARTERY.
   A. Right coronary
   B. Marginal
   C. Posterior descending
   D. Anterior descending
   E. Circumflex

   A B C D E

3. WHICH OF THE FOLLOWING STATEMENTS CONCERNING THE MYOCARDIUM IS CORRECT?
   A. Consists of strong connective tissues
   B. Forms the middle layer of the heart
   C. Produces a serous-like fluid
   D. Often is the thinnest of the three layers of the heart
   E. None of the above

   A B C D E

4. THE FUNCTION OF THE CHORDAE TENDINEAE IS TO:
   A. Prevent eversion of the atrioventricular valves
   B. Support the papillary muscles
   C. Aid in the contraction of the atria
   D. Strengthen the ventricular walls
   E. None of the above

   A B C D E

5. THE CARDIAC CENTER, CONTROLLING THE CONTRACTIONS OF THE HEART, IS FOUND IN THE _______ OF THE BRAIN.
   A. Pons
   B. Medulla oblongata
   C. Cerebellum
   D. Cerebral cortex
   E. None of the above

   A B C D E

SELECT THE SINGLE BEST ANSWER

6. THE A-V NODE IS INDICATED BY _____.

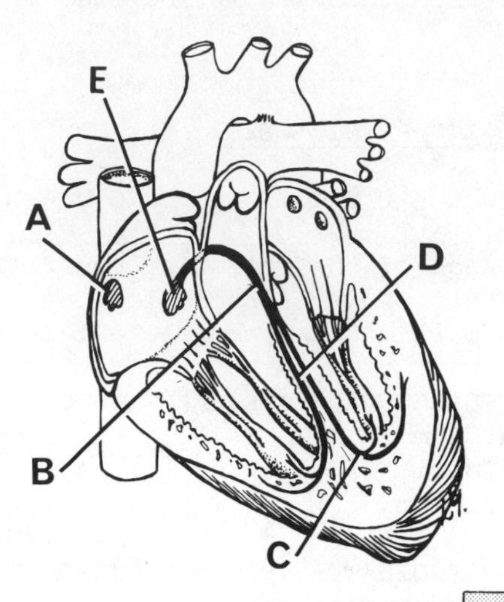

A B C D E

7. THE P WAVE OF THE ELECTROCARDIOGRAM REPRESENTS:
   A. The depolarization of the ventricles
   B. The spread of depolarization through the atria
   C. The repolarization of the atria
   D. The stage of ventricular repolarization
   E. None of the above

A B C D E

8. THE LEFT PULMONARY ARTERY IS INDICATED BY _________.

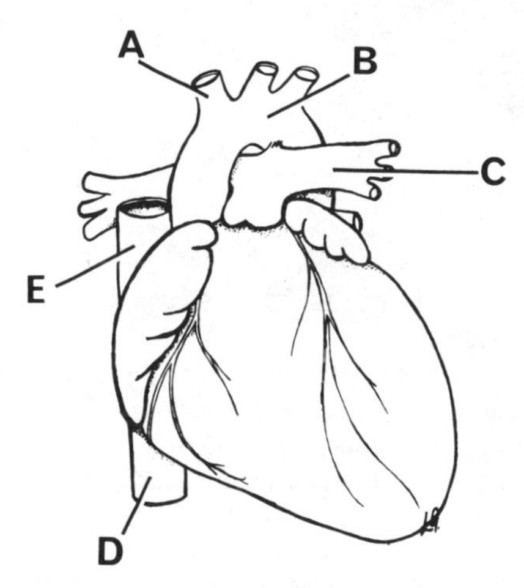

A B C D E

9. WHICH OF THE FOLLOWING STATEMENTS IS INCORRECT WITH REGARD TO THE HEART RATE?
   A. The larger the size, the lower the heart rate
   B. The older a person, the lower the heart rate
   C. Heart rate is faster, when the body temperature is high
   D. The female heart rate is lower than that of the male
   E. Increased thyroxine levels increase the heart rate

A B C D E

SELECT THE SINGLE BEST ANSWER

10. THE ARROW IN THE DIAGRAM INDICATES THE:
   A. Pulmonary artery
   B. Azygos vein
   C. Pulmonary vein
   D. Brachiocephalic artery
   E. Superior vena cava

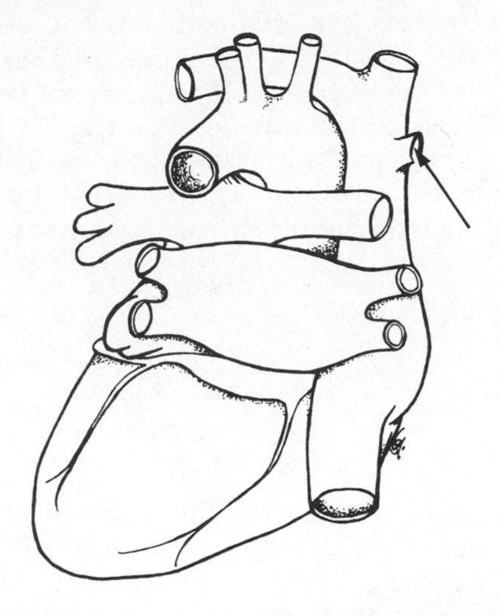

A B C D E

11. THE CARDIAC OUTPUT IS EQUAL TO THE:
   A. Stroke volume
   B. Heart rate multiplied by a factor of two
   C. Stroke volume multiplied by the heart rate
   D. Stroke volume divided by the heart rate
   E. None of the above

A B C D E

12. THE STROKE VOLUME IS APPROXIMATELY:
   A. 70 ml
   B. 80 ml
   C. 90 ml
   D. 100 ml
   E. 110 ml

A B C D E

------------------------- ANSWERS, NOTES AND EXPLANATIONS -------------------------

1. D. The sac surrounding the heart is called the pericardium and it consists of an outer fibrous layer and an inner serous layer. The fibrous pericardium is a tough fibrous (connective tissue) membrane that is attached to the great vessels entering and leaving the heart, the diaphragm, and the posterior surface of the sternum. The fibrous layer is lined by the outer or parietal layer of the serous pericardium, whereas the heart itself is covered by the inner visceral layer of the serous pericardium. This inner visceral layer actually forms the outer layer of the heart or epicardium. Between the visceral and the parietal layers of the serous pericardium there is a potential space which contains a small amount of fluid. This fluid acts as a lubricant to reduce the friction between the two layers and allows the heart to move freely within the pericardium. The middle layer of the heart wall consists of cardiac muscle and is called the myocardium, whereas the inner layer of the heart consists of a single layer of endothelial cells and is called the endocardium.

2. C. The marginal artery (B in the diagram) is a branch of the right coronary artery (A). The right coronary artery also gives rise to the posterior (interventricular) descending artery (C). The left coronary artery (unlabelled) gives rise to the circumflex artery (E) and the anterior descending (interventricular) artery (D). The arteries, the first branches of the aorta, provide the cardiac muscle of the heart with oxygenated blood.

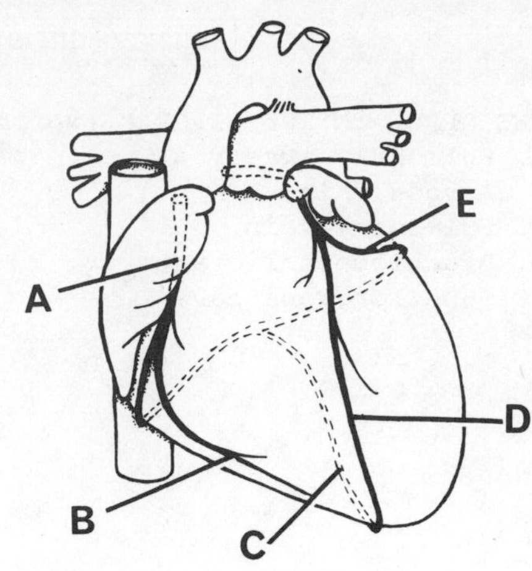

3. B. The myocardium is the middle of three distinct layers, forming the walls of the heart. The myocardium is formed by interlacing bundles of cardiac muscle fibers. This interlacing of the cardiac muscle cells is such that their contraction results in a wringing or twisting of the heart which physically squeezes the blood out of the heart with each beat. The myocardium is the thickest layer of the heart wall. The inner layer of the heart, the endocardium, consists of a single layer of endothelial cells. This endothelial layer continues on to form the outer layer of the valves of the heart and the inner lining of the blood vessels. The outer layer of the heart wall is formed by the visceral serous pericardium and is called the epicardium.

4. A. The atrioventricular valve on the right and left sides of the heart is attached to an apparatus that prevents eversion, gives support, and lends strength to the cusps of the valve. This apparatus consists of chordae tendineae and papillary muscles (labelled in the diagram). When the atrioventricular valve closes after the ventricle is filled with blood this apparatus prevents the valves from everting on contraction of the ventricle. In this way the blood is not returned to the atrium but instead is directed out of the ventricle to either the aorta or the pulmonary artery.

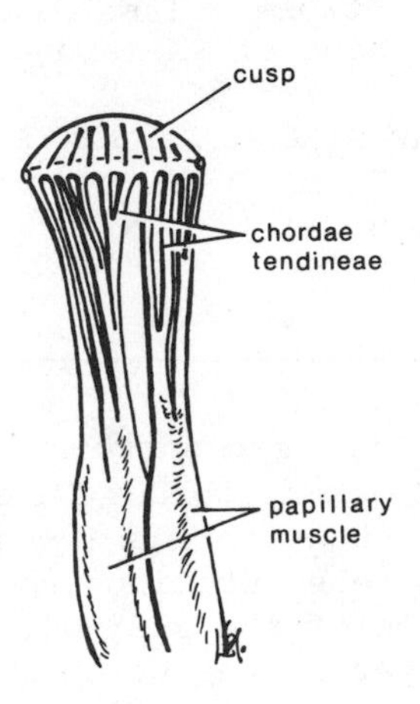

5. B. The cardiac centers of the brain are located in the medulla oblongata. Stimulation of the ventral reticular area of the medulla produces a depressor effect on the circulatory system as a whole, with a slowing of the heart rate and a lowering of the blood pressure. On the other hand a stimulation of the lateral reticular area of the medulla has an opposite or pressor effect.

6. E. The intrinsic, rhythmic contractions of the heart are regulated by pacemakers, which are specialized cardiac muscle fibers. The sinoatrial (S-A) node (A in the diagram), one of these pacemakers, lies beneath the epicardium at the junction of the superior vena cava and the right atrium. The S-A node initiates the normal cardiac impulse which then travels to the atrioventricular (A-V) node (E). The A-V bundle (B) conducts the impulse from the atria into the ventricles. A delay, for a fraction of a second, in the conduction of the impulse through the A-V node causes the ventricles to contract shortly after the atria have contracted. This delay allows the blood to enter the ventricles before they contract. The A-V bundle splits into left (C) and right (D) bundles of Purkinje fibers, which conduct the cardiac impulses to all parts of the ventricles so they can contract as a unit.

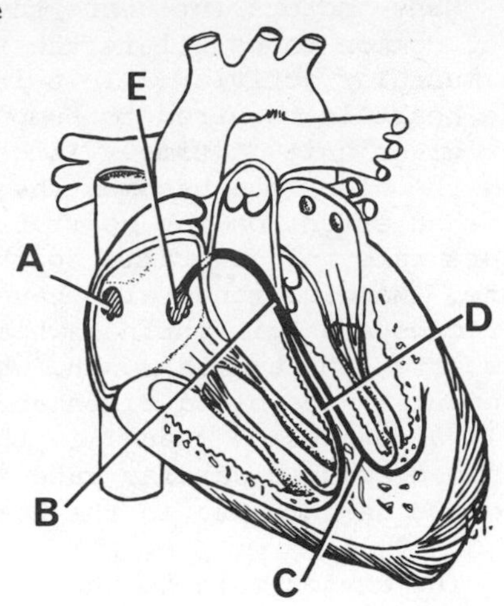

7. B. Each beat of the heart is accompanied by an electrical change or action potential that can be detected by an electrocardiograph tracing. A normal tracing for one cardiac cycle is seen in the diagram on the right. The P wave represents the electrical activity with the depolarization of the atria. The P-R interval gives the time for the original impulse to reach the ventricles. The QRS wave represents the electrical activity which is associated with depolarization of the ventricles or the time required for the impulse to spread through the A-V bundle and its branches. The S-T interval represents the period between the completion of ventricular depolarization and the start of repolarization. The T-wave represents the recovery phase which follows the contraction.

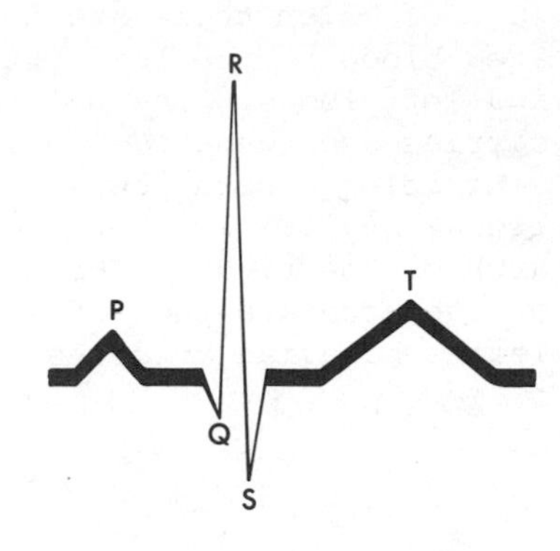

8. C. The left pulmonary artery is indicated by C in the diagram and the right pulmonary artery is not labelled. The aortic arch (B) has three major vessels coming off it: the brachiocephalic artery (A) on the right, the left common carotid artery, and the left subclavian artery (both of which are not labelled). The diagram also shows the inferior (D) and superior (E) venae cavae entering the right atrium.

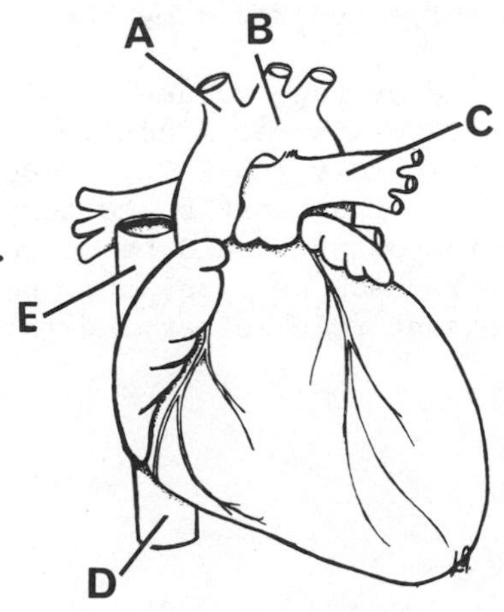

9. D. Many factors influence the rate at which the heart beats, perhaps the most common to us all is the acceleration of the heart rate with an increase in muscular activity. This increased activity requires more oxygen so that the heart is required to pump more blood through the lungs and to the body per given unit of time. In general, men have a slower heart beat than women and the older one becomes the slower the heart beat becomes, so sex and age have an effect on the heart rate. A high body temperature produces a higher heart rate and generally speaking the larger the person the lower the heart rate. Many hormones also have an effect on the rate of the heart. Two such hormones are epinephrine, which increases both the rate of the heart beat and its strength, and thyroxine which increases the heart rate when the amount present in the blood rises above a certain level. A change in blood pressure will also cause a change in the heart rate. An increased blood pressure produces a reduced cardiac rate, whereas a decrease in the blood pressure will produce an increase in the heart rate.

10. B. The azygos vein is indicated by B (in the diagram). It drains the azygos system of veins located in the thoracic cage and along with the superior (A) and inferior (not labelled in the diagram) vena cava, drains into the right atrium. The pulmonary veins (C), of which there are four, bring oxygenated blood to the left atria from the right and left lungs. The pulmonary artery (D) carries unoxygenated blood from the right ventricle to both lungs. The brachiocephalic artery (E) is a branch of the aortic arch on the right. The other two branches of the arch are the left common carotid and left subclavian arteries.

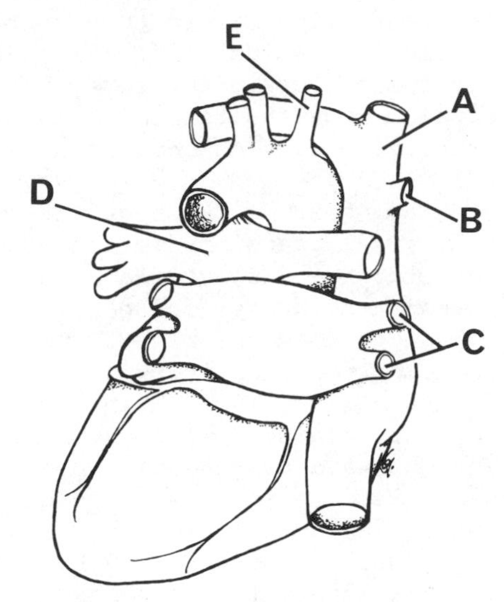

11. C. The cardiac output is equal to the stroke volume multiplied by the heart rate. The stroke volume is the volume of blood forced from the heart per beat. This volume measures about 60-70 ml per beat. The cardiac output depends on the frequency of the heart beat, the amount of venous return, and the force generated by the contractions of the heart.

12. A. The average volume of blood ejected by the heart per beat is 60-70 ml. With 72 beats per minute, as a normal figure, the cardiac output would be approximately (65 x 72) 4,680 ml per minute. Of course during strenuous exercise the heart beats three to four times faster than normal and the cardiac output would be correspondingly greater. Besides exercise other factors such as size, age, sex, body temperature, and blood pressure play an important role in varying the heart rate.

MULTI-COMPLETION QUESTIONS

DIRECTIONS: In each of the following questions or incomplete statements, ONE OR MORE of the completions given is correct. At the lower right of each question circle A if 1, 2, and 3 are correct; B if 1 and 3 are correct; C if 2 and 4 are correct; D if only 4 is correct; and E if all are correct.

1. WHICH OF THE FOLLOWING STATEMENTS IS(ARE) CORRECT FOR CARDIAC MUSCLE FIBERS?
   1. The all-or-nothing principle is applicable
   2. When stretched, cardiac muscle contracts more forcibly
   3. Its velocity of conduction is slower than skeletal muscle
   4. The intercalated discs offer a great electrical resistance

   A B C D E

2. THE SEROUS PERICARDIUM:
   1. Has an inner fibrous layer
   2. Lines the fibrous pericardium
   3. Is a strong connective tissue layer
   4. In part forms the epicardium of the heart

   A B C D E

3. THE LEFT VENTRICLE:
   1. Forms the apex of the heart
   2. Has larger papillary muscles than the right
   3. Is formed by an extremely thick layer of cardiac muscle
   4. Pumps blood to the systemic circulatory system

   A B C D E

4. THE CORONARY SINUS:
   1. Is located anterior to the aorta
   2. Forms a large portion of the right atrial wall
   3. Receives oxygenated blood from the left ventricle
   4. Drains into the right atrium

   A B C D E

5. THE CARDIAC CYCLE:
   1. Consists of a long series of contractions
   2. Is initiated by a spontaneous generation of an action potential in the S-A node
   3. Takes about 1.6 seconds to complete
   4. Initiates in the atria and spreads to the ventricles

   A B C D E

6. THE HEART IS:
   1. Responsible for pumping blood
   2. A cone-shaped muscular organ
   3. Located within the mediastinum
   4. Found, for the most part, to the right of the midsternal line

   A B C D E

7. TRUE STATEMENTS ABOUT THE CORONARY ARTERIES INCLUDE:
   1. Are branches of the ascending aorta
   2. The right coronary artery supplies blood to both ventricles
   3. Lie within grooves on the outer surface of the heart
   4. Each atrium receives blood from its corresponding coronary artery

   A B C D E

8. THE HEART IS CONTROLLED IN PART BY:
   1. The vagus nerve
   2. Sympathetic nerve fibers
   3. Visceral afferent nerve fibers
   4. The cardiac center in the midbrain

   A B C D E

A	B	C	D	E
1,2,3	1,3	2,4	Only 4	All correct

9. WHICH OF THE FOLLOWING STATEMENTS IS(ARE) TRUE FOR THE CONDUCTING SYSTEM OF THE HEART?
   1. Made up of specialized heart cells
   2. Consists of an A-V node in the upper right atrial wall
   3. Communicates into the walls of the atria and ventricles
   4. Has Purkinje fibers spreading into the atrial walls

   A B C D E

10. WHICH OF THE FOLLOWING IS(ARE) CORRECT FOR THE LEFT ATRIUM?
    1. Has thinner walls than the right atrium
    2. Receives blood from the pulmonary veins
    3. Blood leaves this chamber via a three-leafed valve
    4. Is smaller than the right atrium

    A B C D E

11. THE FIRST HEART SOUND IS PRODUCED BY THE:
    1. Closing of the A-V valves
    2. Closing of the aortic valve
    3. Contraction of the ventricular walls
    4. Simultaneous contractions of the atria

    A B C D E

12. THE RIGHT ATRIUM RECEIVES BLOOD VIA THE:
    1. Inferior vena cava
    2. Coronary sinus
    3. Superior vena cava
    4. Pulmonary veins

    A B C D E

ANSWERS, NOTES AND EXPLANATIONS

1. A, <u>1, 2, and 3 are correct</u>. Cardiac muscle has the ability to generate spontaneous and rhythmic impulses independent of any external nerve stimuli. The intercalated discs of cardiac muscle offer less resistance to an electrical stimuli than does the cell membrane, so that the action potential travels from one muscle cell to another with little interference. So cardiac muscle acts as a functional syncytium and because of this, when one cell becomes excited the impulse spreads to all other cells very rapidly. This syncytial nature of cardiac muscle and the fact that the whole atrial or ventricular muscle mass becomes excited together is called the all-or-nothing principle. The velocity of conduction of the action potential in both the atrial and ventricular muscle fibers is about 0.3-0.5 meter per second or about one tenth the speed of skeletal muscle. There is a direct relationship between the volume of blood entering the heart and the force of contraction of the heart walls; this is Starling's Law. When cardiac muscle is stretched by a large volume of blood, the resultant contractions of the cardiac muscle fibers are stronger. The smaller the volume of blood entering the heart, the weaker the contractions are.

2. C, <u>2 and 4 are correct</u>. The serous pericardium consists of two layers separated by a potential space, the pericardial space. The outer or parietal layer of the serous pericardium lines the fibrous pericardium, whereas the inner or visceral layer of the serous pericardium is closely attached to the heart and forms the epicardium. These two layers of the serous pericardium are separated by a thin film of fluid which reduces the friction between these two layers as the heart beats.

3. E, <u>All are correct</u>. The left ventricle forms the apex of the heart and contains two papillary muscles which are larger than their counterparts in the right ventricle. The wall of the left ventricle is up to two times as thick as that of the right ventricle. This increased thickness is because of a higher arterial blood pressure in the systemic system over the pulmonary system

and because the left ventricle has more work to do than the right, as it pumps oxygenated blood to the systemic system.

4. D, Only 4 is correct. The walls of the heart are drained partly by veins that empty into the coronary sinus (A) and partly by small veins that empty directly into the chambers of the heart. The coronary sinus lies in the coronary groove, between the left atrium and left ventricle and drains into the right atrium.

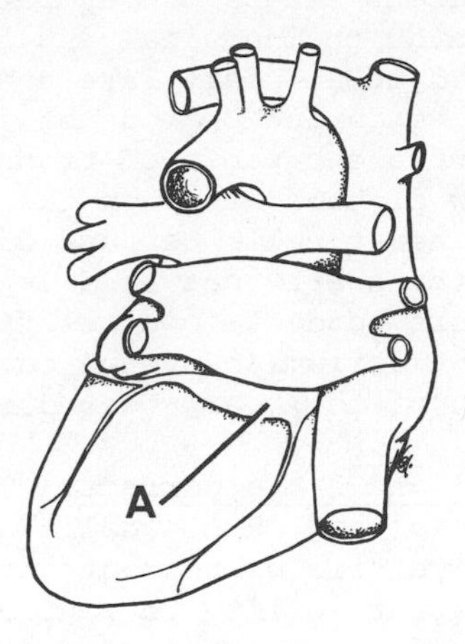

5. C, 2 and 4 are correct. The cardiac cycle is about 0.8 second in duration. The cycle consists of one heart beat which consists of a single contraction (systole) followed by a period of relaxation (diastole). The cycle is initiated by the S-A node and spreads through the atria to the A-V node and its associated A-V bundle. The conduction of the impulse is slowed by the A-V bundle so that the ventricles will contract a split second later than the atria. This slight delay allows the blood to flow into the ventricular chambers before they contract. Atrial systole lasts 0.1 second and the atrial diastole lasts 0.7 second, therefore the atria contract during one-eighth of the cycle and relax seven-eighths of a second. The ventricular systole lasts 0.3 second and diastole lasts 0.5 second.

6. A, 1, 2, and 3 are correct. The sole function of the heart is to pump blood through the pulmonary (lungs) and systemic (remaining portion of the body) vascular systems. The heart is well constructed for its job as it is a durable muscular organ made up of special striated muscle fibers called cardiac muscle. The heart works as the cardiac muscle contracts upon itself in a wringing-out fashion as one would wring water out of a towel. The heart is located in the mediastinum of the thoracic cage, between the right and left lungs. Within the mediastinum about two thirds of the heart lies to the left of the midsternal line.

7. E, All are correct. Both the right and left coronary arteries are the first branches of the aorta. The right coronary artery supplies the sinoatrial node, the right atrium, and both of the ventricles. A branch from the posterior descending artery usually supplies the atrioventricular node. The left coronary artery supplies the left atrium and the left ventricle. These arteries are intimately involved with the epicardium of the heart, lying in small grooves and partially covered byvarying amounts of adipose tissue. These arteries are vital to the functioning of the cardiac muscle, which forms much of the thickness of the heart wall.

8. A, 1, 2, and 3 are correct. Cardiac muscle has a special ability to contract rhythmically and in doing so pumps blood through the chambers of the heart to the pulmonary and systemic arteries. There is however, out of necessity, an external modulating system of control that can change both the rate and strength of the heart beat. This control consists of a double efferent nerve supply that transmits impulses from the cardiac centers in the medulla oblongata of the brain to the heart. The first group of efferent fibers are inhibitory, in that they cause the heart beat to slow down and become weaker. These efferent or parasympathetic fibers reach the heart by way of the vagus

nerve. These parasympathetic vagal fibers are in tonic activity and are continually suppressing the rate and strength of the heart beat. The other efferent fibers are sympathetic or accelerator fibers, that reach the heart through the cervical and thoracic cardiac nerves. Stimuli from these fibers speed up the heart rate and increase its strength. Such stimuli are quite active during periods of emotional stress. Certain afferent fibers from the heart are capable of transmitting pain when there is an interference in the flow of blood through the heart. The resultant decrease in oxygen reaching the heart cells because of the restricted flow stimulates the pain receptors in the heart that send impulses to the medulla of the brain. Other afferent fibers conduct impulses from pressure-sensitive receptors of the heart (i.e., right atrium and venae cavae) that are reflexively involved with the acceleration of the heart beat during exercise.

9. B, 1 and 3 are correct. The conducting system of the heart, consisting of specialized muscle cells, enables the separate muscle masses of the atria and the ventricles to contract in an orderly sequence. Cardiac muscle has an inherent ability to contract rhythmically in the absence of an external stimulus. The S-A (sinoatrial) node, located in the upper right atrial wall, contains cells that have the most rapid inherent rhythm and it is these cells that initiate the normal heart beat. Another mass of specialized muscle cells is the A-V (atrioventricular) node which is found in the wall of the atrium near the coronary sinus. This node carries the stimuli from the S-A node into the ventricles by way of the A-V bundle. This bundle divides and becomes continuous with the Purkinje fibers that stimulate the muscle masses of the right and left ventricles.

10. C, 2 and 4 are correct. The left atrium receives oxygenated blood from the lungs by the two pair of pulmonary veins. Once in the atrium the blood is pumped into the left ventricle through the left atrioventricular (mitral, bicuspid) valve which is a two-leafed valve. The walls of the two atria are similar in thickness and are separated from each other by the interatrial septum. The right atrium is slightly larger than its counterpart on the left. The oxygenated blood passing through the left atrium and left ventricle leaves the heart by way of the aorta to the systemic circulation.

11. B, 1 and 3 are correct. Each beat of the heart produces two sounds. The first heart sound is produced by the closing of the A-V valves (tricuspid and bicuspid) and the contraction of the ventricular walls. Typically the sound is a loud, long *lubb*. The second heart sound is a softer, shorter sound like *dup*. This second heart sound is produced by the closing of the semilunar valves.

12. A, 1, 2, and 3 are correct. The right atrium of the heart receives venous blood from the coronary system of veins and from the systemic venous system. The coronary veins draining the walls of the heart all drain into the coronary sinus, which in turn drains into the right atrium. The systemic system of veins drains into the right atrium by the superior and inferior venae cavae and the azygos system of veins. The pulmonary veins bring oxygenated blood from the lungs to the left atria of the heart.

FIVE-CHOICE ASSOCIATION QUESTIONS

DIRECTIONS: Each group of questions below consists of a numbered list of descriptive words or phrases accompanied by a diagram with certain parts indicated by letters, or by a list of lettered headings. For each numbered word or phrase, SELECT THE LETTERED PART OR HEADING that matches it correctly. Then insert the letter in the space to the right of the appropriate number. Sometimes more than one numbered word may be correctly matched to the same lettered part or heading.

1. _____ Left pulmonary artery

2. _____ Left brachiocephalic vein

3. _____ Pulmonary veins

4. _____ Inferior vena cava

5. _____ Coronary sinus

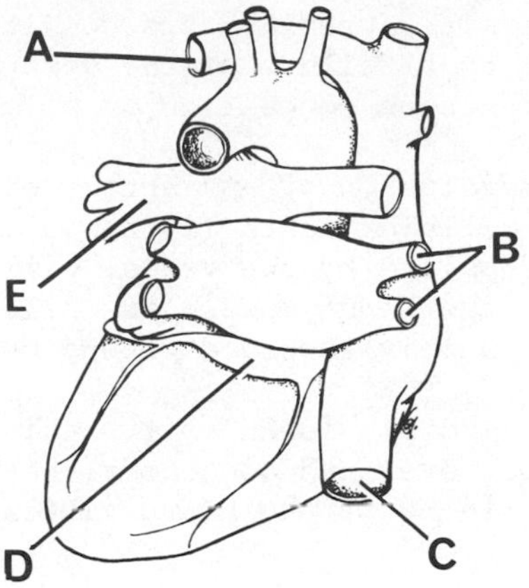

6. _____ The visceral pericardium

7. _____ Inflammation of the heart

8. _____ A coronary artery occlusion produces

9. _____ Inflammation of the lining of the heart

10. _____ The failure of a valve to fully open

A. Infarct

B. Myocarditis

C. Stenosis

D. Epicardium

E. Endocarditis

------------------------ ANSWERS, NOTES AND EXPLANATIONS ------------------------

1. E. The left pulmonary artery carries unoxygenated blood from the right ventricle to the left lung.

2. A. The left brachiocephalic vein is formed by the union of the left subclavian and left internal jugular veins. The right brachiocephalic vein is formed by the same veins on the opposite side. The left brachiocephalic vein joins up with its counterpart on the right and forms the superior vena cava. The azygos vein also drains into the superior vena cava.

3. B. The pulmonary veins, usually four in number, return oxygenated blood from the lungs to the left atrium of the heart. From this chamber the blood passes into the left ventricle where it is pumped via the aorta to the systemic system of arteries.

4. C. The inferior vena cava opens into the right atrium of the heart, along with the openings for the superior vena cava and the coronary sinus. The inferior vena cava receives venous blood from the portal system via the hepatic veins, the kidneys, the posterior abdominal and pelvic walls, and the lower limbs.

5. D. The coronary sinus lies on the posterior surface of the heart, in the groove marking the limits of the left atrium and ventricle. It drains into

the base of the right atrium below the opening for the inferior vena cava. The sinus drains the blood from the coronary veins.

6. D. The visceral pericardium is the inner layer of the serous pericardium of the heart which forms the outer layer of the heart wall or epicardium. The outer (parietal) layer of the serous pericardium lines the fibrous pericardium. Between the two layers, making up the serous pericardium, is a potential space containing a small amount of fluid which allows the two layers to move on each other with a minimal amount of friction.

7. B. Myocarditis is an inflammation of the cardiac muscle or the middle layer of the heart. Usually this condition is due to some systemic disorder and may lead to a failure of the ventricles to function properly which may produce systemic venous congestion or pulmonary congestion, or both.

8. A. The occlusion of any artery or vein will produce an infarct or a localized area of cellular death (ischemia). In a coronary artery occlusion the cardiac muscle supplied by the vessel will die because of a lack of oxygen. An infarct of the heart muscle is called a myocardial infarct. Prolonged chest pain is usually recorded with a myocardial infarct.

9. E. Endocarditis is an inflammation of the endothelial lining of the heart. Rheumatic fever and gonococcal bacteria are two diseases that attack the endothelial lining of the heart valves. They may cause the valves to function improperly.

10. C. The leaflets of the different valves of the heart may fail to open wide enough for the large volume of blood to pass through; such a condition is called stenosis. The fibrous ring to which the leaflets are attached becomes stiff, thickened and fails to open.

THE BLOOD VESSELS AND THEIR PHYSIOLOGY

FIVE-CHOICE COMPLETION QUESTIONS

DIRECTIONS: Each of the following questions or incomplete statements is followed by five suggested answers or completions. SELECT THE SINGLE BEST ANSWER in each case and then circle the appropriate letter at the lower right of each question.

1. THE ARROW INDICATES THE _____ ARTERY.
   A. Celiac
   B. Common hepatic
   C. Right gastric
   D. Splenic
   E. Left gastric

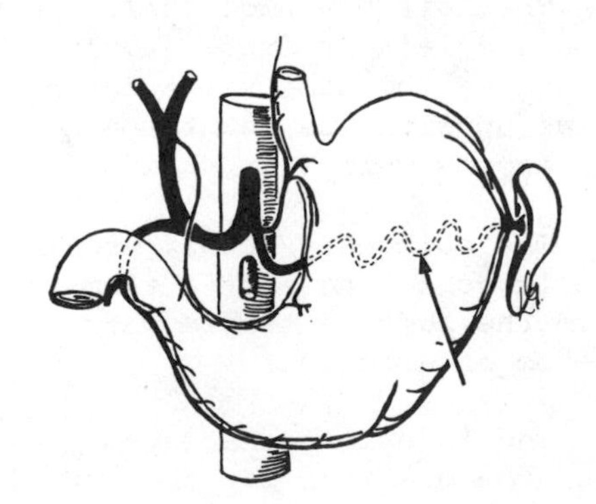

A B C D E

SELECT THE SINGLE BEST ANSWER

2. THE BLOOD VESSEL WHICH ACTS AS A CONTROL VALVE IS THE ______.
   A. Vein
   B. Venules
   C. Artery
   D. Capillary
   E. Arteriole

A B C D E

3. THE _______ ARTERY SUPPLIES MOST OF THE SMALL INTESTINE.
   A. Aorta
   B. Splenic
   C. Hepatic
   D. Inferior mesenteric
   E. Superior mesenteric

A B C D E

4. ARTERIAL DIASTOLIC PRESSURE IS PRODUCED BY THE:
   A. Contractions of the ventricles
   B. Recoil of the elastic arterial walls
   C. Contractions of the atria
   D. Pressure gradient between the arteries and veins
   E. None of the above

A B C D E

5. CHANGES IN BLOOD PRESSURE ARE DETECTED BY SPECIAL END ORGANS CALLED ______.
   A. Corpuscles of Ruffini
   B. Pacinian corpuscles
   C. Baroreceptors
   D. End bulb of Kraus
   E. None of the above

A B C D E

6. THE RENAL ARTERY IS INDICATED BY ________.

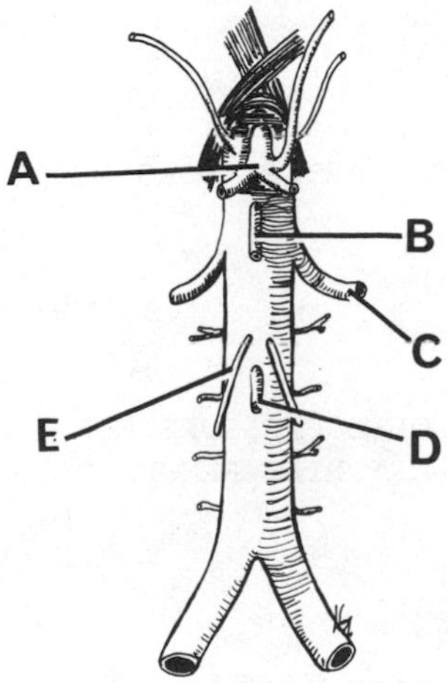

A
B
C
D
E

A B C D E

7. AN ARTERY SUPPLYING THE ANTERIOR ASPECT OF THE SCALP IS THE _______ ARTERY.
   A. Facial
   B. Occipital
   C. Maxillary
   D. Posterior auricular
   E. None of the above

A B C D E

8. THE ______ ARTERY PASSES BEHIND THE KNEE JOINT.
   A. Femoral
   B. Popliteal
   C. Profunda femoris
   D. Posterior tibial
   E. None of the above

A B C D E

SELECT THE SINGLE BEST ANSWER

9. THE GREATEST VOLUME OF BLOOD IS FOUND IN THE:
   A. Capillaries
   B. Systemic arteries
   C. Heart
   D. Pulmonary vessels
   E. Systemic veins

   A B C D E

10. IN THE UPPER LIMB THE SUBCLAVIAN ARTERY BECOMES THE _____ ARTERY.
   A. Axillary
   B. Brachial
   C. Radial
   D. Cephalic
   E. Ulnar

   A B C D E

11. THE ARROW INDICATES THE _____ ARTERY.
   A. Ulnar
   B. Median
   C. Radial
   D. Brachial
   E. Anterior interosseous

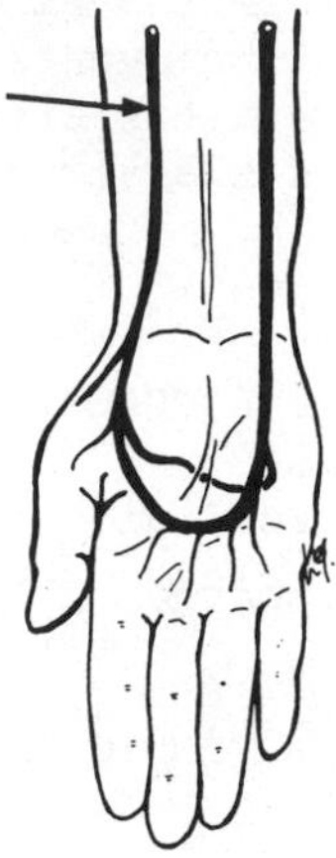

   A B C D E

12. THE INTERNAL ILIAC ARTERY DOES NOT SUPPLY WHICH OF THE FOLLOWING STRUCTURES?
   A. The ovary
   B. The bladder
   C. The uterus
   D. The vagina
   E. The floor of the pelvis

   A B C D E

13. THE AVERAGE SYSTOLIC PRESSURE FOR A NORMAL YOUNG ADULT, AT REST, IN THE SYSTEMIC ARTERIES RANGES BETWEEN ________ mm Hg.
   A. 75-80
   B. 95-100
   C. 115-120
   D. 135-140
   E. None of the above

   A B C D E

------------------------ ANSWERS, NOTES AND EXPLANATIONS ------------------------

1. D. The splenic artery (D) is one of three branches of the celiac (artery) trunk (A). The other two branches of the trunk are the left gastric artery (E) and the common hepatic artery (B). The right gastric artery (C) is a branch of the left hepatic artery.

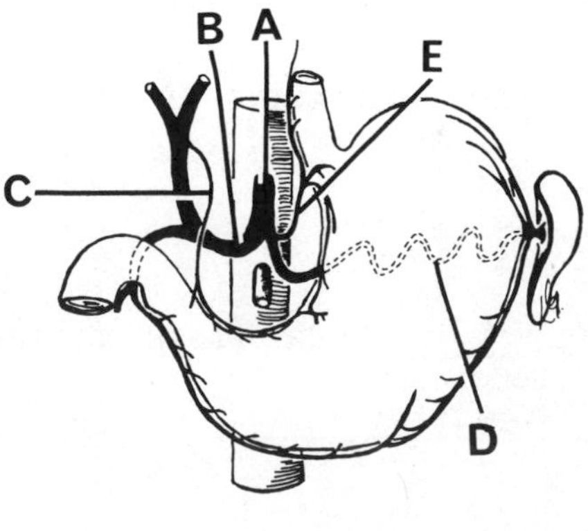

2. E. The arteriole, the last small branch of the arterial system, acts as a control valve in that it releases a given volume of blood into the capillaries. The arteriole is capable of constricting its lumen because of the strong smooth muscle fibers in the tunica media of its wall.

3. E. The superior mesenteric artery, a branch of the abdominal aorta, supplies the lower part of the duodenum, jejunum, ileum, cecum, appendix, ascending colon, and the right half of the transverse colon. The remaining half of the transverse colon, the descending colon, and the upper portion of the rectum are supplied by the inferior mesenteric artery. The upper portion of the duodenum, stomach, and esophagus are supplied by branches of the celiac trunk.

4. B. There are two separate pressure phases in the arteries: a so-called systolic pressure which is produced by the contraction (systole) of the ventricles and a diastolic pressure which is produced by the recoil characteristics of the stretched elastic tissue in the arterial walls during ventricular relaxation (diastole). These two pressure phases keep the blood flowing to the capillaries at an even pace, even though blood is leaving the ventricles only during systole. The recoil (diastolic pressure) of the stretched arterial walls keeps the blood flowing during the period of ventricular relaxation (diastole).

5. C. Lying within the walls of the aortic arch and the carotid sinus of the carotid artery are special pressure sensitive receptors called baroreceptors (pressoreceptors). These end organs, sensitive to changes in blood pressure, inform the medullary centers and they can decrease or increase the constriction of blood vessels and the rate at which the heart beats. So these pressoreceptors are of fundamental importance in regulating the activity of the heart and the blood vessels so as to keep the blood pressure within normal limits.

6. C. The renal artery (C in the diagram) is one of the major paired branches of the abdominal aorta. These arteries supply the kidneys with blood. Other arteries labelled in this diagram are the celiac trunk (A), the superior mesenteric artery (B), the inferior mesenteric artery (D), and the gonadal (testicular or ovarian) artery (E).

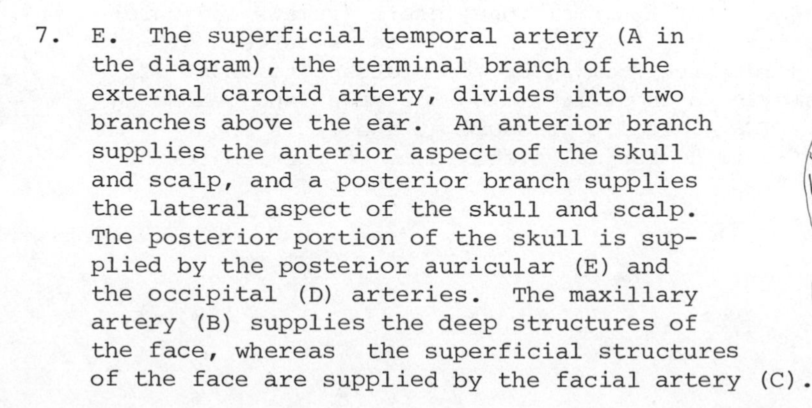

7. E. The superficial temporal artery (A in the diagram), the terminal branch of the external carotid artery, divides into two branches above the ear. An anterior branch supplies the anterior aspect of the skull and scalp, and a posterior branch supplies the lateral aspect of the skull and scalp. The posterior portion of the skull is supplied by the posterior auricular (E) and the occipital (D) arteries. The maxillary artery (B) supplies the deep structures of the face, whereas the superficial structures of the face are supplied by the facial artery (C).

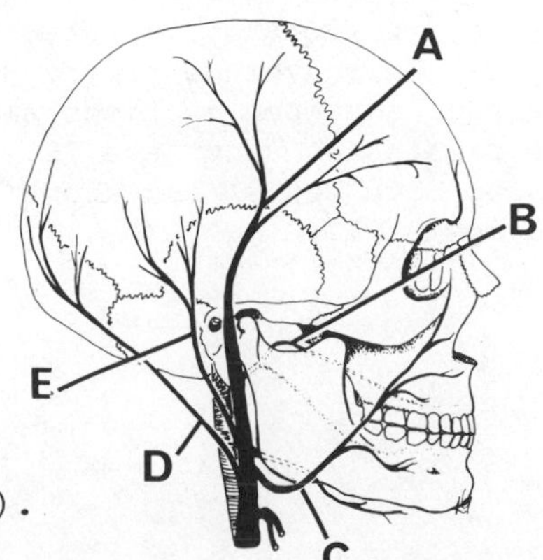

8. B. The popliteal artery is a continuation of the femoral artery lying behind the knee joint. Just below the knee joint the popliteal artery divides into anterior and posterior tibial arteries. The anterior tibial artery supplies the anterior aspect of the leg and the posterior tibial gives off a peroneal branch and together these two arteries supply the posterior aspect of the leg. In the thigh region the femoral artery gives off a profunda femoris branch which supplies the muscles of the thigh region.

9. E. The greatest volume of blood is found in the systemic veins and they contain about 59 per cent of the total blood volume. The remaining volume of blood reaches about 15 per cent in the arteries, five per cent in the capillaries, nine per cent in the heart, and 12 per cent in the pulmonary vessels. So we see that at the important functional site, the capillaries, there is only five per cent of the total volume of blood at any one time.

10. A. The subclavian artery crosses over the top of the first rib and at the lateral margin of the first rib it becomes the axillary artery. The axillary artery is found inthe axilla (arm pit) of the upper limb and it becomes the brachial artery in the arm. Anterior to the elbow joint the brachial artery branches into the radial artery, on the radial (lateral) side of the forearm, and the ulnar artery that extends down the ulnar (medial) side of the forearm. They both enter the palm of the hand and form the superficial and deep palmar arches.

11. C. The arrow in the diagram indicates the radial artery. The other artery passing down the ulnar or medial aspect of the forearm is the ulnar artery (A). Both these arteries are branches of the brachial artery of the arm and both contribute to form the deep (B) and superficial arches (C) of the palm.

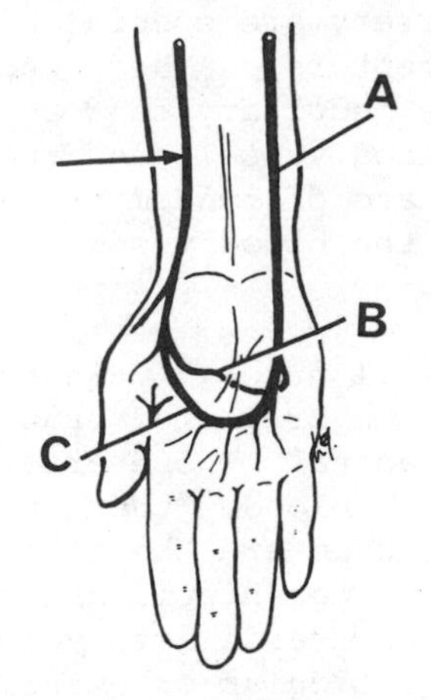

12. A. The internal iliac artery supplies the bladder, uterus, vagina, and the floor of the pelvis. It does not supply the ovary which is supplied mainly by the ovarian artery, a branch of the abdominal aorta which comes off the aorta just below the renal arteries. It should be remembered, however, that there may be anastomoses between the arteries supplying the ovary, the uterus, and the vagina.

13. C. The average systolic pressure for a normal young adult (female and male) at rest in the systemic arteries, is 115-120 mm Hg and ranges between 115-150 mm Hg. This average figure holds true from about the midteens to about 32 years of age when it rises slowly to about 160 mm Hg for males and 175 mm Hg for females at about age 72. The diastolic pressure at age 32 is about 63 mm Hg for both sexes, and about 85 mm Hg for males and 95 mm Hg for females at the age of 72.

MULTI-COMPLETION QUESTIONS

DIRECTIONS: In each of the following questions or incomplete statements, ONE OR MORE of the completions given is correct. At the lower right of each question, circle A if 1, 2, and 3 are correct; B if 1 and 3 are correct; C if 2 and 4 are correct; D if only 4 is correct; and E if all are correct.

1. THE TUNICA MEDIA OF A SMALL ARTERY:
   1. Is lined by an endothelial layer
   2. Consists of smooth muscle fibers
   3. Contains some elastic fibers
   4. Forms the thickest of the three layers of the wall

   A B C D E

2. FLOW THROUGH A BLOOD VESSEL DEPENDS ON THE:
   1. Pressure difference between the two ends of the vessel
   2. Volume of blood present
   3. Vascular resistance
   4. Length of the vessel

   A B C D E

3. SYMPATHETIC VASOCONSTRICTOR FIBERS:
   1. Are found in great numbers in cardiac muscle
   2. Innervate all blood vessels
   3. Are controlled by the vasomotor center of the pons
   4. Maintain the arterioles in a state of partial contraction

   A B C D E

4. WHICH OF THE FOLLOWING VEINS FORM(S) THE PORTAL VEIN?
   1. Left gastric
   2. Splenic
   3. Inferior mesenteric
   4. Superior mesenteric

   A B C D E

5. THE PULMONARY TRUNK:
   1. Divides to form two major arteries
   2. Carries oxygenated blood
   3. Supplies three arterial branches on the right
   4. Arises from the left ventricle

   A B C D E

6. THE MAINTENANCE OF ARTERIAL BLOOD PRESSURE INCLUDES THE:
   1. Force of the heart beat
   2. Viscosity of blood
   3. Amount of peripheral resistance
   4. Elasticity of large arteries

   A B C D E

7. THE ARTERIAL CAPILLARIES OF THE CIRCULATORY SYSTEM:
   1. Are formed by a single layer of cells
   2. Communicate with the venous system
   3. Anastomose freely between the tissue cells
   4. Are the sites of fluid and molecular exchange

   A B C D E

8. WHICH OF THE FOLLOWING STATEMENTS IS(ARE) CORRECT FOR BLOOD PRESSURE?
   1. Capillary pressure is greater than arterial pressure
   2. Is the force exerted by the blood against the vessel wall
   3. Is lower than atmospheric pressure
   4. A pressure gradient is established

   A B C D E

9. ARTERIES OF THE SYSTEMIC SYSTEM:
   1. Carry blood away from the heart
   2. Have strong vascular walls
   3. Transport blood under high pressure
   4. Carry blood to the lungs

   A B C D E

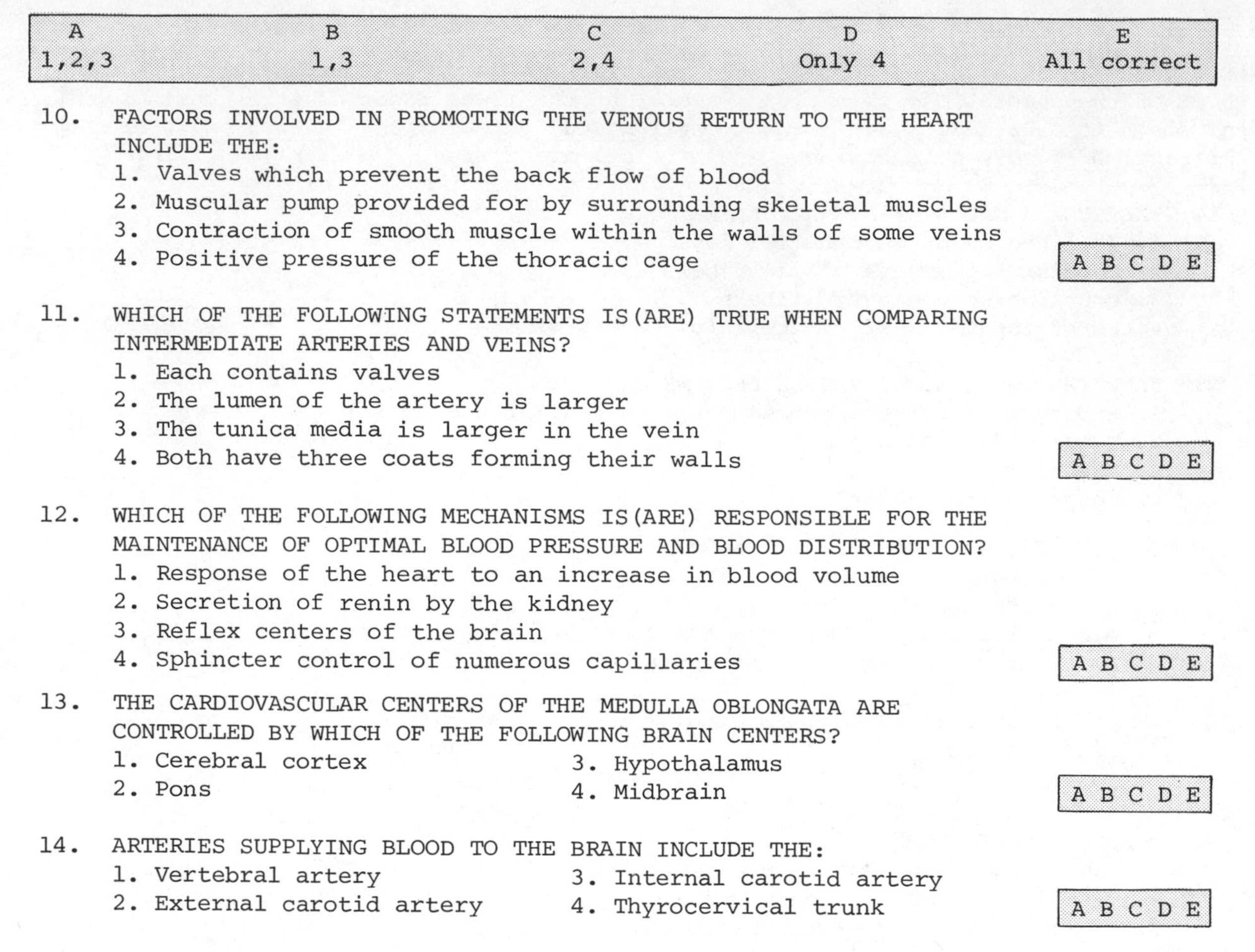

A	B	C	D	E
1,2,3	1,3	2,4	Only 4	All correct

10. FACTORS INVOLVED IN PROMOTING THE VENOUS RETURN TO THE HEART INCLUDE THE:
    1. Valves which prevent the back flow of blood
    2. Muscular pump provided for by surrounding skeletal muscles
    3. Contraction of smooth muscle within the walls of some veins
    4. Positive pressure of the thoracic cage

    A B C D E

11. WHICH OF THE FOLLOWING STATEMENTS IS(ARE) TRUE WHEN COMPARING INTERMEDIATE ARTERIES AND VEINS?
    1. Each contains valves
    2. The lumen of the artery is larger
    3. The tunica media is larger in the vein
    4. Both have three coats forming their walls

    A B C D E

12. WHICH OF THE FOLLOWING MECHANISMS IS(ARE) RESPONSIBLE FOR THE MAINTENANCE OF OPTIMAL BLOOD PRESSURE AND BLOOD DISTRIBUTION?
    1. Response of the heart to an increase in blood volume
    2. Secretion of renin by the kidney
    3. Reflex centers of the brain
    4. Sphincter control of numerous capillaries

    A B C D E

13. THE CARDIOVASCULAR CENTERS OF THE MEDULLA OBLONGATA ARE CONTROLLED BY WHICH OF THE FOLLOWING BRAIN CENTERS?
    1. Cerebral cortex
    2. Pons
    3. Hypothalamus
    4. Midbrain

    A B C D E

14. ARTERIES SUPPLYING BLOOD TO THE BRAIN INCLUDE THE:
    1. Vertebral artery
    2. External carotid artery
    3. Internal carotid artery
    4. Thyrocervical trunk

    A B C D E

---------------------- ANSWERS, NOTES, AND EXPLANATIONS ----------------------

1. E, All are correct. The tunica media of a small artery consists of large quantities of smooth muscle cells and some elastic fibers. Large arteries contain large quantities of elastic connective tissue and some smooth muscle fibers. The tunica media is the thickest of the three layers. The smooth muscle of the wall changes the size of the lumen by either contracting or dilating (relaxing).

2. B, 1 and 3 are correct. The rate of blood flow through a vessel depends on two factors: a) the pressure difference between the two ends of the vessel, and b) the impediment of blood flow through the vessel (resistance). If both ends of the vessel have the same pressure then there will be no flow of blood, the pressure difference between the two ends determines the flow. Resistance impedes the flow of blood. Slight increases in the diameter of a vessel greatly increase the flow and it is this factor which has the greatest role in the determination of the flow rate. Neither the volume of blood nor the length of the vessel has anything to do with the rate of flow.

3. D, Only 4 is correct. The sympathetic vasoconstrictor fibers innervate all blood vessels except capillaries. However, there are a lot more fibers to the blood vessels of the skin and the abdominal viscera when compared to those vessels found in cardiac and skeletal muscle. These nerve fibers are controlled by the vasomotor center of the medulla in the maintenance of peripheral resistance. The smooth muscle fibers are in a state of vasomotor tone

or a state of partial contraction by continual nerve stimulation. A marked increase in stimulation produces a constriction of the arterioles and an increase in the blood pressure.

4. C, 2 and 4 are correct. The superior mesenteric (C) and splenic veins (A) form the portal vein (D). The inferior mesenteric vein (B) joins the splenic vein. All the veins draining the gall bladder, stomach, pancreas, small and large intestines, and spleen enter the portal vein. This vein then takes the blood through the liver rather than directly to the inferior vena cava. The blood from these organs for the most part contains nutrients or products of digestion that are utilized by the liver cells. The venous blood is drained from the liver by the two hepatic veins which empty into the inferior vena cava posterior to the liver.

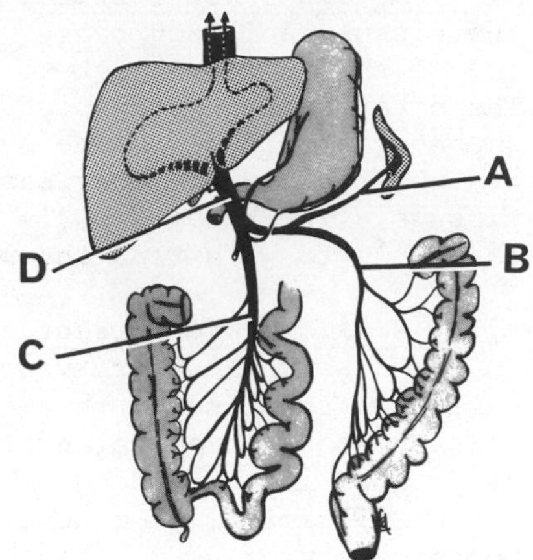

5. B, 1 and 3 are correct. The pulmonary trunk arises from the right ventricle and above the heart it divides into right and left pulmonary arteries. The right pulmonary artery divides into three branches, one to each of the three major lobes of the right lung, whereas the left pulmonary artery divides into two branches each supplying a major lobe of the left lung. These arteries carry unoxygenated blood to the capillaries of the lung, where carbon dioxide is exchanged for oxygen in the capillaries. The pulmonary veins carry oxygenated blood back to the left atrium of the heart. These pulmonary arteries, capillaries, and veins form the pulmonary circulation.

6. E, All are correct. The maintenance of the arterial blood pressure depends on a number of factors. One factor is the force and rate of the ventricular contractions. The stronger and faster the ventricles contract the higher the arterial blood pressure will be. Another factor directly involved is the elasticity of the large arteries. When the elastic recoil of these large arteries is high then the systolic pressure will also be high and if the elastic recoil is poor then the systolic pressure will be low. The amount of resistance to flow also plays a role in the maintenance of arterial blood pressure. If the lumina of the arterioles are wide there is less resistance and the pressure will drop, however, if the lumina are constricted the pressure will rise. An increase in blood viscosity will also increase the arterial blood pressure. Another factor in maintaining the arterial blood pressure is the total blood volume. If the blood volume drops because of a hemorrhage or a severed blood vessel the arterial blood pressure will also drop.

7. E, All are correct. The capillaries are the sites of functional exchange for the cardiovascular system. It is here that the exchange of fluid and nutrients between the blood and the interstitial spaces takes place. The capillary wall consists of a single layer of endothelial cells that are highly permeable. The arterial capillaries anastomose with the venous capillaries and drain the blood into the venous system.

8. C, 2 and 4 are correct. Blood pressure can be defined as the force exerted by the blood against the vessel wall in which it is contained. It is expressed in millimeters of mercury above the atmospheric pressure (760 mm Hg). There are three types of blood pressure: arterial, capillary, and venous. These three vascular components form a pressure gradient decreasing from a high arterial system, through the capillary system to a low venous system. The establishment of this pressure gradient depends on a high head of pressure which is generated during contraction of the ventricles of the heart. Once

blood leaves the heart it must pass from an area of higher pressure to an area of low pressure in order for it to flow.

9. A, <u>1, 2, and 3 are correct</u>. All arteries carry blood away from the heart to the tissue capillary beds. The arteries that transport blood to the capillary bed of the lung and back to the heart are components of the pulmonary system. The arteries that transport blood to all other tissues of the body and back to the heart again form the systemic system. The function of the arteries is to transport blood (in most cases oxygenated blood) under high pressure to the tissues. The blood flow is rapid in the arteries and they have strong vascular walls to withstand the pressure.

10. A, <u>1, 2, and 3 are correct</u>. Gravity works against the return of venous blood to the heart. However, venous blood is moved towards the heart by the so-called muscular pump, that is the skeletal muscles of the body. Any movement forces blood to move along the veins towards the heart. The valves of these veins work in only one direction so that the blood is moved from one valve to the next, similar to a number of separate elevators moving cargo up to the top of a skyscraper. Also helping the movement of venous blood towards the heart is the constriction of the smooth muscle found in the walls of the large veins. The pressure in the thoracic cage is negative and on inspiration becomes even more so. This negative pressure produces a suction effect so that the venous blood is literally pulled into the thorax veins and the systemic veins leading to the heart.

11. D, <u>Only 4 is correct</u>. Medium-sized arteries and veins have the same three layers that go to form the walls of the larger arteries and veins. However, when compared to the artery the layers in the wall of the vein are thinner. Thicker walls in the veins are not necessary because the blood pressure is lower in the veins than in the arteries. The lumina of arteries are smaller than that of veins. Veins contain valves spaced throughout their length and function to prevent the back flow of venous blood. Valves are most abundant in the veins of the extremities, especially the lower limb.

12. A, <u>1, 2, and 3 are correct</u>. The maintenance of optimal blood pressure and the distribution of blood throughout the circulatory system is controlled by both intrinsic and extrinsic factors. The intrinsic factors include: the ability of the heart muscle to respond to an increase in blood volume by stretching the ventricular walls, thus producing a greater contraction force; the regulation of fluid volume and arterial pressure by the renin-angiotensin system (low pressure causes the kidney to secrete renin which stimulates angiotensin II formation, which is a very potent vasoconstrictor); and the ability of precapillary arterioles to constrict and to adjust the blood flow to the tissues and to help maintain an adequate blood pressure. An extrinsic factor involved in maintaining optimal blood pressure and blood distribution is the nervous input to the cardiac and vasomotor reflex centers of the brain from receptors throughout the body.

13. B, <u>1 and 3 are correct</u>. The cardiovascular centers of the medulla are controlled by higher centers of the brain. These centers include many diffuse areas of the cerebral cortex and the hypothalamus. These higher centers are continually being stimulated, by afferent impulses via the cord and cranial nerves, which in turn send impulses to the cardiovascular centers of the medulla to modulate cardiac and vasomotor activities.

14. B, 1 and 3 are correct. The brain receives its blood by way of two important pairs of arteries: the vertebral arteries (A in the diagram), which are branches of the subclavian artery, and the internal carotid arteries (B), branches of the common carotid arteries. These four arteries enter the base of the skull and set up an anastomosis of arteries at the base of the brain. This anastomosis is called the circle of Willis and as seen in the diagram it provides three major arterial branches to the tissues of the brain (C).

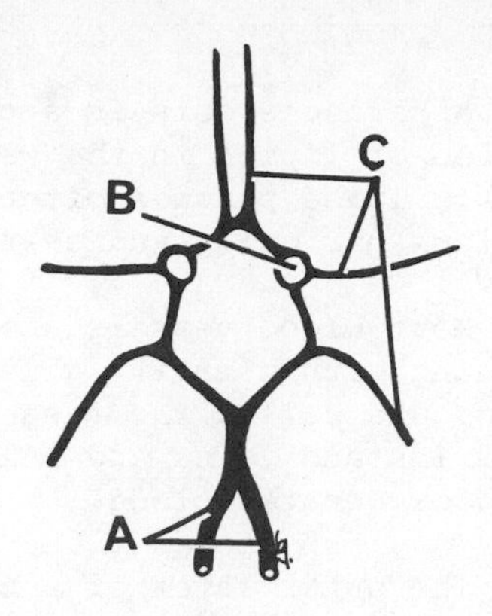

FIVE-CHOICE ASSOCIATION QUESTIONS

DIRECTIONS: Each group of questions below consists of a numbered list of descriptive words or phrases accompanied by a diagram with certain parts indicated by letters, or by a list of lettered headings. For each numbered word or phrase, SELECT THE LETTERED PART OR HEADING that matches it correctly. Then insert the letter in the space to the right of the appropriate number. Sometimes more than one numbered word may be correctly matched to the same lettered part or heading.

1. _____ An accumulation of lipids in arterial walls
2. _____ The small nutrient blood vessels of vessels
3. _____ The outer layer of a blood vessel
4. _____ A dilated vessel
5. _____ A layer consisting only of endothelium
6. _____ A hyaline thickening of arterioles

A. Adventitia
B. Aneurysm
C. Tunica intima
D. Vasa vasorum
E. None of the above

7. _____ The vertebral artery
8. _____ The thyrocervical trunk
9. _____ Arteries carrying unoxygenated blood
10. _____ The subclavian artery
11. _____ The common carotid artery

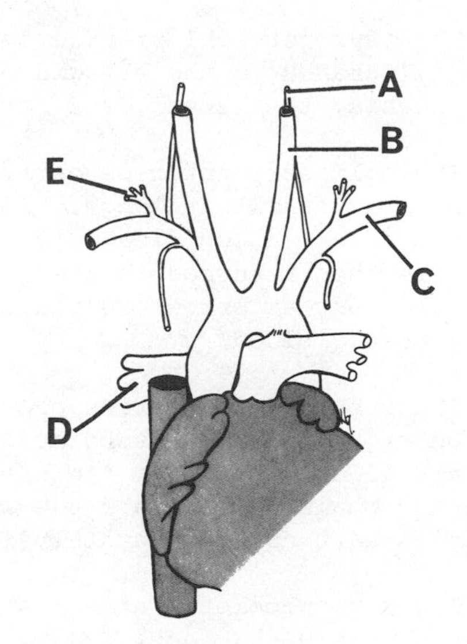

1. E. Atherosclerosis is a common disease of arteries and consists of an accumulation of lipids in the walls which form elevated lipid-laden plaques (atheromas). These plaques often encroach on the lumen of the vessels and may cause serious narrowing and eventually may completely occlude the vessel.

2. D. Most blood vessels above the calibre of small arteries have small vessels running within their walls. These small networks of nutrient arteries and veins are called the vasa vasorum. These vasa vasorum are found in both arteries and veins and help to supply the thickened walls of these vessels with oxygenated blood.

3. A. The outer layer of a blood vessel is called the tunica externa or adventitia. It consists primarily of fibrous connective tissues which blend with the surrounding connective tissue of the organs in which they are located.

4. B. A dilated portion of an artery wall is called an aneurysm. It is caused by some weakening of the arterial wall and the blood pressure of the artery is sufficiently high to cause the weakened area to bulge out and form a dilated sac.

5. C. The inner layer (tunica intima) of the wall of a blood vessel consists of a single layer of endothelial cells. These cells are a continuation of those forming the inner lining of the heart and they sit on an elastic membrane. In capillaries, the endothelial cells form the entire thickness of the wall.

6. E. Arteriosclerosis is a thickening of arteriole walls by the deposition of hyaline. It is fairly common in the elderly of our population and is called hardening of the arteries. This progressive disease process decreases the elasticity and the diameter of the lumen of the affected arterioles.

7. A. The vertebral artery is a branch of the subclavian artery and it passes up through the transverse foramen of the upper six cervical vertebrae on its way to the brain. This artery along with the internal carotid artery forms the circle of Willis and supplies blood to the brain.

8. E. The thyrocervical trunk arises from the subclavian artery and supplies an arterial branch to the thyroid gland, branches to the prevertebral muscles, the trachea, the larynx, and structures of the shoulder and scapular regions.

9. D. The pulmonary arteries carry unoxygenated blood from the right ventricle, to the right and left lungs. In the lungs the hemoglobin of the red blood cell releases carbon dioxide and then binds with oxygen. This oxygenated blood is then returned to the left atrium of the heart by the pulmonary veins. The pulmonary arteries are the only arteries of the adult body that transport unoxygenated blood.

10. C. The subclavian arteries provide the major blood supply to the upper limbs. On the right side, the subclavian and common carotid arteries arise not from the aortic arch but from the brachiocephalic artery. On the left there is no brachiocephalic artery and the left common carotid and the left subclavian arteries each arise separately from the aortic arch.

11. B. The common carotid arteries and their branches supply most of the structures of the neck, face, scalp, and brain. The left common carotid artery is a branch of the aortic arch, however on the right it is a branch of the brachiocephalic artery. The common carotid artery on each side divides into the external carotid artery, that supplies most of the structures of the neck, face, and scalp, and the internal carotid artery, that supplies the brain and certain structures within and near the orbit (eye).

12. THE LYMPHATIC SYSTEM

LEARNING OBJECTIVES

Be Able To:

- Define the components of the lymphatic system and give the chemical composition of lymph.
- Describe the functions of the lymphatic system.
- Describe the superficial (cutaneous) and deep lymphatic drainage patterns of the body.
- Discuss the formation of lymph and the factors influencing its rate of flow.
- Using a simple diagram illustrate the anatomical components of a lymph node.
- Describe the general area that the following clinically important groups of lymph nodes drain: axillary, mental, inguinal, superficial cervical and superficial cubital.
- For each of the following organs describe their location, structure, and function: spleen, tonsils, and thymus.

RELEVANT READINGS

Anthony & Kolthoff - Chapter 13.
Chaffee & Greisheimer - Chapters 11 and 13.
Crouch & McClintic - Chapters 22 and 25.
Jacob, Francone & Lossow - Chapter 11.
Landau - Chapters 16 and 17.
Tortora - Chapter 11.

DIRECTIONS: Each of the following questions or incomplete statements is followed by five suggested answers or completions. SELECT THE SINGLE BEST ANSWER in each case and then circle the appropriate letter at the lower right of each question.

1. WHICH OF THE FOLLOWING IS NOT A FUNCTION OF THE LYMPHATIC SYSTEM?
   A. An important site of RBC production
   B. Maintains a low protein concentration in the tissue fluids
   C. Filters out foreign bacteria from the tissue spaces
   D. Absorbs lipids from the gut lumen
   E. Drains excess intercellular fluid

   A B C D E

2. WHICH OF THE FOLLOWING IS NOT A FACTOR IN THE MOVEMENT OF LYMPH?
   A. Arterial pulsations
   B. The lymph node
   C. Interstitial fluid pressure
   D. Smooth muscle in the walls of the trunks
   E. Negative pressure in the brachiocephalic veins

   A B C D E

3. AN AREA OF SKIN IN THE RIGHT HYPOCHONDRIUM IS DRAINED BY THE _________ OF GROUP OF NODES.
   A. Axillary
   B. Deep cervical
   C. Inguinal
   D. Submental
   E. None of the above

   A B C D E

4. LYMPH CAPILLARIES ARE ABSENT FROM THE:
   A. Heart
   B. Lungs
   C. Dermis
   D. Large intestine
   E. Spinal cord

   A B C D E

5. THE SMALL LYMPHATIC VESSELS OF THE SMALL INTESTINE ARE CALLED _________.
   A. Lacunae
   B. Laminae
   C. Lamellae
   D. Lacteals
   E. None of the above

   A B C D E

6. THE PRESENCE OF EXCESSIVE AMOUNTS OF INTERSTITIAL FLUID IN THE BODY TISSUES IS CALLED _______.
   A. Obesity
   B. Hypertrophy
   C. Edema
   D. Lymphadenitis
   E. None of the above

   A B C D E

7. AN INFECTION IN THE FOOT WOULD LIKELY CAUSE A SWELLING OF THE _______ LYMPH NODES.
   A. Axillary
   B. Inguinal
   C. Superficial cubital
   D. Superficial cervical
   E. Submental

   A B C D E

------------------------ ANSWERS, NOTES AND EXPLANATIONS ------------------------

1. A. The lymphyatic system does not produce RBC (red blood corpuscles) but produces large numbers of lymphocytes in the lymph nodes. These lymphocytes produce antibodies and thereby contribute a significant protective function to the body. The lymph nodes are also responsible for filtering out foreign bacteria from the tissue fluids. Once filtered, the foreign bacteria are phagocytosed by large macrophages lining the sinuses of the nodes. The lymphatic system drains excess body (intercellular, interstitial) fluid and proteins from the

intercellular spaces and in this way helps maintain a low tissue colloidal osmotic pressure. The lymphatic system also absorbs lipids from the gut (small intestine) by small lymphatic vessels called lacteals. These lacteals then drain into the larger lymphatic vessels draining the gut.

2. B. The movement and rate of lymph flow is determined by a product of the interstitial fluid pressure and the activity of the 'lymphatic pump'. The interstitial fluid pressure in the tissue spaces is generated by the filtration of fluid from the blood capillaries, which in turn forces the interstitial fluid into the lymphatic capillaries. The 'lymphatic pump' functions in part because of the pressure applied to the lymphatic vessels by: the arterial pulsations of accompanying arteries, the contractions and movements of adjacent skeletal muscle masses, and respiratory movements. Another factor involved in moving lymph is that the brachiocephalic veins have a negative pressure, so that the lymph flows into the venous system unimpeded. The smooth muscle cells found in the walls of the large lymphatic trunks are stimulated by the sympathetic nerves to contract the lymphatic vessel walls which forces the lymph forward. The back flow of lymph is prevented by the valves located in the large lymph vessels and trunks.

3. A. An area of skin in the right hypochondrium is drained by the right axillary group of lymph nodes (A in the diagram). The lymphatic drainage of the skin follows a very definite pattern as illustrated. There are three areas of lymph drainage on each half of the body. These areas are bound by lines drawn through the clavicles and the umbilicus (see the diagram). The axillary group of nodes drains the upper limb and the upper portions of the front and back of the body limited by the lines drawn through the clavicles and the umbilicus. The cervical group of nodes (B), drains the skin of the appropriate half of the head and neck above the clavicles. The inguinal group of nodes (C) drains the skin area below the umbilicus including the lower limb. Each of these three superficial groups of lymph nodes then drain into deeper groups which in turn then drain into the right lymphatic duct or the thoracic duct.

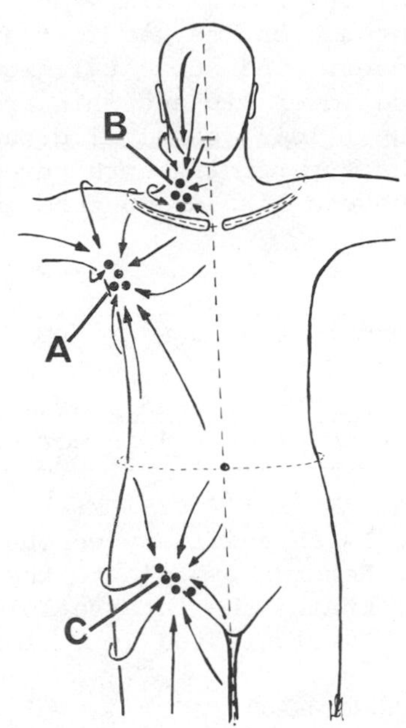

4. E. The lymph capillaries are present in most of the tissues of the body, but are absent from avascular structures such as the epidermis and its derivatives, the cornea, and the articular cartilages. Lymphatics are also not found in the brain, spinal cord, or bone marrow.

5. D. The small lymphatic vessels located within the villi of the small intestine are called lacteals. These small vessels are involved in the absorption of lipid material from the lumen of the small intestine. These lymphatic vessels will ultimately drain their lymph into the thoracic duct and return it to the venous system. Following a meal high in fat, the lymph found in the lacteals contains a large quantity of absorbed lipids and appears a milky color. This milky fluid is called chyle.

6. C. The presence of excessive amounts of interstitial fluid in the tissues of the body is called edema. Normally the interstitial fluid pressure is in the negative range, so that any factor that increases the interstitial fluid pressure high enough may cause an excess interstitial fluid volume and thereby cause edema. Among the causes of edema are: increased capillary pressure and

permeability, decreased plasma protein, lymphatic obstruction, and the failure of the kidneys to excrete adequate quantities of urine. Lymphadenitis is the inflammation of a lymph node or nodes.

7. B. As in other parts of the body, the lymphatics of the lower limb begin as lymphatic capillaries in the tissue spaces. These capillaries empty into lymphatic vessels which drain the superficial (skin and subcutaneous tissues) and deeper structures (deep fascia, muscles, joints, and periosteum). The lymphatic vessels draining the lower limb eventually all pass through the inguinal group of nodes (B in the diagram). These nodes also drain the gluteal and lower abdominal areas as well. The axillary group of nodes (A) drain the upper limb and the anterior and posterior thoracic walls. The superficial cubital nodes (C) are intermediary nodes which lymph passes through on its way to the axillary group of nodes. The submental group of nodes (E) drains the lower lip and chin area of the face. The superficial cervical group of nodes (D) drain the superficial structures of the head and neck above a line drawn through the clavicles.

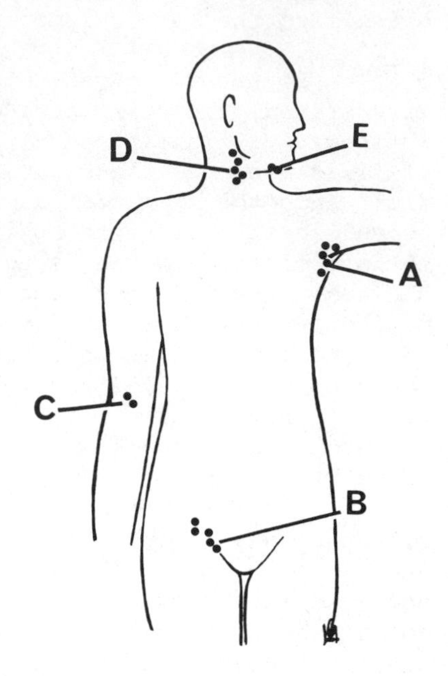

M U L T I - C O M P L E T I O N   Q U E S T I O N S

DIRECTIONS: In each of the following questions or incomplete statements, ONE OR MORE of the completions given is correct. At the lower right of each question, circle A if 1, 2, and 3 are correct; B if 1 and 3 are correct; C if 2 and 4 are correct; D if only 4 is correct; and E if all are correct.

1. THE LYMPHATIC SYSTEM:
   1. Is an auxiliary venous system
   2. Returns protein to the circulatory system
   3. Drains the intercellular spaces
   4. Is classified as a closed system

A B C D E

2. THE THORACIC DUCT:
   1. Has a dilated portion lying anterior to the second lumbar vertebra
   2. Empties into the junction of the right internal jugular and subclavian veins
   3. Drains the whole body below the level of the diaphragm
   4. Contains no valves similar to other lymphatic vessels

A B C D E

3. LYMPHATIC CAPILLARIES:
   1. Are more permeable than blood capillaries
   2. Begin as blind tubular structures
   3. Are formed by endothelial cells that overlap
   4. Have large numbers of valves

A B C D E

4. LYMPHATIC NODULES:
   1. Contain a highly organized germinal center
   2. Appear in the cortex of lymph nodes
   3. Are located beneath wet epithelial surfaces
   4. Aggregate to form components of the tonsillar ring

A B C D E

A	B	C	D	E
1,2,3	1,3	2,4	Only 4	All correct

5. THE WHITE PULP OF THE SPLEEN:
   1. Acts as a site for red blood cell filtration
   2. Is surrounded by a reticular network
   3. Contains stem cells of the erythrocyte line
   4. Provides the site for plasma cell production

   A B C D E

6. THE KNOWN FUNCTIONS OF THE THYMUS INCLUDE THE:
   1. Stimulation of plasma cell development
   2. Selected destruction of worn-out red blood cells
   3. "Seeding" of lymphatic tissue with lymphocytes
   4. Production of lymphocytes in old age

   A B C D E

7. WHICH OF THE FOLLOWING STATEMENTS IS(ARE) CORRECT FOR THE SPLEEN?
   1. Contains reticuloendothelial cells within its sinusoids
   2. Found in the upper left quadrant of the abdominal cavity
   3. Stores large quantities of red and white blood cells
   4. Destroys large numbers of red blood cells

   A B C D E

8. WHICH OF THE FOLLOWING STATEMENTS IS(ARE) CORRECT FOR THE PALATINE TONSILS?
   1. Are partially encapsulated organs of lymphatic tissue
   2. Participate in the formation of new lymphocytes
   3. Have epithelial lined crypts
   4. Contain will defined sinuses

   A B C D E

------------------------ ANSWERS, NOTES AND EXPLANATIONS ------------------------

1. A, 1, 2, and 3 are correct. The lymphatic system is not a true closed circulatory system but rather a blind system of tubes that drains the intercellular spaces of the body. It is considered auxiliary to the venous system in that it has many openings or connections with the venous system. The intercellular (tissue) fluid forms when more fluid leaves the capillaries by filtration than enters by osmosis. The excess of tissue fluid and the small amounts of protein, within the tissue spaces, enter the lymphatic vessels, and return thence to the circulatory system. The tissue fluid is called lymph once it has entered the lymphatic vessels.

2. B, 1 and 3 are correct. The thoracic duct is the larger of the two major lymphatic vessels. It begins as a dilated portion (cysterna chyli) lying anterior to the second lumbar vertebra and passes upward where it drains into the junction of the left internal jugular and left subclavian veins. The other major vessel of the lymphatic system, the right lymphatic duct, drains into the junction of the right internal jugular and subclavian veins. The thoracic duct drains the whole body below the level of the diaphragm and the left halves of the thorax, head, and neck regions. The thoracic duct has numerous valves placed throughout its length and has a pair at its juncture with the veins in the neck which prevents the back flow of lymph and venous blood.

3. A, 1, 2, and 3 are correct. The lymphatic capillaries are tiny thin-walled tubes that begin blindly and anastomose freely throughout the intercellular (interstitial) spaces of the body. Their walls are formed by overlapping endothelial cells and are much more permeable than blood capillary walls. The lymph capillaries have no valves.

4. E, All are correct. Lymphatic nodules are compact accumulations of lymphoid cells within diffuse lymphoid tissue. They are found in the cortex of lymph nodes, at the periphery of the white pulp regions of the spleen and beneath wet epithelial surfaces such as are found in the digestive tract, the respiratory passages, and the genitourinary ducts. Nodules are also very numerous in the tonsils, Peyer's patches of the ileum and the appendix. The germinal centers of the nodules have been shown to be the site of active lymphocyte proliferation. Whether the lymphocytes are in a diffuse or aggregated (nodular) form their function is to defend the body against bacteria or other disease-inducing agents that break through the mucous membrane barriers of the body.

5. C, 2 and 4 are correct. The spleen consists of two components: small islands of white pulp and surrounding areas of red pulp. The white pulp consists of lymph nodules that produce both lymphocytes and plasma cells. The plasma cells are stimulated into production by the presence of antigens in the circulation. The red pulp is found surrounding the small clusters of white pulp and consists of an extensive network of reticular fibers which contain both fixed and free macrophages, red blood cells, and lymphocytes. Monocytes and granular leukocytes are also found in the red pulp but in fewer numbers. The blood percolates through the red pulp and it is here that foreign bacteria, cell debris, and red blood cells are phagocytosed.

6. B, 1 and 3 are correct. An insight into the possible functions of the thymus has been recently derived from experimental evidence, however, there still remains many questions about its exact function. It appears that the thymus plays an important role in developing the immunological competence (ability to recognize and attack foreign substances) of the body. The immunologically competent lymphocytes are called T- (thymus dependent) cells and they migrate via the blood stream to populate the lymphatic tissue throughout the body (i.e., spleen, lymph nodes, tonsils) with immunologically competent lymphocytes. Some evidence indicates that the thymus may secrete a hormone which stimulates the proliferation of lymphocytes. Other evidence suggests that lymphocytes may evolve into antibody producing plasma cells. These plasma cells are thought to secrete their antibodies into the blood and lymph, where they inactivate most foreign bacteria and some viruses. The thymus does not phagocytose dead or dying red blood cells as does the spleen. In fact, the thymus, once puberty is reached, begins to involute and gradually ceases to produce lymphocytes. By mid-adulthood thymic tissue is hard to find and appears to be replaced by adipose tissue.

7. E, All are correct. The oval spleen is located in the upper left quadrant (hypochondrium) of the abdominal cavity. It is formed by numerous areas of lymphatic nodules surrounded by tortuous venous sinuses. These sinuses contain large numbers of reticuloendothelial cells (macrophages) which form a defense mechanism against foreign microorganisms. Besides defense the spleen plays a role in hemopoiesis and produces large numbers of nongranular leukocytes (monocytes and lymphocytes). The spleen also forms plasma cells. Before birth the spleen forms red blood cells but after birth it is more actively involved in destroying red blood cells. The spleen also destroys platelets. The spleen normally stores large numbers of blood cells and on sympathetic stimulation empties large quantities of blood cells into the circulation.

8. A, 1, 2, and 3 are correct. The tonsils are a group of well defined organs of lymphatic tissue that are located in the region of the oral cavity, where it communicates with the pharynx. There are three groups of tonsils: 1) the palatine tonsils, which are paired structures lying on either side of the oral cavity between the glossopalatine and the pharyngopalatine arches; 2) the unpaired lingual tonsil, found on the dorsum of the tongue above its root; and 3) the unpaired pharyngeal tonsil in the roof and posterior wall of the nasopharynx. The palatine tonsils are ovoid masses of lymphoid tissue devoid of

sinuses, as found in the lymph nodes. However, the oral epithelium invaginates into the lymphoid tissue forming deep crypts. These crypts may become infected by microorganisms which may then spread and cause a general infection in other parts of the body. The tonsilar tissue is backed by a connective tissue capsule. The tonsils as a group are involved in the formation of new lymphocytes and it is generally believed that they have something to do with the protection of the body against an invasion by microorganisms.

FIVE-CHOICE ASSOCIATION QUESTIONS

DIRECTIONS: Each group of questions below consists of a numbered list of descriptive words or phrases accompanied by a diagram with certain parts indicated by letters, or by a list of lettered headings. For each numbered word or phrase, SELECT THE LETTERED PART OR HEADING that matches it correctly. Then insert the letter in the space to the right of the appropriate number. Sometimes more than one numbered word may be correctly matched to the same lettered part or heading.

1. _____ A trabecula
2. _____ A cortical follicle
3. _____ The efferent vessels
4. _____ The capsule
5. _____ The sinuses

6. _____ The right lymphatic duct
7. _____ The lumbar trunk
8. _____ The cisterna chyli
9. _____ The subclavian trunk
10. _____ The thoracic duct

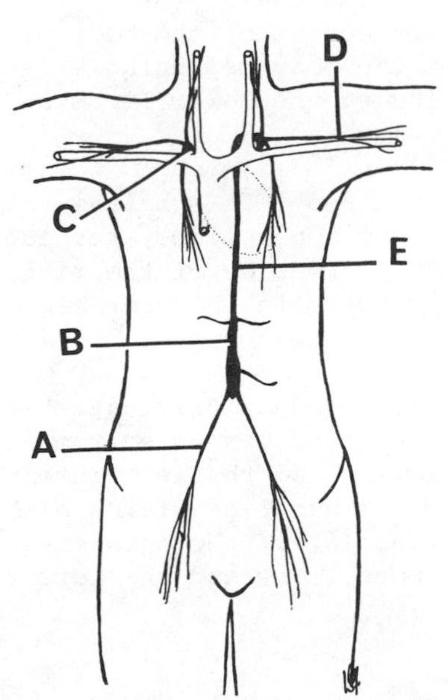

---------------------- ANSWERS, NOTES AND EXPLANATIONS ----------------------

1. E. The many trabeculae found in a lymph node are inward projections of the connective tissue capsule (C). They serve to support the lymphatic nodules and the sinuses.

2. B. The cortical follicle is the outer region of the nodules forming a lymph node. The follicular or outer region of the nodule contains lymphocytes which have been produced in the center of the nodule (germinal center). The germinal center is the site of lymphocyte production.

3. D. The lymph which percolates through the sinuses (A) of the node exits the node by the efferent lymphatic vessels. The efferent lymphatics leave the node at its hilum along with arteries and veins. These efferent vessels eventually will lead to larger lymphatic collecting vessels. The lymph enters the node by the afferent lymphatic vessels (not labelled).

4. C. The lymph node is encapsulated by a connective tissue layer or capsule. This capsule projects into the node forming trabeculae. The capsule of the node distinguishes it from a nodule, which has no capsule. So a lymph node is an encapsulated aggregation of lymph nodules.

5. A. Surrounding the lymphatic nodules of the lymph node are the lymph filled sinuses. These sinuses besides containing lymph contain many free and fixed macrophages that are active in removing foreign bacteria. These macrophages form part of the reticuloendothelial system.

6. C. The right lymphatic duct is a short segment that is formed by the junction of lymphatic trunks draining the right side of the head (jugular trunk), the right upper limb (subclavian trunk), and the right thoracic cavity (broncho-mediastinal trunk). The right lymphatic duct drains into the venous system at the superior junction of the right subclavian vein with the right internal jugular vein.

7. A. The lumbar trunks (right and left) drain the lymphatic vessels from the lower abdomen, the pelvis and the lower limbs. These trunks drain into the cisterna chyli (B), the dilated beginning of the thoracic duct (E).

8. B. The cisterna chyli, the dilated beginning of the thoracic duct, lies on the anterior aspect of the body of the second lumbar vertebra. It drains lymph from the body beginning at and including all areas below the umbilical line, including the lower limbs and the abdominal cavity (especially the abdominal viscera).

9. D. The right and left subclavian trunks drain the lymphatic vessels of the right and left upper limbs, respectively. The right subclavian trunk contributes to the formation of the right lymphatic duct and the left subclavian trunk contributes to the thoracic duct just before each drains into the venous system.

10. E. The thoracic duct originates in the cisterna chyli (B) and terminates as it drains into the venous system at the superior aspect of the junction of the left subclavian and the left internal jugular veins. The thoracic duct through its tributaries drains all the areas of the body below the level of the umbilicus (i.e., the area drained by the cisterna chyli), the left trunk, the left upper limb, and the left side of the head and neck.

REVIEW EXAMINATION OF PART 3

INTRODUCTORY NOTE: This review examination consists of 100 multiple-choice questions based on the learning objectives listed for the following chapters of this Study Guide and Review Manual: 8. THE NERVOUS SYSTEM; 9. SPECIAL SENSES; 10. THE ENDOCRINE SYSTEM; 11. THE CARDIOVASCULAR SYSTEM; and 12. THE LYMPHATIC SYSTEM. Before beginning tear out an answer sheet from the back of the book and read the directions on how to use it. The key to the correct responses is on page 339.

FIVE-CHOICE COMPLETION QUESTIONS

DIRECTIONS: Each of the following questions or incomplete statements is followed by five suggested answers or completions. SELECT THE SINGLE BEST ANSWER in each case and then circle the appropriate letter at the lower right of each question.

1. MOST OF THE OXYGEN TRANSPORTED IN THE BLOOD IS CARRIED:
   A. Chemically bound with hemoglobin
   B. Combined with Na^+ ions
   C. In simple solution
   D. As small bubbles
   E. None of the above

2. WHICH OF THE FOLLOWING STATEMENTS CONCERNING THE POSTGANGLIONIC PARASYMPATHETIC NEURONS IS CORRECT?
   A. Usually has a long neuron
   B. Originates in the sacral spinal cord
   C. Elicits a response in a single effector
   D. Uses norepinephrine as its neurotransmitter substance
   E. None of the above

3. ALL BLOOD CELLS ORIGINATE FROM A(AN):
   A. Hemocytoblast
   B. Erythrocyte
   C. Megakaryocyte
   D. Reticulocyte
   E. Normoblast

4. THE PARATHYROID GLANDS ARE NORMALLY LOCATED:
   A. In the substance of another gland
   B. Deep to the sternum of the chest
   C. On the superior surface of the kidneys
   D. Lateral to the hyoid bone
   E. None of the above

SELECT THE SINGLE BEST ANSWER

5. THE HEMATOCRIT REFERS TO THE:
   A. Normal pH range of blood
   B. Specific gravity of whole blood
   C. Ratio of plasma volume to the cellular volume
   D. Cellular volume of blood
   E. None of the above

6. WHICH OF THE FOLLOWING GANGLIA CAN BE CONSIDERED TO BE ASSOCIATED WITH THE SYMPATHETIC DIVISION OF THE AUTONOMIC NERVOUS SYSTEM (ANS)?
   A. Otic
   B. Ciliary
   C. Submandibular
   D. Sphenopalatine
   E. None of the above

7. THE PIA MATER IS INDICATED BY ________.

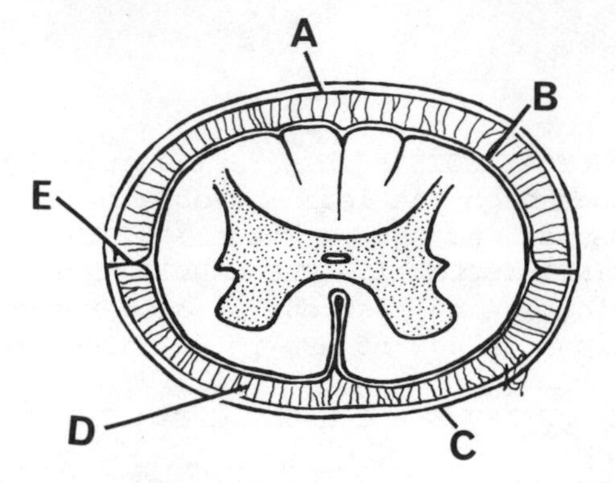

8. WHICH OF THE FOLLOWING IS NOT TRUE FOR THE RIGHT VENTRICLE?
   A. Has a thinner wall than the left ventricle
   B. Blood is prevented from returning to it by semilunar valves
   C. Contains papillary muscles attached to the tricuspid valve
   D. Receives oxygenated blood
   E. Pumps blood to the lungs

9. THE AQUEDUCT OF SYLVIUS IS INDICATED BY _________.

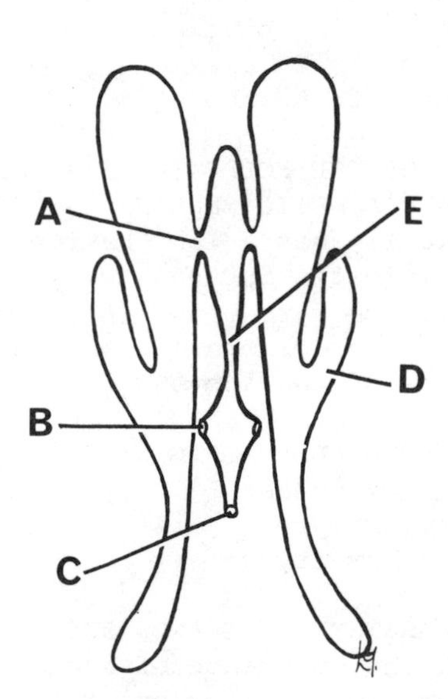

10. CALCITONIN IS RESPONSIBLE FOR:
   A. Increasing the rate of cellular diffusion of carbohydrates
   B. Increasing the level of calcium in the blood
   C. Decreasing the activity of osteoclasts
   D. Stimulating thyroxine secretion
   E. None of the above

SELECT THE SINGLE BEST ANSWER

11. THE ATRIOVENTRICULAR VALVE ON THE LEFT SIDE OF THE HEART IS CALLED THE _________.
    A. Mitral
    B. Tricuspid
    C. Left semilunar
    D. Left tricuspid
    E. None of the above

12. SYSTOLE RELATES TO THE:
    A. Relaxation of heart muscle
    B. Venous pressure gradient
    C. Ventricular relaxation
    D. Contraction of heart muscle
    E. None of the above

13. THE SUPPORT CELLS OF THE PERIPHERAL NERVOUS SYSTEM ARE CALLED:
    A. Neurons
    B. Schwann cells
    C. Fibrocytes
    D. Neuroglial cells
    E. Satellite cells

14. WHICH OF THE FOLLOWING IS CLASSIFIED AS A GENERAL SENSE OF THE BODY?
    A. Pain
    B. Smell
    C. Taste
    D. Hearing
    E. Equilibrium

15. THE VALVE BETWEEN THE RIGHT ATRIUM AND VENTRICLE IS CALLED THE _______.
    A. Aortic
    B. Mitral
    C. Pulmonary
    D. Tricuspid
    E. None of the above

16. THE RECEPTOR INVOLVED IN THE RELAYING TO THE BRAIN THE SENSE OF JOINT POSITIONING IS CALLED:
    A. Pain
    B. Touch
    C. Pressure
    D. Proprioception
    E. None of the above

17. THE TECTORIAL MEMBRANE IS INDICATED BY _________.

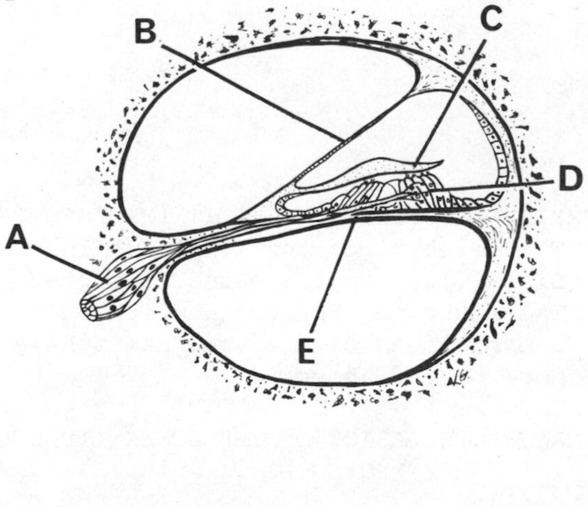

18. THE PACEMAKER OF THE HEART IS THE:
    A. A-V node
    B. S-A node
    C. A-V bundle
    D. Purkinje fibers
    E. None of the above

19. THE NERVE SUPPLYING THE MUSCLES AND THE OVERLYING SKIN BETWEEN TWO RIBS IS CALLED THE _______ NERVE.
    A. Femoral
    B. Median
    C. Musculocutaneous
    D. Pudendal
    E. None of the above

SELECT THE SINGLE BEST ANSWER

20. THE SUPRARENAL MEDULLA SECRETES:
    A. Prolactin
    B. Insulin
    C. Antidiuretic hormone
    D. Epinephrine
    E. Thyroxine

21. THE RIGHT LYMPHATIC DUCT IS INDICATED BY __________.

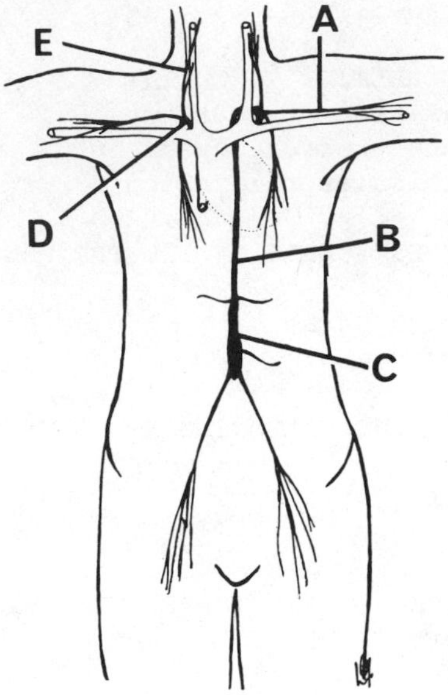

22. THE FOURTH VENTRICLE IS INDICATED BY ____________.

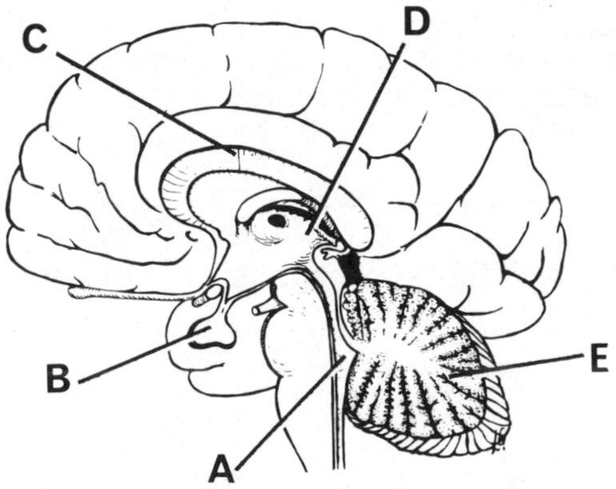

23. COMMISSURAL FIBERS TRANSMIT IMPULSES FROM:
    A. The cerebral cortex to distal loci
    B. Distal loci to the cerebral cortex
    C. The right to the left hemisphere
    D. Convolutions of the same hemisphere
    E. None of the above

24. A CANCEROUS LESION OF THE LOWER LIP WOULD FIRST RELEASE METASTASES TO THE __________ LYMPH NODES.
    A. Axillary
    B. Inguinal
    C. Submandibular
    D. Superficial cervical
    E. None of the above

25. THE SPINAL CORD HAS ______ SEGMENTS.
    A. 29
    B. 30
    C. 31
    D. 32
    E. None of the above

SELECT THE SINGLE BEST ANSWER

26. THE OPTIC RADIATION IS INDICATED BY ____________.

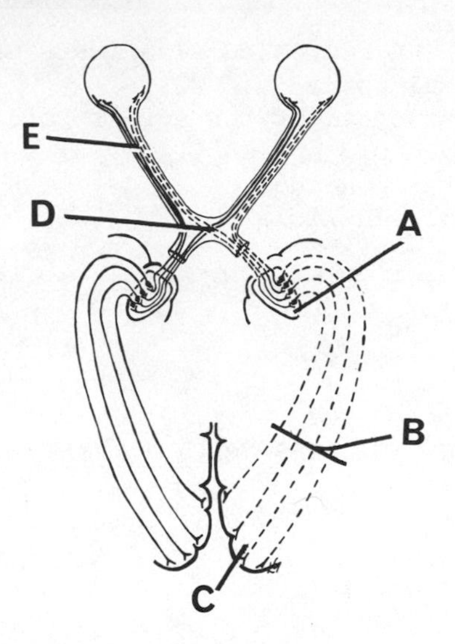

27. THE BRACHIOCEPHALIC ARTERY IS INDICATED BY __________.

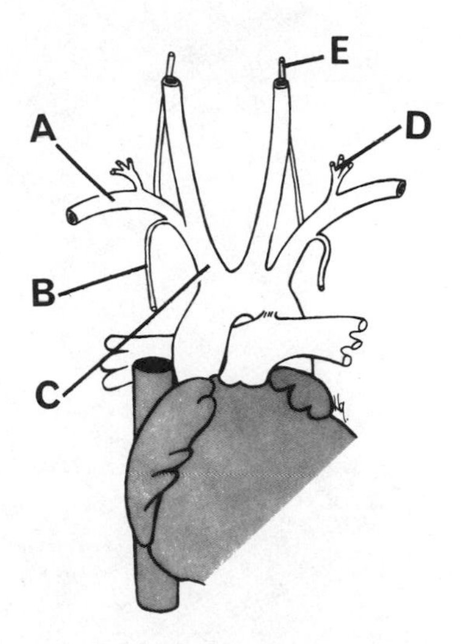

28. THE ARROW INDICATES THE:
A. Coronary sinus
B. Great cardiac vein
C. Small cardiac vein
D. Anterior cardiac vein
E. None of the above

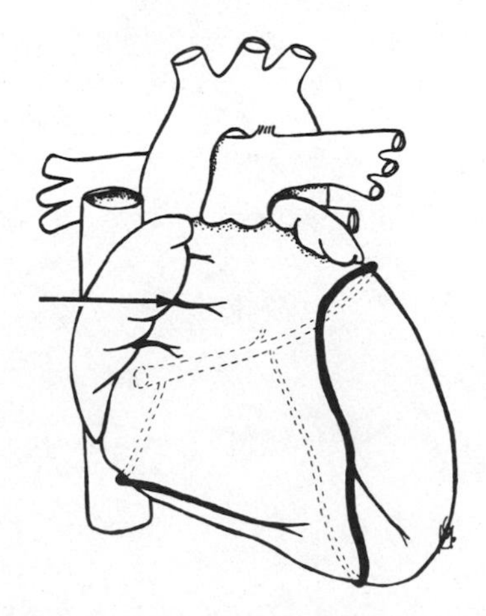

SELECT THE SINGLE BEST ANSWER

29. WHICH OF THE FOLLOWING STRUCTURES OR SPACES DOES NOT CONTAIN CSF?
    A. Subdural space
    B. Central canal of the spinal cord
    C. Subarachnoid space
    D. Cerebral aqueduct
    E. Lateral ventricle

30. Which of the following centers is not located in the medulla oblongata?
    A. Cardiac
    B. Swallowing
    C. Vasomotor
    D. Equilibrium
    E. Vomiting

31. THE ARROW INDICATES THE ______ VEIN.
    A. Basilic
    B. Cephalic
    C. Subclavian
    D. Median cubital
    E. None of the above

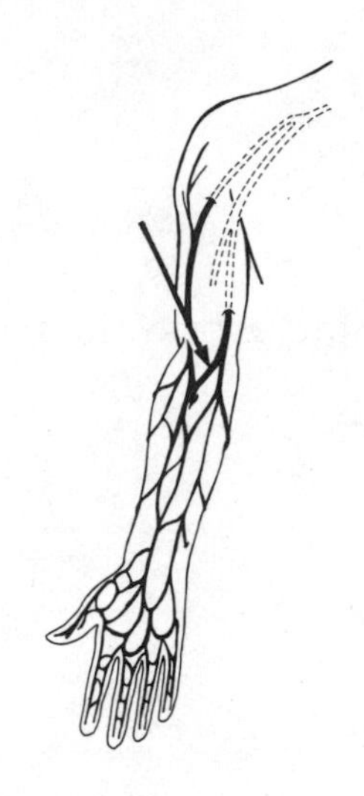

32. THE HORMONE PRODUCING A HYPERGLYCEMIC EFFECT IN THE BODY IS FOUND IN THE:
    A. Adrenal cortex
    B. Thymus
    C. Parathyroid
    D. Pancreas
    E. None of the above

33. THE CENTRAL SULCUS IS INDICATED BY __________.

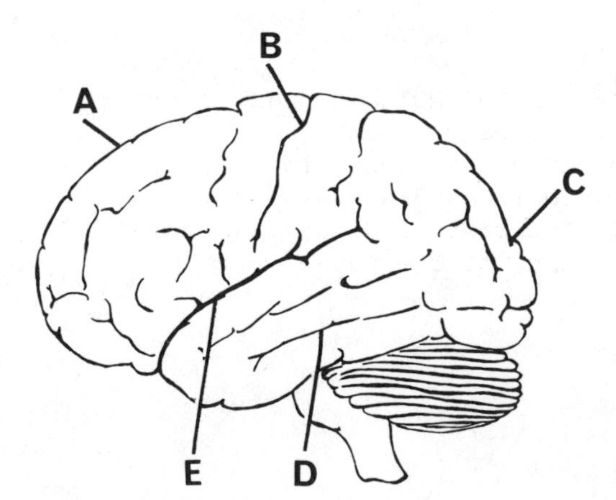

34. WHICH OF THE FOLLOWING IS NOT A COMPONENT OF THE RETINA?
    A. Rods
    B. Cones
    C. Motor neurons
    D. Ganglion neurons
    E. Pigment epithelium

SELECT THE SINGLE BEST ANSWER

35. THE HORMONE STIMULATING THE SECRETION OF TESTOSTERONE IS THE _______ HORMONE.
    A. Luteinizing
    B. Follicle-stimulating
    C. Glucocorticoids
    D. Aldosterone
    E. None of the above

36. THE MAXILLARY ARTERY IS INDICATED BY __________.

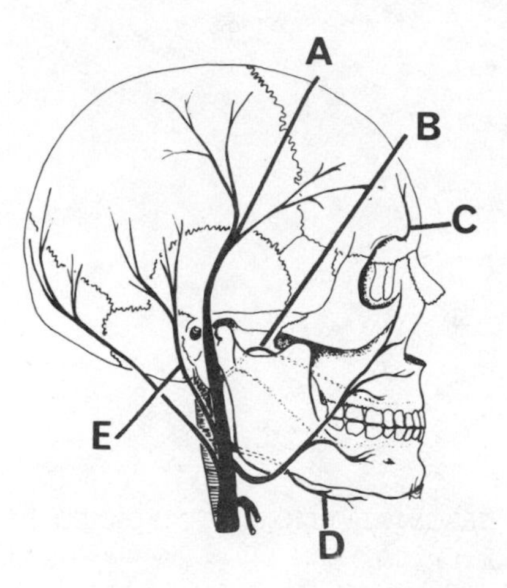

37. WHICH OF THE FOLLOWING IS NOT A FACTOR IN INCREASING THE FLOW OF LYMPH?
    A. Contraction of smooth muscle
    B. Decreased capillary pressure
    C. Increased interstitial fluid protein
    D. Increased capillary permeability
    E. Decreased plasma colloid osmotic pressure

38. THE ARROW INDICATES WHICH OF THE FOLLOWING STRUCTURES?
    A. Dorsal root ganglion
    B. Dorsal primary ramus
    C. Ventral root
    D. Ventral primary ramus
    E. Dorsal root

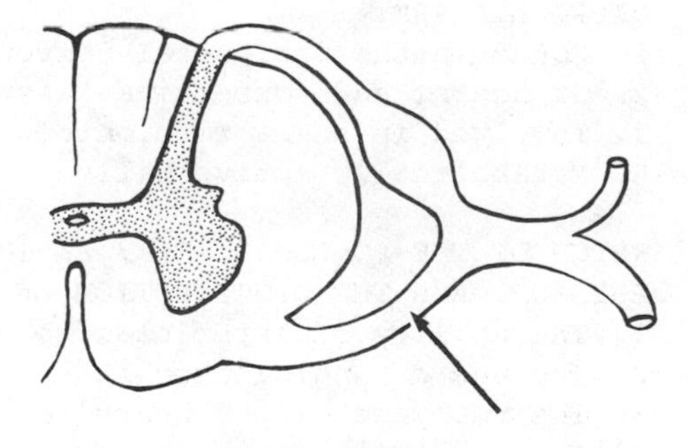

39. THE ARROW INDICATES THE _____.
    A. Left brachiocephalic vein
    B. Superior vena cava
    C. Hemiazygous vein
    D. Azygous vein
    E. Inferior vena cava

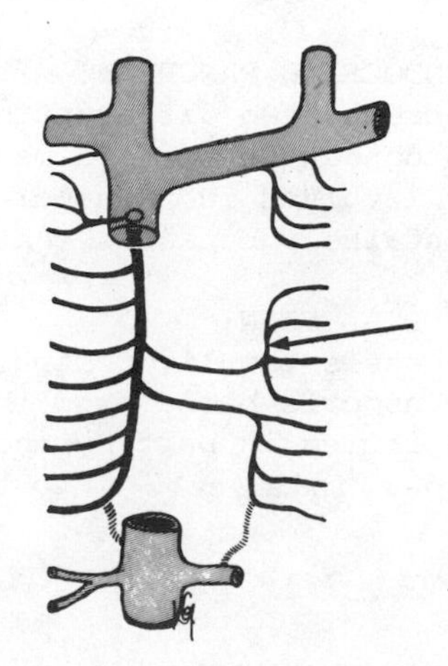

SELECT THE SINGLE BEST ANSWER

40. THE ARROW INDICATES THE ______.
A. Gonadal vein
B. Renal vein
C. Inferior vena cava
D. Superior mesenteric artery
E. None of the above

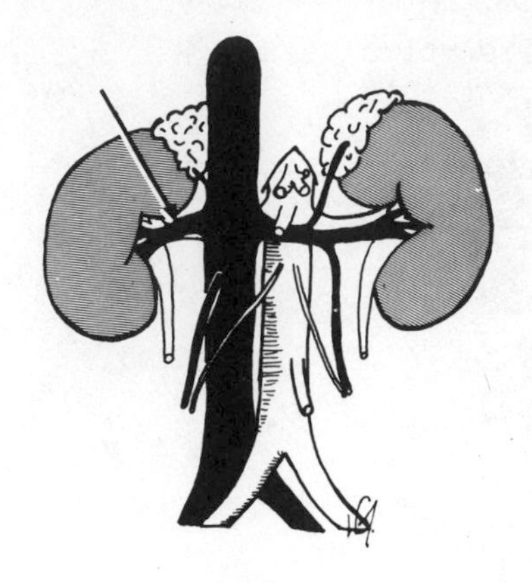

41. THE BRACHIAL PLEXUS IS FORMED BY THE PRIMARY RAMI OF SPINAL NERVES ______.
A. C_4-C_8
B. C_5-C_8
C. C_5-T_1
D. C_4-T_2
E. None of the above

MULTI-COMPLETION QUESTIONS

DIRECTIONS: In each of the following questions or incomplete statements, ONE OR MORE of the completions given is correct. At the lower right of each question, circle A if 1, 2, and 3 are correct; B if 1 and 3 are correct; C if 2 and 4 are correct; D if only 4 is correct; and E if all are correct.

42. NEUROGLIA ARE:
1. Found in the peripheral nervous system
2. Of neural and connective tissue origin
3. Involved in the formation of synapses
4. Metabolically active cells

43. WHICH OF THE FOLLOWING PLAY AN IMPORTANT ROLE IN THE MAINTENANCE OF BLOOD PRESSURE AND THE DISTRIBUTION OF BLOOD?
1. The ability of arterioles to constrict their lumen
2. The use of angiotensin II
3. The response of the heart to an increased blood volume
4. The input into the cardiovascular centers of the brain

44. THE ENDOCRINE PANCREAS:
1. Secretes two different hormones
2. Is concentrated as a mass of cells surrounding a capillary
3. Is involved in transporting glucose into a cell
4. Contains substances that are involved with digestion

45. THE PORTAL VEIN:
1. Bypasses the liver sinusoids
2. Transports blood from the small intestine
3. Is formed in part by the renal vein
4. Flows indirectly into the inferior vena cava

A	B	C	D	E
1,2,3	1,3	2,4	Only 4	All correct

46. VASOCONSTRICTOR FIBERS OF THE SYMPATHETIC SYSTEM:
    1. Are controlled by a center in the medulla oblongata
    2. Innervate the muscles of the heart
    3. Maintain arterioles in a state of partial contraction
    4. Innervate the capillaries of most tissues

47. THE AUTONOMIC NERVOUS SYSTEM:
    1. Has two divisions
    2. Depends on control from the hypothalamus
    3. Is a two neuronal system
    4. Is considered to be an afferent system

48. THE LYMPHATIC SYSTEM
    1. Contains many macrophages
    2. Drains into the venous system
    3. Acts as a blood filter
    4. Returns protein to the circulatory system

49. WHICH OF THE FOLLOWING STATEMENTS IS(ARE) CORRECT FOR THE DIENCEPHALON?
    1. Includes the thalamus
    2. Located between the cerebrum and the mesencephalon
    3. Lies about the third ventricle
    4. Contains the basal (nuclei) ganglia

50. THE GRANULAR LEUKOCYTES INCLUDE:
    1. Lymphocytes
    2. Platelets
    3. Monocytes
    4. Basophils

51. THE PLACENTA OF THE PREGNANT FEMALE SECRETES:
    1. A hormone that relaxes the pelvic ligaments
    2. Hormones essential to the continuance of pregnancy
    3. A hormone that stimulates the secretion of male fetal testosterone
    4. Large quantities of estrogen during the latter part of pregnancy

52. FUNCTIONS OF THE BLOOD INCLUDE:
    1. Carrying products of digestion to all parts of the body
    2. Distributing heat produced by active muscles
    3. Transporting oxygen to tissues of the body
    4. Maintaining the acid-base balance

53. THE CEREBRUM:
    1. Consists of two hemispheres
    2. Contains basal (nuclei) ganglia
    3. Is the largest component of the brain
    4. Has five lobes

54. WHICH OF THE FOLLOWING IS(ARE) NOT TRUE FOR THE MATURE ERYTHROCYTE?
    1. Its number is closely regulated
    2. Its life span is about 120 days
    3. Has about 34 per cent hemoglobin
    4. Usually has a diameter of 10 microns

55. WHICH OF THE FOLLOWING STATEMENTS IS(ARE) TRUE FOR THE RED PULP OF THE SPLEEN?
    1. Produces large numbers of lymphocytes
    2. Contains free and fixed macrophages
    3. Made up of small lymphatic nodules
    4. Surrounds the small masses of white pulp

A	B	C	D	E
1,2,3	1,3	2,4	Only 4	All correct

56. THE STRUCTURE(S) SUPPORTING THE SPINAL CORD WITHIN THE VERTEBRAL CANAL IS(ARE):
    1. Falx cerebelli
    2. Tentorium cerebelli
    3. Falx cerebri
    4. Denticulate ligament

57. WHICH OF THE FOLLOWING STATEMENTS CONCERNING THE IRIS IS(ARE) CORRECT?
    1. Forms part of the sclera
    2. Is the colored portion of the eye
    3. Attaches to the cornea
    4. Composed of smooth muscle cells

58. THE FIBROUS PERICARDIUM:
    1. Is a tough membrane of connective tissue
    2. Produces a serous-like fluid
    3. Attaches to the diaphragm
    4. Is a rather inflexible sac

59. WHICH OF THE FOLLOWING STATEMENTS IS(ARE) CORRECT OF THE POSTERIOR RAMI?
    1. Is the smaller of the two rami
    2. Involved in the formation of plexuses
    3. Supplies the muscles of the back and neck
    4. Give rise to white rami in the thoracic region

60. WHICH OF THE FOLLOWING STATEMENTS IS(ARE) CORRECT FOR THE SPINAL NERVES?
    1. Are either sensory or motor
    2. There is a total of 31 pairs
    3. Posterior rami gives rise to plexuses
    4. The lower nerve roots form the cauda equina

61. THE HORMONE(S) RESPONSIBLE FOR SPERMATOGENESIS IS(ARE) THE:
    1. Chorionic gonadotropin
    2. Follicle-stimulating
    3. Corticotropin
    4. Luteinizing

62. WHICH OF THE FOLLOWING STATEMENTS CONCERNING THE MYOCARDIUM IS(ARE) CORRECT?
    1. Contracts to expel the blood from the ventricles
    2. Contributes to the formation of the valves of the heart
    3. Forms the middle layer of the heart wall
    4. Consists of stratified endothelial cells

63. THE SPINAL CORD:
    1. Ends at the lower border of the first lumbar vertebra
    2. Is bathed in CSF
    3. Has an expanded area in the cervical region
    4. Does not completely fill the vertebral canal

64. THE VENOUS RETURN TO THE HEART DEPENDS ON WHICH OF THE FOLLOWING?
    1. The negative pressure within the thoracic cage
    2. Smooth muscle contraction of the venous walls
    3. Extravascular movement of body tissues
    4. Valves located along the length of the large veins

65. THE EXTRINSIC MUSCLES OF THE EYE ARE INNERVATED BY WHICH OF THE FOLLOWING NERVES?
    1. Oculomotor
    2. Facial
    3. Trochlear
    4. Trigeminal

A	B	C	D	E
1,2,3	1,3	2,4	Only 4	All correct

66. BLOOD FLOW DEPENDS ON THE:
    1. Volume of blood present
    2. Thickness of the tunica media
    3. Length of the vessel involved
    4. Amount of vascular resistance

67. SOMATIC PAIN MAY RESULT FROM:
    1. The stimulation of the skin
    2. Excitation of proprioceptive nerve endings
    3. Receptors in a tendon
    4. The excitation of the stomach wall

68. THE RIGHT ATRIUM:
    1. Is larger than the left
    2. Receives blood from the azygos system of veins
    3. Drains blood from the coronary system of veins
    4. Communicates with the right ventricle by the tricuspid valve

69. THE MIDDLE EAR:
    1. Opens into the oropharynx by the auditory tube
    2. Is located in the temporal bone
    3. Opens into the external ear by the fenestra vestibuli
    4. Communicates with the mastoid air cells

70. WHICH OF THE FOLLOWING NERVES ARE DERIVED FROM THE BRACHIAL PLEXUS?
    1. Sciatic
    2. Pudendal
    3. Phrenic
    4. Median

71. WHICH OF THE FOLLOWING STATEMENTS IS(ARE) TRUE FOR CARDIAC MUSCLE?
    1. Forms a functional syncytium
    2. Contains both actin and myosin
    3. The intercalated disc has a low electrical resistance
    4. Has an inherent pulse independent of any nerve stimuli

72. WHICH OF THE FOLLOWING GLANDS IS(ARE) NOT UNDER CONTROL OF THE PITUITARY?
    1. Thyroid
    2. Ovary
    3. Islet cells of the pancreas
    4. Adrenal medulla

73. THE CRANIAL NERVE(S) EXITING THE BRAIN BETWEEN THE PONS AND THE MEDULLA OBLONGATA IS(ARE) THE:
    1. Facial
    2. Acoustic
    3. Abducens
    4. Vagus

74. THE RETICULOSPINAL TRACT:
    1. Forms part of the pyramidal system
    2. Originates within the brain stem
    3. Consists solely of inhibitory fibers
    4. Facilitates repetitious movements

75. THE ATRIA DIFFER FROM THE VENTRICLES IN THAT THE ATRIA:
    1. Are more muscular
    2. Are not regulated by the S-A node
    3. Have papillary muscles
    4. Have thinner walls

A	B	C	D	E
1,2,3	1,3	2,4	Only 4	All correct

76. THE PALATINE TONSILS:
    1. Are located lateral to the base of the tongue
    2. Filter blood as it passes through the sinuses
    3. Actively produce new quantities of lymphocytes
    4. Are important sites of new red blood cell formation

77. THE BRAIN STEM INCLUDES THE:
    1. Pons
    2. Medulla oblongata
    3. Midbrain
    4. Cerebellum

78. MEDIUM-SIZE VEINS:
    1. Have relatively large lumens when compared to medium arteries
    2. In the lower limbs have numerous valves
    3. Have smaller walls than do companion arteries
    4. Usually do not have a tunica media

79. THE ACIDOPHILIC EPITHELIAL CELLS OF THE ADENOHYPOPHYSIS SECRETES WHICH OF THE FOLLOWING:
    1. Oxycytocin
    2. Prolactin
    3. Luteinizing hormone
    4. Growth hormone

80. TASTE FIBERS ARE FOUND IN WHICH OF THE FOLLOWING CRANIAL NERVES?
    1. Vagus
    2. Facial
    3. Glossopharyngeal
    4. Hypoglossal

81. WHICH OF THE FOLLOWING STATEMENTS IS(ARE) TRUE FOR BLOOD PLATELETS?
    1. Are small cytoplasmic fragments
    2. Produced in the white pulp of the spleen
    3. Release a factor involved in blood clotting
    4. Derived from hemocytoblasts

82. THE SYMPATHETIC DIVISION OF THE ANS:
    1. Produces dilation of the pupil
    2. Prevents peristalsis of the gut
    3. Increases the rate of the heart beat
    4. Constricts the smooth muscle of blood vessels

83. WHICH OF THE FOLLOWING IS(ARE) FEATURES OF AN ENDOCRINE GLAND?
    1. Are ductless in their mature form
    2. Helps control metabolic functions
    3. Produces substances called hormones
    4. Secrete their products directly into capillaries

84. THE CONDUCTING SYSTEM OF THE HEART:
    1. Consists of specialized cardiac muscle cells
    2. Coordinates the contraction of the atria and ventricles
    3. Is controlled by a center in the medulla
    4. Plays a role in initiating the heart beat

85. WHICH OF THE FOLLOWING STATEMENTS IS(ARE) CORRECT FOR THE SOMATIC SENSORY AREA OF THE BRAIN?
    1. Receives information from the opposite side of the body
    2. Located in the postcentral gyrus of the cerebral cortex
    3. Involved in detecting changes in body position
    4. Receives neurons that have passed through the internal capsule

DIRECTIONS: Each group of questions below consists of a numbered list of descriptive words or phrases accompanied by a diagram with certain parts indicated by letters, or by a list of lettered headings. For each numbered word or phrase, SELECT THE LETTERED PART OR HEADING that matches it correctly. Then blacken the appropriate space on the answer sheet. Sometimes more than one numbered word or phrase may be correctly matched to the same lettered part or heading.

86. _____ Parafollicular cells

87. _____ Interstitial cells

88. _____ Alpha cells

89. _____ Corpus luteum

90. _____ Juxtaglomerular cells

A. Glucagon

B. Testosterone

C. Estrogen

D. Renin

E. Calcitonin

91. _____ An artery supplying the ovary

92. _____ An artery supplying the stomach

93. _____ An artery which supplies the diaphragm

94. _____ An artery which supplies the posterior abdominal wall

95. _____ An artery supplying the sigmoid colon

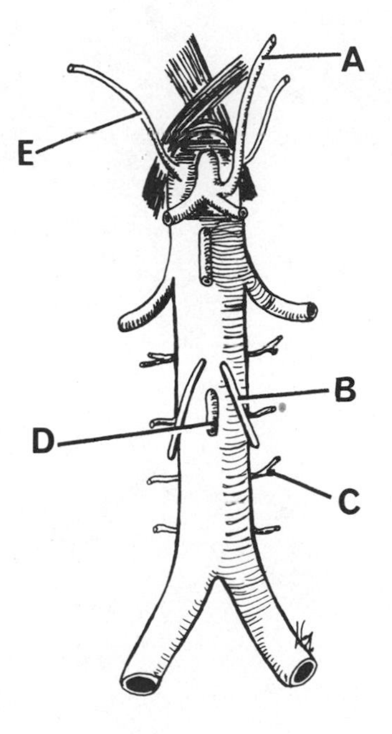

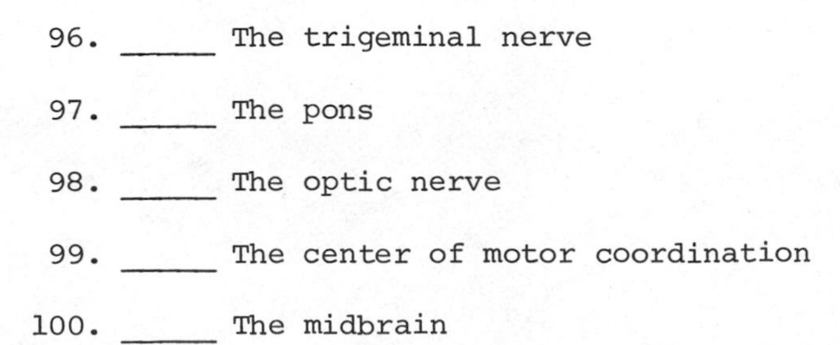

96. _____ The trigeminal nerve

97. _____ The pons

98. _____ The optic nerve

99. _____ The center of motor coordination

100. _____ The midbrain

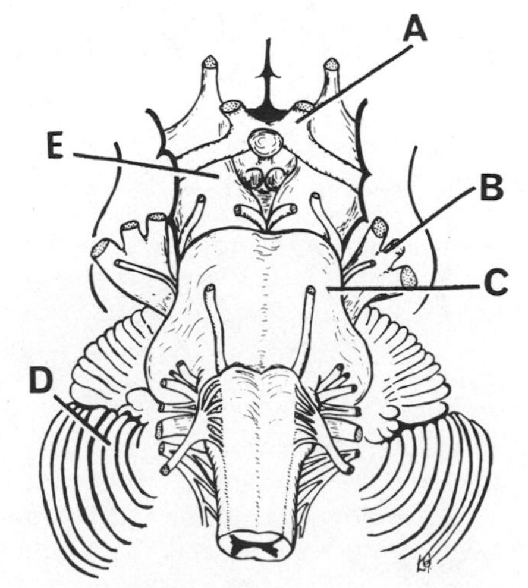

PART FOUR

THE RESPIRATORY SYSTEM

THE DIGESTIVE SYSTEM

THE URINARY SYSTEM

BODY FLUIDS AND ACID-BASE BALANCE

THE REPRODUCTIVE SYSTEM

13. THE RESPIRATORY SYSTEM

LEARNING OBJECTIVES

Be Able To:

- List the organs involved with respiration and give the general functions of each.
- Describe the anatomy and functions of the following components of the nose: septum, nares, conchae, mucosa, meatuses, paranasal sinuses, and nasolacrimal ducts.
- Describe the components of the pharynx and larynx, with special reference to the structures of the larynx which are involved in speech.
- Discuss the anatomy of the trachea, primary bronchi, secondary bronchi, respiratory bronchioles, alveolar ducts, alveolar sacs, and alveoli.
- Describe the components of the respiratory unit and the respiratory membrane.
- Describe the anatomy of the lungs with reference to their location, pleura, root, number of lobes, fissures, blood supply, venous drainage, and innervation.
- Describe the location and structure of the diaphragm and list the muscles responsible for moving the thoracic cage in quiet and forced respiration.
- Define the following terms: atmospheric, intrapulmonic, and intrathoracic (interpleural) pressures.
- Describe the events and pressure changes found in the respiratory cycle.
- Define the following terms: tidal volume, inspiratory reserve, expiratory reserve, vital capacity, minute respiratory residual volume, functional residual capacity, total lung capacity, and compliance.
- Describe the components of the respiratory membrane and discuss the respiratory gases with reference to: their concentrations in inspired, alveolar, and expired air; their pressures in inspired and alveolar air, and in blood; and their transport in external and internal respiration.
- Define the term hypoxia and discuss the common possible causes of hypoxia.

- Describe the mechanisms involved in controlling respiration with special reference to: respiratory centers and pathways, voluntary control, and the effects of chemical and physical factors.

- Define each of the following clinical conditions of the lung: asthma, bronchiectasis, bronchitis, emphysema, hiccough, pleurisy, pneumonia, pneumothorax, and tuberculosis.

RELEVANT READINGS

Anthony & Kolthoff - Chapter 12.
Chaffee & Greisheimer - Chapter 15.
Crouch & McClintic - Chapter 26.
Jacob, Francone & Lossow - Chapter 12.
Landau - Chapters 19, 20, and 21.
Tortora - Chapter 19.

FIVE-CHOICE COMPLETION QUESTIONS

DIRECTIONS: Each of the following questions or incomplete statements is followed by five suggested answers or completions. SELECT THE SINGLE BEST ANSWER in each case and then circle the appropriate letter at the lower right of each question.

1. THE ARROW INDICATES THE:
   A. Alveolus
   B. Alveolar sac
   C. Alveolar duct
   D. Respiratory bronchiole
   E. None of the above

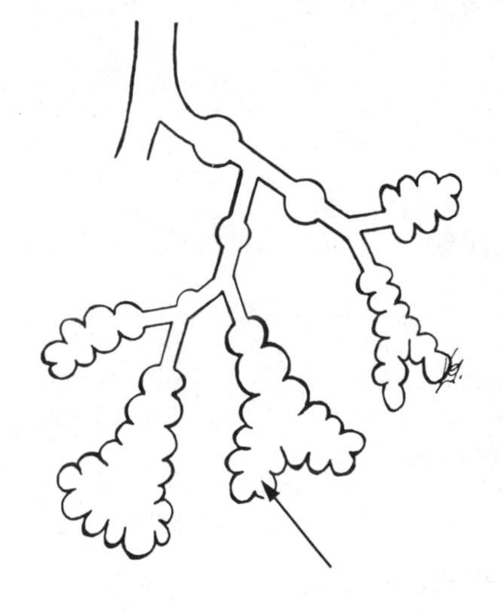

A B C D E

2. THE AVERAGE INSPIRATORY RESERVE VOLUME EQUALS:
   A. 1500 ml
   B. 2000 ml
   C. 2500 ml
   D. 3000 ml
   E. 3500 ml

A B C D E

3. THE WALLS OF THE TRACHEA ARE SUPPORTED BY ________.
   A. A continuous cartilagenous tube
   B. A series of complete bony rings
   C. Large quantities of elastin fibers
   D. Loose connective tissue
   E. Incomplete rings of cartilage

A B C D E

SELECT THE SINGLE BEST ANSWER

4. THE ARROW IN THIS VIEW OF THE ENTRANCE OF THE LARYNX INDICATES THE ________.
   A. Trachea
   B. Epiglottis
   C. Vocal fold
   D. Ventricular fold
   E. Corinculate cartilage

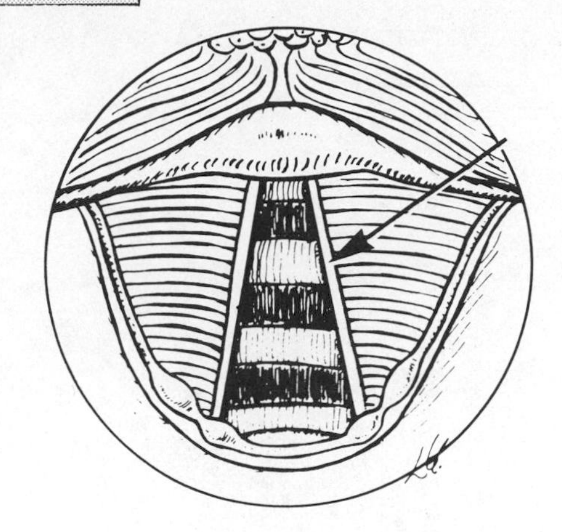

A B C D E

5. IN THIS POSTERIOR VIEW OF THE LARYNX, THE ARYTENOID CARTILAGE IS INDICATED BY _______.

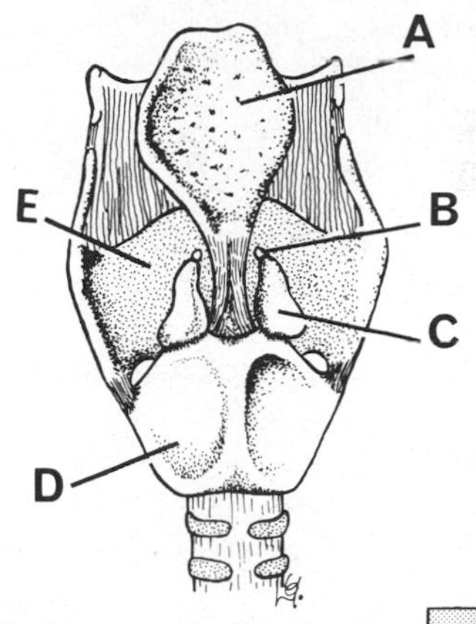

A B C D E

6. WHICH OF THE FOLLOWING IS NOT A CAUSE OF HYPOXIA?
   A. Production of erythropoietin by the kidney
   B. Slow or decreased heart beat
   C. High altitude mountain climbing
   D. Decreased or low numbers of red blood cells
   E. A substance combining chemically with hemoglobin

A B C D E

7. THE TIDAL VOLUME EQUALS THE VOLUME OF AIR:
   A. Passing in and out of the lungs with each respiratory movement
   B. Filling the bronchial tree, trachea, and larynx
   C. Exhaled by the deepest possible expiration
   D. Remaining in the lungs after they collapse
   E. None of the above

A B C D E

8. THE PRESSURE OF AIR WITHIN THE BRONCHIAL TREE IS CALLED THE_____ PRESSURE.
   A. Atmospheric
   B. Hydrostatic
   C. Intrapulmonic
   D. Intrathoracic
   E. None of the above

A B C D E

SELECT THE SINGLE BEST ANSWER

9. THE ETHMOIDAL PARANASAL SINUS IS INDICATED BY __________.

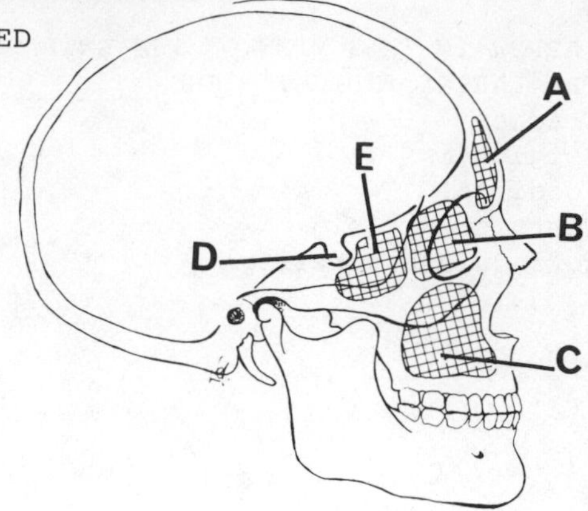

A B C D E

10. WHICH OF THE FOLLOWING IS A SINGLE CARTILAGE OF THE LARYNX?
A. Arytenoid
B. Corniculate
C. Cuneiform
D. Cricoid
E. None of the above

A B C D E

11. WHICH OF THE FOLLOWING STATEMENTS IS NOT TRUE FOR THE RESPIRATORY MUSCLES?
A. The diaphragm contracts and increases diameters of the thoracic cavity
B. Inspiratory muscles elevate the ribs
C. The internal intercostal muscles are primary expiratory muscles
D. Relaxation of the diaphragm expels air from the lungs
E. Contraction of the abdominal muscles increases the thoracic cavity

A B C D E

12. THE OBLIQUE FISSURE IS INDICATED BY __________.

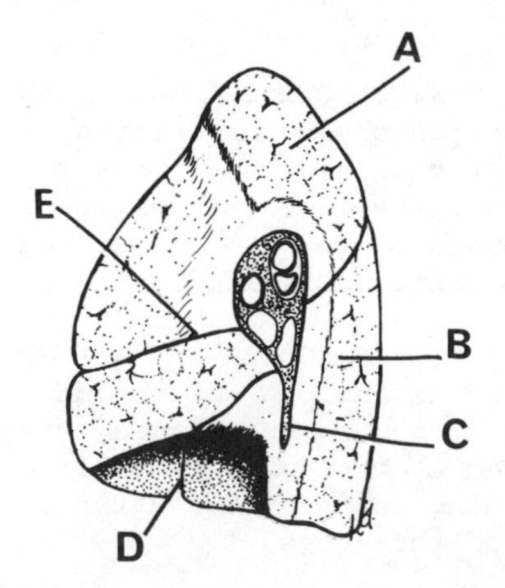

A B C D E

13. THE MAJOR RESPIRATORY CENTER IS LOCATED IN THE ________.
A. Pons
B. Medulla oblongata
C. Midbrain
D. Cerebellum
E. None of the above

A B C D E

SELECT THE SINGLE BEST ANSWER

14. INFLAMMATION OF THE TISSUES OF THE LUNG IS CALLED ________.
A. Asthma
B. Pleurisy
C. Emphysema
D. Pneumonia
E. Tuberculosis

A B C D E

------------------------ ANSWERS, NOTES AND EXPLANATIONS ------------------------

1. A. This diagram represents a functional unit or primary lobule of the lung. This unit includes the respiratory bronchiole (A in the diagram), the alveolar duct (B), the atrium (E), the alveolar sac (C), and the alveolus (D). Also included in this unit are the associated blood vessels, lymphatics, nerves, and connective tissue (not seen in the diagram). The respiratory bronchioles divide into a number of radiating alveolar ducts which terminate into an area called the atrium. The atrium opens into the alveolar sacs which consists of two or more alveoli. The walls of the alveoli contain very conspicuous blood capillaries and the combined walls of these two structures form the respiratory membrane.

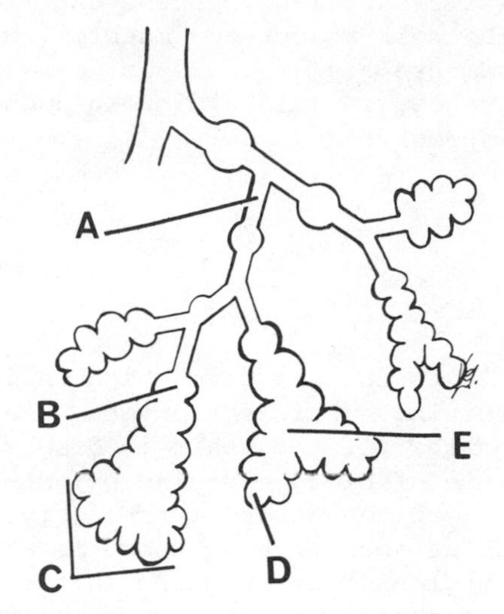

2. D. The average inspiratory reserve volume is about 3000 ml and represents the volume of air inspired over and beyond the normal tidal volume (500 ml). These two volumes together form the inspiratory capacity of the lungs. This is the volume (about 3500 ml) of air that a person can breath beginning at the normal expiratory level (500 ml) and distending his lungs to the maximum amount.

3. E. The trachea is continuous above with the larynx and inferiorly it divides into two primary bronchi; one passes into each lung. The trachea is a major component of the duct system carrying inspired and expired air to and from the lungs. The trachea is kept open by a series of incomplete rings of cartilage, shaped like a horseshoe. The posterior wall of the trachea is free of cartilage but both its anterior and lateral walls are supported by hyaline cartilage. Therefore the lumen of the trachea never collapses but is always kept open by its cartilage skeleton.

4. C. The vocal fold or "true vocal cord" is indicated by B in the two diagrams. This fold lies below and medial to the ventricular fold (A). The two vocal folds are separated by a space termed the rami glottidis (F) which is the narrowest part of the laryngeal cavity. The cuneiform (C) and arytenoid (D) cartilages along with the aryepiglottic fold (E) are indicated near the posterior wall of the pharynx. The epiglottis (G) is seen anteriorly.

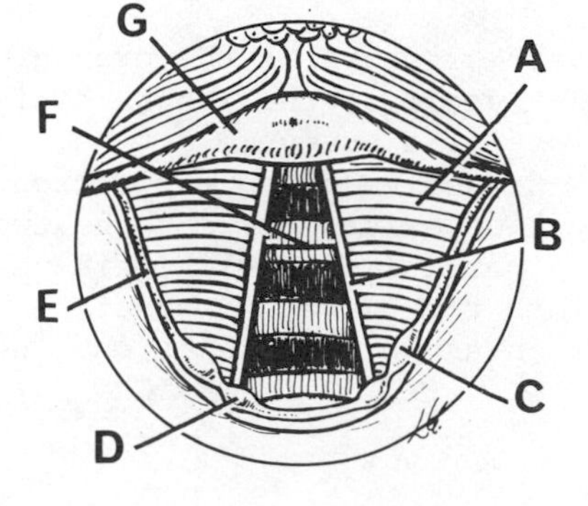

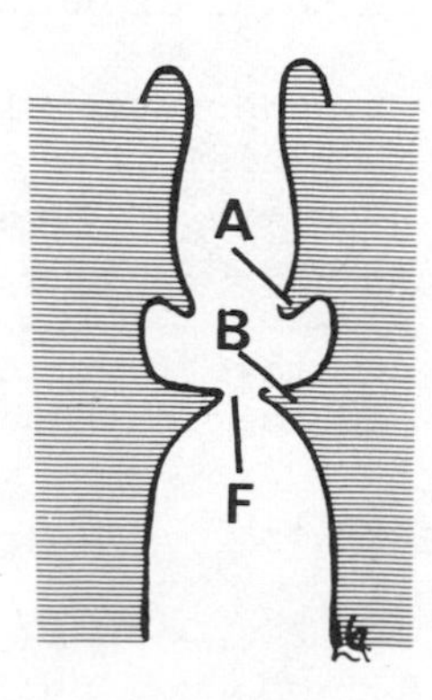

Coronal Section

5. C. The larynx is formed by six cartilages. They are the single thyroid (E, in the diagram), epiglottic (A), and cricoid (D) cartilages and the paired arytenoid (C), corniculate (B), and cuneiform (not shown) cartilages. The thyroid, arytenoid and cricoid cartilages consist of hyaline cartilage and the remaining cartilages consist of elastic cartilage. These cartilages along with their associated muscles, membranes, and ligaments are involved in guarding the air passages, in maintaining an open airway, and in speech.

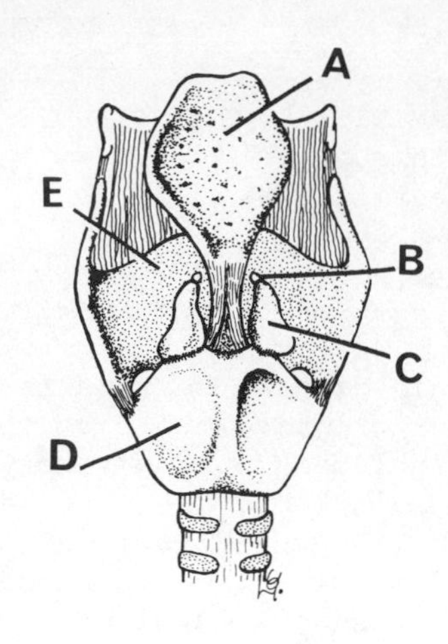

6. A. Hypoxia is a term which indicates that the cells of the body are not receiving sufficient oxygen. Many factors may be responsible for producing hypoxia. If the heart is diseased and cannot pump properly then an inadequate volume of blood carrying oxygen fails to reach the tissues and therefore does not meet the needs of the active cells of the body. Similarly, conditions such as anemia, where there is a decrease in the number of red blood cells (hemoglobin) circulating in the blood, will also lead to hypoxia. There are substances like cyanide, a poison, which forms complexes with hemoglobin and prevents oxygen from combining with it and subsequently will cause hypoxia or even death. Some physical activities that take place in high altitudes, where the oxygen content of the air is low may also produce hypoxia. Low levels of oxygen in the blood are known to stimulate the kidney to produce erythropoietin which in turn stimulates the production of new red blood cells (hemoglobin) in the body. These new red blood cells will carry more oxygen to the oxygen starved tissues of the body. So erythropoietin does not produce hypoxia but is a method by which the body combats against hypoxia.

7. A. The tidal volume is equal to the amount of air inspired and expired with each normal breath. It equals approximately 500 ml in the average healthy male and about 400 ml in the average healthy female. Other lung volumes are: the inspiratory reserve which is the volume of air inspired (3000 ml in the male) over and beyond the normal tidal volume, the expiratory reserve which is the volume of air that can be forcefully expired after the end of a normal tidal expiration (1100 ml in the male), and the residual volume equalling about 1200 ml (in the male) and is the volume remaining in the lungs after the most forceful expiration.

8. C. The intrapulmonic pressure is the pressure of the air located within the bronchial tree and alveoli of the lungs. This pressure fluctuates from a negative or subatmospheric pressure during inspiration to a slightly positive (above atmospheric) pressure during expiration. The intrathoracic (intrapleural) pressure which exists between the two layers of the pleura is also negative (forming a partial vacuum). This negative pressure allows the outside air to rush into the lungs by way of the organs of the external respiratory system. This inrushing air makes the intrapulmonic pressure slightly positive.

9. B. The paranasal sinuses are mucous membrane lined cavities found in the interior of certain bones of the skull. They are named for the bone in which they are found: frontal (A), ethmoidal (B), maxillary (C), and sphenoidal (E) sinuses. The lining of these sinuses (mucoendosteum) is continuous with the respiratory mucosa of the nasal cavity and contains glands which secrete mucus. The paranasal sinuses all drain into the nasal cavities. The hypophyseal fossa which contains the pituitary is indicated by D in the diagram.

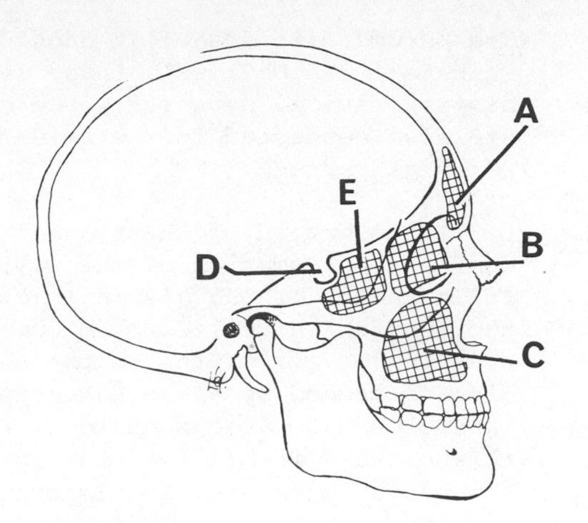

10. D. The larynx has six cartilages which are involved in preventing food from entering the air passages, maintaining open air passages, and the production of speech. The cartilages which are paired include the arytenoid, corniculate, and cuneiform, whereas the thyroid, cricoid, and epiglottic are single cartilages.

11. E. The diaphragm is considered to be the basic muscle of respiration and on contraction it increases the vertical, transverse, and anteroposterior diameters of the thorax. The diaphragm along with the external intercostal and the sternocostalis muscles elevate the ribs and are therefore the primary inspiratory muscles. The internal intercostal muscles are the primary expiratory muscles but it should be remembered that normal expiration is a purely passive process due to the recoil of the thoracic wall and the relaxation of the diaphragm. The abdominal muscles by contracting decrease the transverse and anteroposterior diameters of the thorax and at the same time increase the intraabdominal pressure. This pressure pushes the diaphragm upward and decreases the vertical diameter of the thorax. So in forced expiration the anterolateral abdominal muscles help to expel the air from the lungs.

12. D. The oblique fissure (D in the diagram) of the right lung separates the lower lobe (B) from the middle (unlabelled) and upper (A) lobes. On the left the oblique fissure separates the lung into an upper and a lower lobe and there is no middle lobe. On the right a transverse fissure (E) separates the lung into upper and middle lobes. The root of the lung is surrounded by pleura which inferiorly forms the pulmonary ligament (C). The root of the lung contains the primary bronchi, pulmonary and bronchiole vessels, nerves, lymph nodes, and lymphatic vessels.

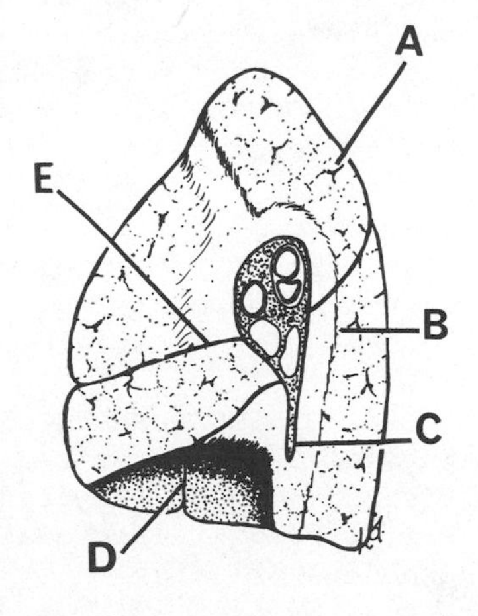

13. B. The respiratory center is a widely dispersed group of neurons found within the medulla oblongata and the pons of the brain stem. The major center is the rhythmicity area which is located in the medulla. This center establishes the basic rhythm of respiration and it is this basic rhythm which is moderated by impulses from the apneustic and pneumotaxic areas of the pons. The apneustic area is involved in changing the ratio of inspiration time to expiration time.

The pneumotaxic area is thought to be involved in regulating the rate of inspiration. Obviously these centers regulate the contractions of the muscles of respiration, especially the diaphragm. The centers of the pons and medulla are also connected to the spinal cord and centers in the hypothalamus and cerebral cortex.

14. D. Inflammation of the tissues of the lung is called lobar pneumonia, whereas inflammation of the walls of the bronchi is called bronchopneumonia. Inflammation of the pleura is called pleurisy. Emphysema is a condition of the lung characterized by bronchiole obstruction, loss of elasticity, and trapping of air distal to the obstructions. Tuberculosis is an infectious disease caused by a mycobacterium and is characterized by the formation of tubercles and caseous necrosis in the tissues. The lung is a major site of this disease in man. Asthma is a spasm of bronchial muscles, with edema of the mucous membrane, and is brought on because of a sensitivity to inhaled or ingested substances. This condition is characterized by frequent attacks of difficult breathing, coughing, wheezing, a mucus sputum, and a feeling of a congested chest.

MULTI-COMPLETION QUESTIONS

DIRECTIONS: In each of the following questions or incomplete statements, ONE OR MORE of the completions given is correct. At the lower right of each question, circle A if 1, 2, and 3 are correct; B if 1 and 3 are correct; C if 2 and 4 are correct; D if only 4 is correct; and E is all are correct.

1. THE PARANASAL SINUSES:
   1. Are lined by a mucous membrane
   2. Give a resonance to the voice
   3. Are encased in bone
   4. Empty into the oral cavity

   A B C D E

2. WHICH OF THE FOLLOWING STATEMENTS IS(ARE) TRUE FOR THE NASAL MUCOSA?
   1. Provides moisture for the inspired air
   2. Removes dust from the incoming air
   3. Secretes large quantities of mucus
   4. Has areas responsible for the detection of odors

   A B C D E

3. FACTORS INVOLVED WITH OXYGEN TRANSPORT INCLUDE:
   1. Low temperatures facilitate the combination of oxygen with hemoglobin
   2. Some oxygen is transported in solution in the blood plasma
   3. High partial pressures of oxygen affect the amount of oxygen in red blood cells
   4. Low P_{CO2} levels are responsible for the release of oxygen

   A B C D E

4. FUNCTIONS OF THE LUNG INCLUDE THE:
   1. Warming of the inspired air
   2. Distribution of inspired air
   3. Filtering of inspired air
   4. Exchange of gases

   A B C D E

5. WHICH OF THE FOLLOWING STRUCTURES IS(ARE) INVOLVED WITH BOTH THE RESPIRATORY AND DIGESTIVE SYSTEMS?
   1. Oropharynx
   2. Nasopharynx
   3. Laryngopharynx
   4. Esophagus

   A B C D E

A	B	C	D	E
1,2,3	1,3	2,4	Only 4	All correct

6. THE LARYNX:
   1. Is maintained open by cartilaginous and ligamentous structures
   2. Receives its innervation from the hypoglossal nerve
   3. Has a narrow opening called the rima glottidis
   4. Continues inferiorly to open into the esophagus

   A B C D E

7. FACTORS INVOLVED IN THE ELASTIC TENDENCY OF THE LUNGS TO COLLAPSE INCLUDE THE:
   1. Epithelial cells lining the alveolar sac
   2. Surface tension of alveoli fluid
   3. Smooth muscle of the interstitial space of the respiratory membrane
   4. Presence of elastic fibers within the walls of the lung

   A B C D E

8. WHICH OF THE FOLLOWING IS(ARE) TRUE FOR THE TRACHEA?
   1. Its wall is supported by U-shaped cartilages
   2. Is lined by a ciliated columnar mucous membrane
   3. Its posterior wall contains smooth muscle
   4. Carries air to the primary bronchi

   A B C D E

9. RESIDUAL AIR:
   1. Remains after the deepest expiration
   2. Is found in the larynx and trachea
   3. Averages about 1200 ml
   4. May be expired after a great deal of voluntary effort

   A B C D E

10. THE RIGHT BRONCHUS:
    1. Has a smaller diameter than the left bronchus
    2. Is more vertical than the left bronchus
    3. Is lined by non-ciliated columnar cells
    4. Has cartilage in its wall

    A B C D E

11. BREATHING IS CONTROLLED BY WHICH OF THE FOLLOWING FACTORS:
    1. Chemoreceptors
    2. Temperature of the circulating blood
    3. Conscious voluntary nerve impulses
    4. Changes in blood pressure

    A B C D E

12. THE LEFT LUNG HAS:
    1. Three primary lobes
    2. No cardiac impression
    3. A rather large horizontal fissure
    4. A primary bronchus with a smaller diameter

    A B C D E

13. EXPIRED AIR CONTAINS:
    1. More carbon dioxide than alveolar air
    2. Less oxygen than alveolar air
    3. Less carbon dioxide than inspired air
    4. Less oxygen than inspired air

    A B C D E

14. WHICH OF THE FOLLOWING STRUCTURES PASSES THROUGH THE ROOT OF THE LUNG?
    1. Pulmonary artery
    2. Bronchi
    3. Bronchial artery
    4. Phrenic nerve

    A B C D E

A	B	C	D	E
1,2,3	1,3	2,4	Only 4	All correct

15. WHICH OF THE FOLLOWING VOLUMES CONTRIBUTES TO THE VITAL CAPACITY OF THE LUNGS?
    1. Tidal volume
    2. Inspiratory reserve volume
    3. Expiratory reserve volume
    4. Maximal breathing capacity volume

A B C D E

16. WHICH OF THE FOLLOWING COMPONENTS FORM THE RESPIRATORY MEMBRANE?
    1. An interstitial space
    2. An endothelial layer
    3. A lipoprotein layer
    4. At least one basement membrane layer

A B C D E

------------------------ ANSWERS, NOTES AND EXPLANATIONS ------------------------

1. A, 1, 2, and 3 are correct. The four paranasal sinuses, located in the frontal, maxilla, ethmoid, and sphenoid bones, are formed by compact bone which is lined by a mucous membrane. This mucous membrane is very similar to the membrane lining the nasal cavities but it contains fewer glands. The four paranasal sinuses all drain into the nasal cavity by small openings. The exact function of the sinuses is not known, however they are thought to lighten the weight of the skull and give resonance to the voice.

2. E, All are correct. The nasal mucosa for the most part consists of mucus-secreting pseudostratified ciliated epithelium. This epithelium contains mucous glands which produce mucus that moistens the inspired air and helps keep the nasal membrane moist. The vestibule of the nose is lined by stratified squamous epithelium (skin) with stiff hairs that help in removing dust particles from the inspired air. Beneath the epithelium of the lower nasal conchae are rich venous plexuses which serve to warm the air passing through the nose. The olfactory epithelium, lining the upper reaches of the nasal cavity, is a highly specialized form of ciliated epithelium responsible for detecting odors.

3. A, 1, 2, and 3 are correct. Oxygen is transported in the blood in two ways: 1) about 0.5 ml of oxygen (per 100 ml of blood) is found in solution, and 2) about 19.5 ml of oxygen (per 100 ml of blood) is carried in combination with hemoglobin. So most of the oxygen being transported to the tissues is done so in combination with hemoglobin. This combination of oxygen with hemoglobin is enhanced by an increased partial pressure of oxygen and low temperatures and is inhibited by a high partial pressure of carbon dioxide. Oxygen is released from the hemoglobin at low partial pressures of oxygen and at high partial pressures of carbon dioxide. Both of which are the normal conditions found at the active cellular tissue sites in the body. Increases in the temperature and in the acid production (lowered pH, because of high levels of carbon dioxide) during active contraction of skeletal muscle both facilitate the unloading of oxygen to the muscle so it will continue to function.

4. D, Only 4 is correct. The function of the lungs is to provide a site for the exchange of gases between the blood and the inspired air. This exchange takes place across the respiratory membrane formed by the pulmonary capillary and alveolar walls. The nose serves as a site where the inspired air is filtered and either warmed or cooled, depending on the condition of the external air. The nose along with the pharynx, larynx, trachea, and its bronchial tree all serve to transport the inspired air to the respiratory membrane.

5. B, 1 and 3 are correct. The pharynx consists of three portions: the nasopharynx (A in the diagram) forming the back portion of the nasal cavities; the oropharynx (C) that portion between the soft palate (B) and the epiglottis (D) which communicates with the oral cavity; and the laryngopharynx (E) extending from the upper border of the epiglottis and continuing below with the larynx anteriorly and the esophagus posteriorly. The oropharynx and the laryngopharynx are involved with both the respiratory and the digestive systems. The nasopharynx is involved with the respiratory system only, whereas the esophagus (F) is only involved with the digestive system. The larynx which is involved with the respiratory system is indicated by G.

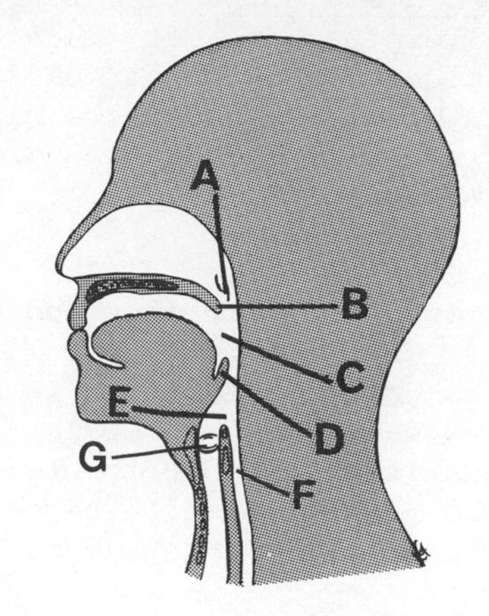

6. B, 1 and 3 are correct. The larynx is the organ that connects the lower part of the pharynx with the trachea. This air passage is maintained open by cartilages and their supporting ligaments. The larynx acts as a valve and guards the air passages during swallowing (deglutition) and it is also involved in speech. The laryngeal cavity has a constricted portion called the rima glottidis which lies between the vocal folds. The intrinsic muscles of the larynx are innervated by branches of the vagus nerve.

7. C, 2 and 4 are correct. The lungs have an elastic tendency to collapse and therefore to recoil away from the thoracic wall. This recoil is because of two factors: 1) the many elastic fibers found within the walls of the lung, and 2) the surface tension of the fluid lining the alveoli has a continual elastic tendency to collapse the lungs. The surface tension accounts for about two-thirds of the recoil tendency and the elastic fibers of the lung account for about one-third. The surface tension in the lungs is reduced by a lipoprotein mixture (dipalmitoyl lecithin) called surfactant. This substance allows the lungs to expand during inspiration. During inspiration the intraalveolar pressure becomes slightly negative (about -1 mm Hg, when compared to atmospheric pressure) and this causes air to flow inward through the respiratory passageways. When surfactant is absent it is very difficult to expand the lungs. In expiration the intraalveolar pressure becomes slightly positive (+1 mm Hg) and this causes the air to flow outward through the respiratory passageways.

8. E, All are correct. The trachea extends downwards from the cricoid cartilage and transports air to the primary bronchus of each lung. The lumen (A) of the trachea is always open due to the presence of U-shaped hyaline cartilages (B) in its wall. The posterior aspect of the tracheal wall contains smooth muscle (C) and this allows the wall of the esophagus (D) to expand in that direction when a bolus of food passes through it. The trachea is lined by a ciliated columnar epithelium (E).

9. B, 1 and 3 are correct. The residual air is that left after the deepest (or forced) expiration. It averages about 1200 ml in the young adult male. The residual air is important because it allows the alveoli to aerate the blood between breaths so that the blood is continually aerated. Therefore, the levels of oxygen and carbon dioxide in the blood are maintained at fairly uniform levels.

10. C, 2 and 4 are correct. The right and left primary bronchi are found at the distal end of the trachea and carry air into the right and left lung, respectively. The right bronchus lies in a more vertical plane than the left bronchus and it also has a larger diameter than the left bronchus. These facts explain the more common presence of foreign objects in the right bronchus than the left. Both bronchi have cartilagenous rings that become plates once the primary bronchi become pulmonary bronchi. The bronchi are lined by ciliated columnar epithelium.

11. E, All are correct. The main factors controlling respiration are voluntary (conscious) control, humoral (chemical) factors, and a number of physical factors. Voluntary or conscious control of breathing is a learned task which is utilized when speaking, singing, or holding ones breath. However, voluntary control in holding ones breath is limited and soon the respiratory center will ignore impulses from the cortex when it is necessary to meet the basic oxygen needs of the body and breathing will resume. There are chemoreceptors in the body that detect the levels of oxygen, carbon dioxide, and hydrogen ions in both the blood and body fluids, which includes the cerebrospinal fluid. When oxygen concentrations are low and carbon dioxide and hydrogen ion levels are high then the chemoreceptors of the brain, aortic body and carotid bodies detect the lack of oxygen and stimulate the respiratory center to increase both the rate and depth of breathing. Pressoreceptors in the aortic and carotid sinuses are affected by sudden changes in the blood pressure. For example, a high blood pressure will inhibit the respiratory centers by means of impulses transmitted by the vagus and glossopharyngeal nerves. In this way the rate and depth of respiration will become slower and more shallow. Conversely, a declining blood pressure will stimulate the respiratory system. The temperature of the blood flowing through the respiratory center also affects respiration. A high temperature stimulates the respiratory system, whereas a low temperature inhibits the respiratory system.

12. D, Only 4 is correct. The left lung has an oblique fissure that separates it into an upper and lower lobe but it has no horizontal fissure as does the right lung. Both lungs have a cardiac impression on their medial surfaces, with the one on the left lung being greater than the one on the right lung. The primary bronchus of the left lung is more horizontal, longer, and has a smaller diameter than the bronchus to the right lung.

13. D, Only 4 is correct. The amounts of oxygen and carbon dioxide in inspired, alveolar, and expired air are physiologically important in understanding the functioning of the lungs. Inspired air contains 20.93 per cent oxygen, 0.04 per cent carbon dioxide and 89.03 per cent water vapor and nitrogen. Expired air contains 15.70 per cent oxygen and 4.40 per cent carbon dioxide, while alveolar air has 14.00 per cent oxygen and 5.50 per cent carbon dioxide. The composition of the alveolar air is fairly constant and only about one-fifth of the air in the lungs is renewed by each inspiration. In this way, sudden and marked changes in the composition of alveolar air are prevented.

14. A, 1, 2, and 3 are correct. The root of the lung is formed by the structures entering and emerging at the hilum of each lung. The main structures of the root are the bronchi and the pulmonary vessels (arteries and veins), which serve to connect the medial surface of each lung to the trachea and the heart,

respectively. Other structures forming the root of the lung are the bronchial vessels, nerves, lymphatic vessels and nodes. The root of the lung is surrounded by the pleura that extends inferiorly to form the pulmonary ligament.

15. A, 1, 2, and 3 are correct. The vital capacity volume equals the sum of the inspiratory reserve volume, the tidal volume, plus the expiratory reserve volume. This is the amount of air that can be expelled from the lungs after first filling the lungs to their maximum extent and then expiring the maximum extent (4600 ml). This measurement is among the most important of all clinical respiratory measurements and used for assessing the normal function of the lungs and assessing the progress of different diseases. Factors affecting the vital capacity are: anatomical build, position of person during the testing, strength of the respiratory muscles, and pulmonary compliance (distensibility of the lungs and thoracic wall).

16. E, All are correct. The respiratory (pulmonary) membrane consists of the combined layers making up the walls of the alveolus and the capillary. The alveolus is lined by a lipoprotein mixture called surfactant (A in the diagram) which reduces the surface tension in the alveolus. The alveolar epithelium (B) and its basement membrane (C) are separated from the capillary wall by an interstitial space (D). The capillary wall consists of an endothelial layer (E) and its basement membrane. It is across this membrane that the gaseous exchange occurs between the alveolar air and the pulmonary blood. The oxygen also has to be transported across the membrane of the red blood cell (F).

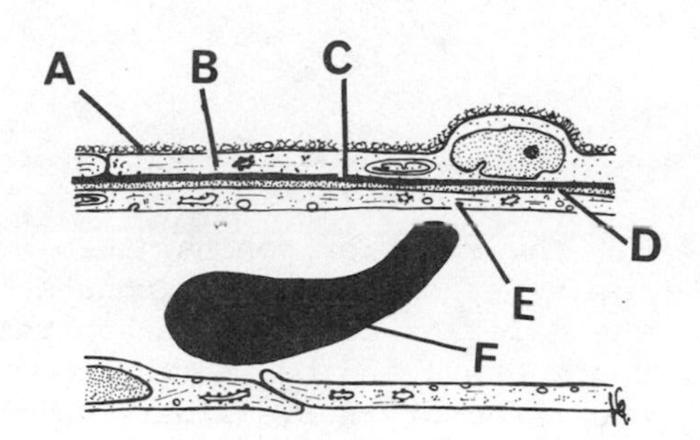

FIVE-CHOICE ASSOCIATION QUESTIONS

DIRECTIONS: Each group of questions below consists of a numbered list of descriptive words or phrases accompanied by a diagram with certain parts indicated by letters, or by a list of lettered headings. For each numbered word or phrase, SELECT THE LETTERED PART OR HEADING that matches it correctly. Then insert the letter in the space to the right of the appropriate number. Sometimes more than one numbered word may be correctly matched to the same lettered part or heading.

1. _____ The inferior concha

2. _____ Is continuous with the middle ear

3. _____ The site of drainage for the lacrimal apparatus

4. _____ The sphenoidal sinus

5. _____ The vestibule

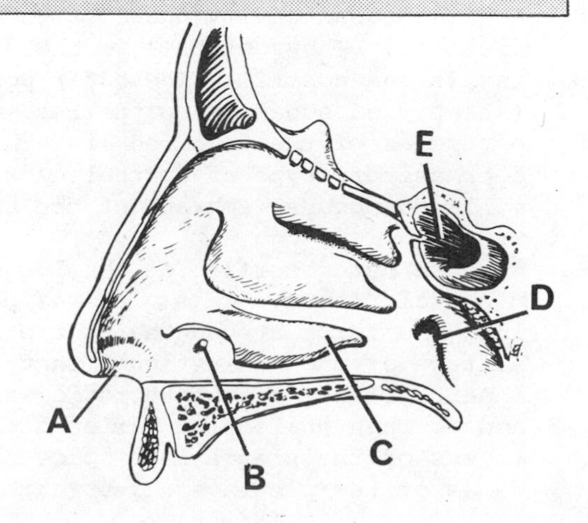

ASSOCIATION QUESTIONS

6. _____ The visceral pleura

7. _____ A fissure separating the lung into upper and middle lobes

8. _____ A muscle of respiration

9. _____ The parietal pleura

10. _____ The oblique fissure

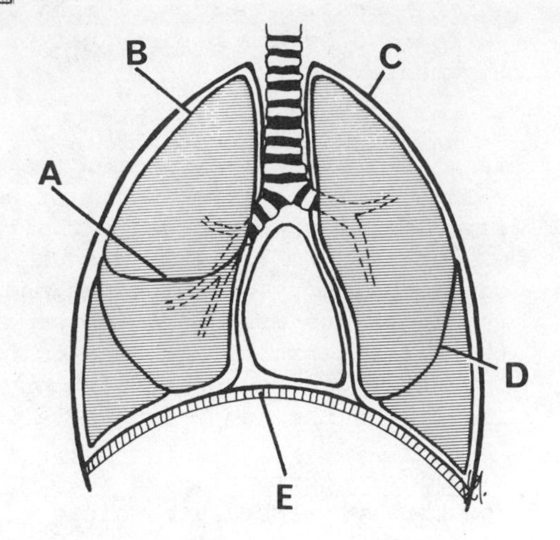

------------------------ ANSWERS, NOTES AND EXPLANATIONS ------------------------

1. C. The inferior concha (turbinate) is the lower of three constant conchae. The superior and middle conchae are projections of the ethmoid bone, whereas the inferior concha is a separate bone. The mucous membrane lining the space above and behind the superior concha (sphenoethmoidal recess) contains the olfactory nerve fibers which are involved in detecting odors.

2. D. The auditory tube opens into the posterior aspect of the nasal cavity (nasopharynx). This tube is normally closed, but when open it communicates with the middle ear. The opening of this tube by muscle action allows the pressure of the middle ear to equalize with the atmospheric pressure. This equalization of pressure allows the tympanic membrane of the ear to function properly.

3. B. The nasolacrimal duct drains the tears from the anterior surface of the eye and empties them into the inferior nasal meatus of the nasal cavity through a small opening.

4. E. The sphenoidal sinus is located within the body of the sphenoid bone. It is lined by a mucous membrane which is continuous with that lining the spheno-ethmoidal recess of the nasal cavity.

5. A. The nasal cavity opens to the outside through the anterior nares or nostrils. The vestibule is the dilated portion of the nasal cavity just inside the nostril. The outer portion of the vestibule is lined with skin (stratified squamous epithelium) bearing coarse hairs which serve to trap dust particles of the inspired air. The inner portion of the vestibule is lined by a transitory type of epithelium that blends with the highly vascularized ciliated mucous membrane of the nasal cavity proper.

6. B. The outer surface of the lungs is covered by a serous membrane called the visceral pleura. This visceral pleura projects down into the fissures of the lung and forms around the root of each lung. It then reflects on the medial surface of the mediastinum (that area between the right and left lungs), the inner surfaces of the thoracic wall, and the upper surface of the diaphragm and is then called the parietal pleura. Between the visceral and parietal layers of the pleura is a space called the pleural cavity. The adjacent surfaces of the pleura are lubricated by a thin film of serous fluid.

7. A. The right lung has its upper lobe divided into two lobes by the horizontal fissure forming a smaller upper lobe and a middle lobe. The oblique fissure of the right lung separates the upper and middle lobes from the lower lobe. The left lung has no horizontal fissure but is divided by the oblique fissure into an upper and lower lobe.

8. E. The diaphragm, a musculomembranous structure, partitions the thoracic and abdominal cavities. Each half of the muscular portion of the diaphragm is innervated by a phrenic nerve ($C_{2,3,4}$), which is a branch of the cervical plexus. The diaphragm contracts on stimulation and increases the internal thoracic volume which permits air to be taken into the lungs. When the diaphragm relaxes its elasticity returns it to its normal shape and expels the air from the lungs.

9. C. The parietal pleura of the lung is continuous with the visceral pleura at the root of each lung. The parietal pleura lines the mediastinal structures, the inner surfaces of the thoracic cage, and the upper surface of the diaphragm. It is separated from the visceral pleura, covering the lung, by the pleural cavity which contains a thin film of serous fluid. This fluid allows the lung to move freely within the pleural cavity with a minimum of friction.

10. D. The oblique fissure of the left lung divides the lung into upper and lower lobes. On the right, the oblique fissure separates the lower lobe from the middle and upper lobes. The middle and upper lobes on the right are separated by a second fissure called the horizontal fissure.

14. THE DIGESTIVE SYSTEM

LEARNING OBJECTIVES

ANATOMY

Be Able To:

- Describe the general functions of the different components of the digestive system.
- List the organs forming the digestive system and describe the four layers that form the wall of its hollow organs.
- Describe the anatomy of the oral cavity with reference to the following: hard and soft palate, uvula, pharyngopalatine and glossopalatine arches, palatine tonsil, vestibule, tongue, and the openings of the salivary glands.
- Describe the anatomy of a tooth and the types and numbers of deciduous and permanent teeth.
- Describe the anatomy, relationships and functions of the pharynx and the esophagus.
- Describe the cell types found in the mucous membrane, the rugae, the cardiac and pyloric sphincters, the curvatures, and the named portions of the stomach.
- Describe the anatomy of the small intestine under the following headings: mucous membrane, length, named parts, duodenal papilla, ileocecal valve, plicae circulares, villi, lacteals, and Peyer's patches.
- Describe the anatomy of the large intestine using the following headings: cecum, appendix, ascending colon, hepatic (right colic) flexure, transverse colon, splenic (left colic) flexure, descending colon, sigmoid colon, taenia coli, appendices epiploicae, and mucous membrane.
- Describe the anatomy of the rectum and the anal canal with reference to the valves, the hemorrhoidal plexuses, and the sphincters.
- Describe the anatomy of the liver with reference to its location, lobes, ligaments, blood supply, venous drainage, and the intrahepatic and extrahepatic biliary systems.

- Describe the gallbladder with reference to its structure, histology, blood supply, and function.

- Describe the anatomy of the pancreas with reference to its position, named parts, ductal sytem, and its exocrine functions.

- Describe the organs supplied by branches of the celiac trunk, and the superior mesenteric, the inferior mesenteric, and internal iliac arteries.

- Define the term portal and describe the veins forming the hepatic portal system, giving its significance.

- Describe the nerves carrying sympathetic and parasympathetic innervation to the organs of the digestive system and functions they perform.

- Describe the functions of mesenteries and define the following terms: parietal peritoneum, visceral peritoneum, retroperitoneal, lesser sac, greater sac, greater omentum, lesser omentum, mesentery, transverse mesocolon, and sigmoid mesocolon.

- Describe each of the following terms: peritonitis, mumps, stenosis, appendicitis, colostomy, hemorrhoids, cirrhosis, cholecystitis, and cholelithiasis.

DIGESTION, ABSORPTION, AND METABOLISM

Be Able To:

- Describe the digestive processes that begin in the oral cavity and the role the tongue, the pharynx, and the esophagus play in swallowing.

- Describe the three phases of digestion in the stomach with emphasis on the enzymes secreted and the factors involved in each.

- Describe the secretions which empty into the small intestine, giving their functions and the factors that control their rate of secretion.

- Describe the functions of the large intestine and the composition of feces and its elimination.

- Compare and contrast the absorption of the products of digestion from the mouth, the stomach, and the small and large intestines.

- Define the terms metabolism, anabolism, catabolism, enzymes, and basal metabolism.

- Discuss the metabolism of carbohydrates under the following headings: sources, anaerobic metabolism, aerobic metabolism, glycogenesis, storage in various forms, blood levels of glucose, energy produced and its utilization.

- Discuss lipid metabolism under the following headings: sources, storage, oxidation, ketone bodies and their formation (metabolic acidosis) and biosynthesis.

- Discuss the metabolism of proteins using the following headings: sources, the synthesis of amino acids, body tissue proteins, the processes of deamination and oxidation, and the utilization of amino acids and plasma proteins.

- Describe the maintenance of body temperatures under the following headings: normal basal level, production, sites of loss, controlling mechanisms, and normal and abnormal variations.

- Explain the meaning of the following terms: monosaccharides, disaccharides, polysaccharides simple and compound lipids, steroids, complete and conjugated proteins, and essential amino acids.

- Describe the normal sources, functions, and the abnormal conditions produced by the lack of sufficient quantities of each of the following vitamins: A, B_{12}, C, D, E, K, thiamine, niacin, and riboflavin.

RELEVANT READINGS

Anthony & Kolthoff - Chapters 14 and 15.
Chaffee & Greisheimer - Chapters 16 and 17.
Crouch & McClintic - Chapters 27 and 28.
Jacob, Francone & Lossow - Chapter 13.
Landau - Chapters 22, 23, 24, and 25.
Tortora - Chapter 20.

ANATOMY

FIVE-CHOICE COMPLETION QUESTIONS

DIRECTIONS: Each of the following questions or incomplete statements is followed by five suggested answers or completions. SELECT THE SINGLE BEST ANSWER in each case and then circle the appropriate letter at the lower right of each question.

1. THE BILE DUCT IS INDICATED BY _______.

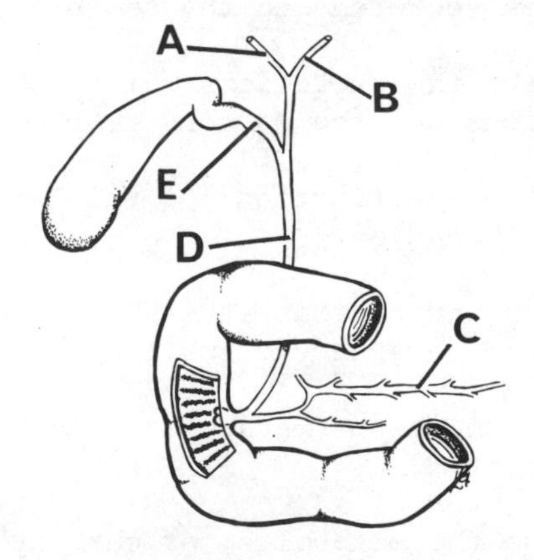

A B C D E

2. WHICH OF THE FOLLOWING STATEMENTS CONCERNING THE TEETH IS INCORRECT?
   A. Have enamel crowns
   B. Dentin forms a component of the gingiva
   C. The pulp cavity contains blood vessels and nerves
   D. Are connected to bone by the periodontal ligament
   E. Permanent teeth begin to appear at about six years

A B C D E

SELECT THE SINGLE BEST ANSWER

3. THE ARROW INDICATES THE ________.
   A. Portal vein
   B. Inferior vena cava
   C. Superior mesenteric vein
   D. Inferior mesenteric vein
   E. Splenic vein

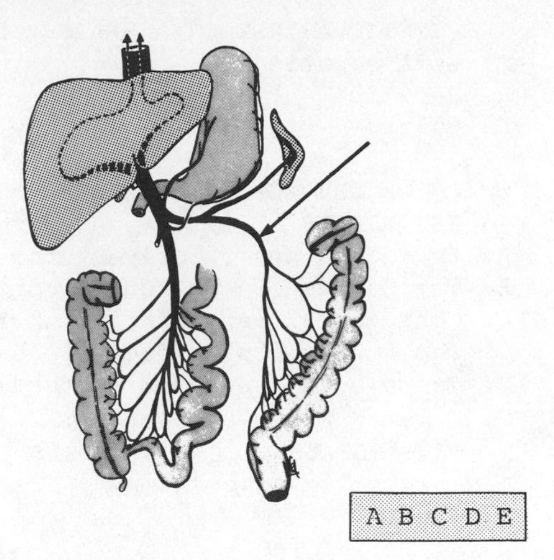

A B C D E

4. A SURGICAL FORMATION OF AN ARTIFICAL ANUS IS CALLED A(AN):
   A. Appendectomy
   B. Cholecystectomy
   C. Colostomy
   D. Hemorrhoidectomy
   E. None of the above

A B C D E

5. THE ARROW INDICATES THE ________ LAYER OF THE WALL OF THE DIGESTIVE TRACT.
   A. Serosa
   B. Submucosa
   C. Muscular layer
   D. Mucous membrane
   E. None of the above

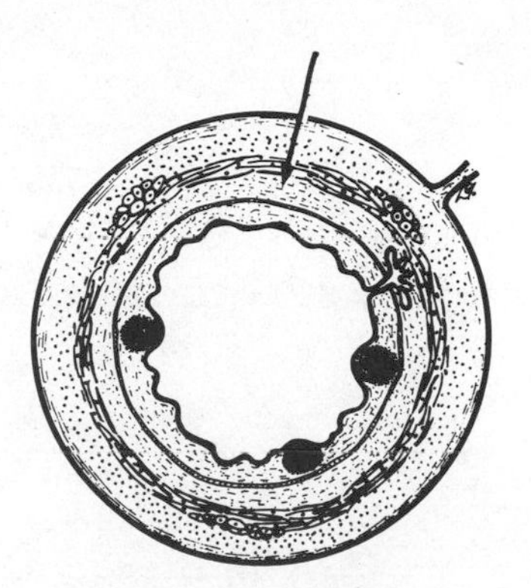

A B C D E

6. THE LESSER SAC IS INDICATED BY ______.

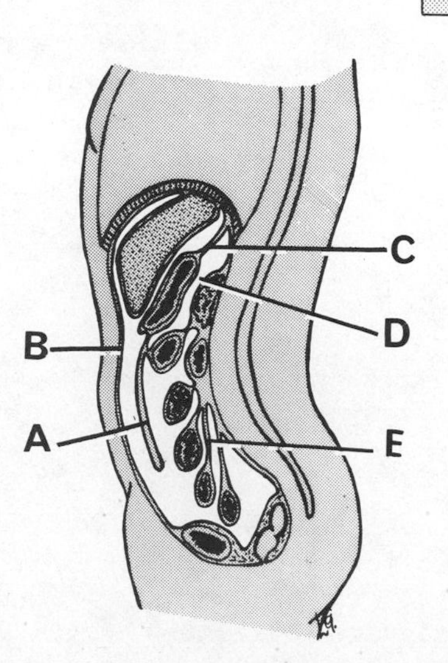

A B C D E

SELECT THE SINGLE BEST ANSWER

7. THE CYSTIC ARTERY IS A BRANCH OF THE _______ ARTERY.
   A. Left hepatic
   B. Hepatic
   C. Celiac
   D. Superior mesenteric
   E. None of the above

A B C D E

8. WHICH OF THE FOLLOWING STATEMENTS CONCERNING THE ORAL CAVITY IS INCORRECT?
   A. Is roofed posteriorly by the hard palate
   B. The dorsum of the tongue contains lymphatic nodules
   C. Contains 32 permanent teeth in the adult
   D. The submandibular duct opens on the floor of the mouth
   E. Taste buds are found on the dorsum of the tongue

A B C D E

9. A NERVE PLEXUS, LOCATED WITHIN THE WALL OF THE DIGESTIVE TUBE IS THE _______ PLEXUS.
   A. Celiac
   B. Choroid
   C. Sacral
   D. Myenteric
   E. Hypogastric

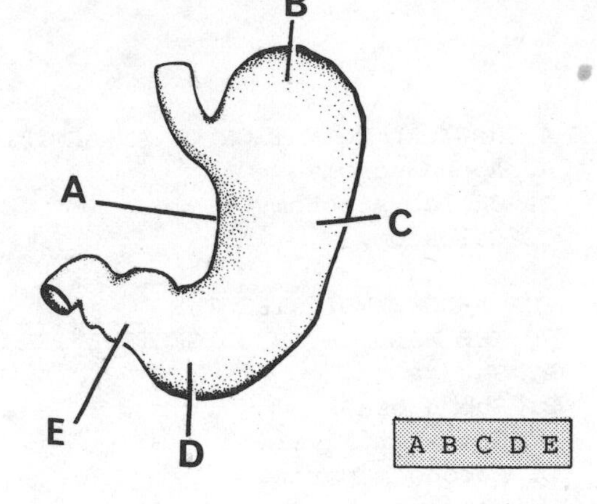

A B C D E

10. THE FUNDUS OF THE STOMACH IS INDICATED BY:

A B C D E

11. THE ARROW INDICATES THE _________.
   A. External anal sphincter
   B. Transverse fold
   C. Anal column
   D. Anal valve
   E. None of the above

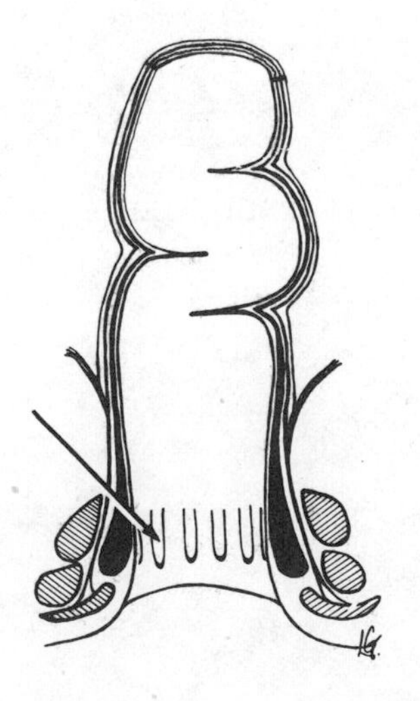

A B C D E

SELECT THE SINGLE BEST ANSWER

12. THE ARROW INDICATES THE _________ ARTERY.
 A. Splenic
 B. Left gastric
 C. Superior mesenteric
 D. Inferior mesenteric
 E. None of the above

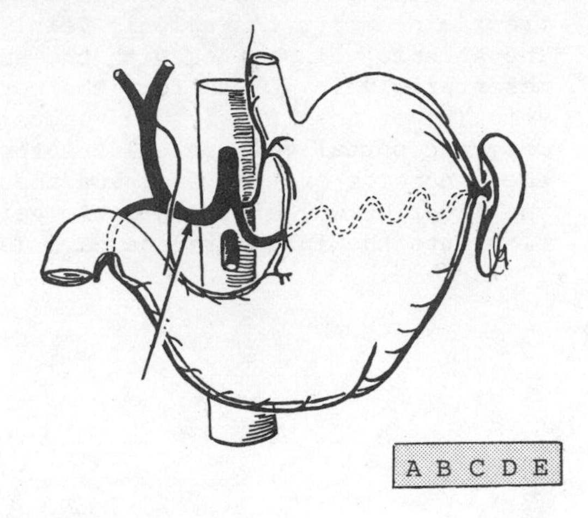

A B C D E

13. THE POUCH-LIKE SACCULATIONS OF THE LARGE INTESTINE ARE CALLED:
 A. Villi
 B. Haustra
 C. Appendices epiploicae
 D. Taeniae coli
 E. None of the above

A B C D E

------------------------ ANSWERS, NOTES AND EXPLANATIONS ------------------------

1. D. The right (A in the diagram) and left (B) hepatic ducts drain bile from their corresponding halves of the liver. They unit to form the common hepatic duct (not labelled in the diagram) which runs downward and joins the cystic duct (E) to form the bile duct (D). The bile duct runs along the free edge of the lesser omentum, crosses the head of the pancreas, and enters the duodenum, usually in conjunction with the pancreatic duct (C) at the hepatopancreatic ampulla.

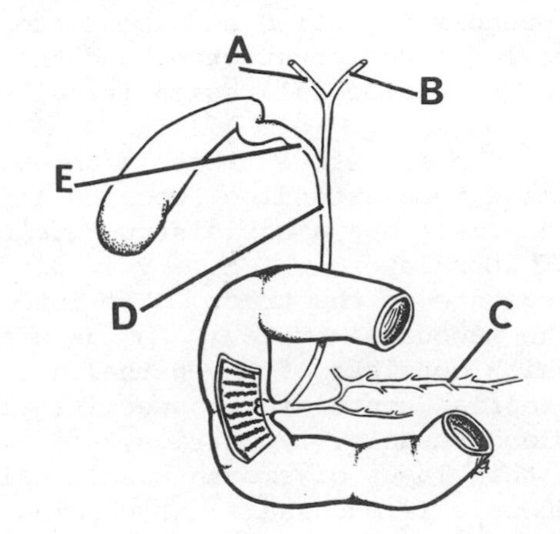

2. B. A tooth consists of a crown which lies above the gum (gingiva) and a root which lies below the gum line and inserts into sockets of the alveolar bone of the mandible and maxillae. A typical tooth consists of an enamel covered crown, an inner portion of calcified connective tissue or dentine, and a central portion or pulp cavity. The pulp cavity contains loose connective tissue, blood vessels, and nerve endings. The roots, which project into the bones of the upper and lower jaw, are covered by cementum and are firmly attached to the bone of the sockets by the periodontal ligament. The permanent teeth begin to appear at about six years of age and except for the wisdom teeth have all erupted by 17 years of age. The deciduous (milk) teeth begin to appear at about six months of age and usually are all gone by 13 years of age.

3. D. The inferior mesenteric vein (D in the diagram) drains venous blood from the distal portion of the gastrointestinal tract and enters the splenic vein (E). The splenic vein unites with the superior mesenteric vein (C) to form the portal vein (A) of the liver. The venous blood from the portal vein percolates through the sinusoids of the liver and then exits the liver by way of the hepatic veins to flow into the inferior vena cava (B).

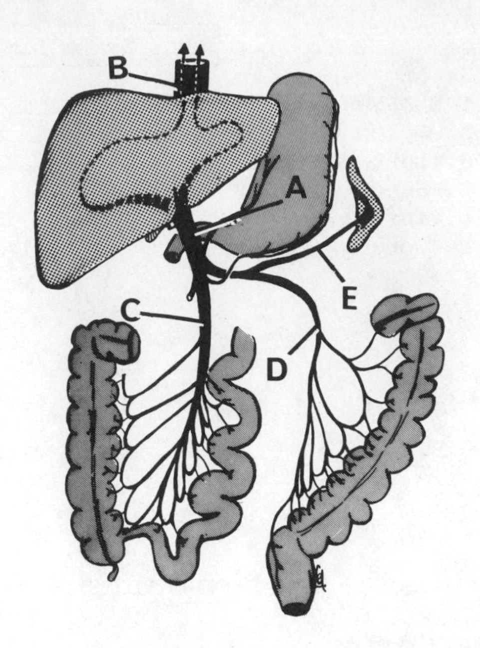

4. C. A colostomy or the surgical procedure of producing a colostomy, establishes an opening between the colon and the skin of the anterior abdominal wall. The more mobile portions of the large bowel (cecum, transverse and sigmoid colons) are used. Colostomies are constructed so as to form a temporary fecal fistula or a permanent artificial anus. The surgical procedure for removing the appendix is called an appendectomy, whereas the removal of the gallbladder is called a cholecystectomy and the removal of the hemorrhoidal plexus of veins in the anal canal region is called a hemorrhoidectomy.

5. B. The digestive tract is an epithelial lined tube extending from the lips to the anus. It has a basic structure consisting of four layers which vary in different segments of the tract. The inner layer is the mucous membrane (A in the diagram) which consists of an epithelium, a lamina propria (containing connective tissues, blood vessels, lymphatics, and glands) and a thin layer of smooth muscle called the muscularis mucosae. Surrounding the mucosa is the submucosa (B) which consists of loose connective tissue and it is bound by the muscularis externa (C). The muscularis externa consists of an inner circular and an outer longitudinal layer of smooth muscle. Between these two layers of smooth muscle are located bundles of nerve fibers or the myenteric plexus (D). The outermost layer of the digestive tract consists of an inner layer of connective tissue and an outer moist serous membrane, which together form the serosa (E). The serous membrane does not surround all segments of the gastrointestinal tract. In certain portions of the tract, the serosa forms a mesentery (F) to suspend portions of the tract within the abdominal cavity and other portions are retroperitoneal.

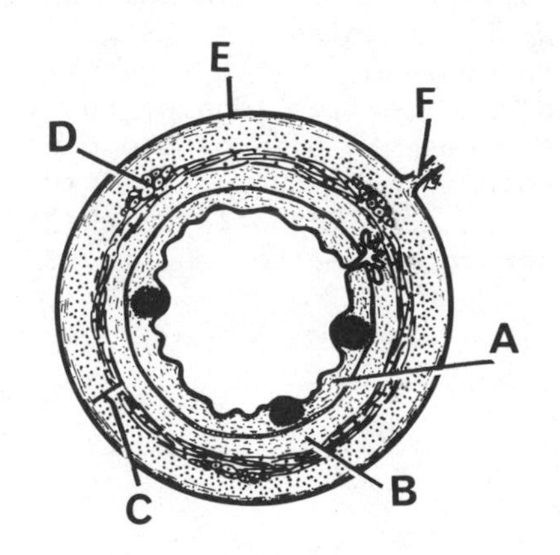

6. D. The peritoneum is a smooth, glistening, serous membrane. When it lines the abdominal wall it is known as parietal peritoneum (B in the diagram). When peritoneum is reflected from the abdominal wall on to the various organs it is then termed the visceral peritoneum (unlabelled in the diagram). Some of the abdominal viscera, the kidneys, for example, lie on the posterior abdominal wall and are covered by peritoneum only on their anterior aspects. Such organs are said to be retroperitoneal in position. Many organs of the abdominal cavity are suspended by folds of peritoneum forming mesenteries (F). The lesser omentum (C), a broad reflection of peritoneum, is seen extending between the liver and the stomach. The greater omentum (A) extends downward from the greater curvature of the stomach and the transverse colon into the greater sac. The greater sac extends from the diaphragm above to the floor of the pelvis below. The lesser sac (D) lies behind the stomach and in front of the retroperitoneally positioned pancreas (E).

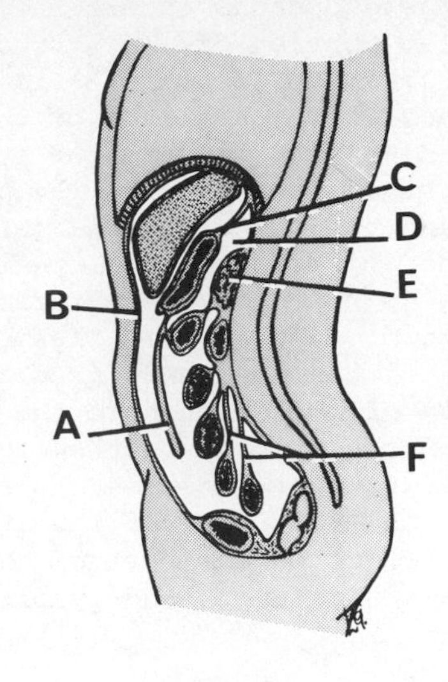

7. E. The cystic artery, which supplies blood to the gallbladder, is a branch of the right hepatic artery. The cystic artery may be doubled and supplies the gallbladder and on occasion may also supply the cystic duct and the upper segments of the bile duct.

8. A. The roof of the oral cavity consists of the anterior hard palate and the posterior soft palate, both of which are covered by a mucous membrane. The dorsum of the root of the tongue contains lymphatic nodules which form the lingual tonsil. The anterior two-thirds (body) of the tongue is studded with papillae. Some of these papillae are involved in roughing the surface of the tongue for use in licking and others contain taste buds which are the end organs of the sense of taste. The submandibular duct, which drains the gland of the same name, opens below the tongue through the floor of the mouth. The numerous ductules from the sublingual gland open into the floor of the mouth along a ridge of tissue. The duct of the parotid gland opens into the mouth through the substance of the cheek opposite the second upper molar tooth. The adult oral cavity has 16 (permanent) teeth in the upper jaw (maxillae) and 16 in the lower jaw (mandible) for a total of 32. Occasionally the last molar in each side of the upper and lower jaws fails to erupt, these are the so-called wisdom teeth. In children there are 20 deciduous teeth which are usually all replaced by permanent teeth by the time the child is 13 years of age.

9. D. The myenteric plexus is located between the inner circular and outer longitudinal layers of smooth muscle forming the muscularis externa of the gut wall. This plexus supplies these smooth muscle layers, whereas the submucosal plexus supplies the muscularis mucosae (smooth muscle) and their projections into the intestinal villi. These plexuses contain both parasympathetic and sympathetic fibers which locally control the peristaltic movements of the tube. The extrinsic nerves to the gut merely moderate the local autonomic reflexes.

10. B. The esophagus enters the stomach at the cardiac portion of the stomach (shaded in the diagram). The fundus (B) is the part of the stomach above the level of the entrance of the esophagus, whereas the pyloric part of the stomach consists of the pyloric antrum (D) and the pyloric canal (E). The pyloric sphincter surrounds the opening between the distal portion of the pyloric canal and the duodenum. The shorter, lesser curvature (A) of the stomach extends from the cardiac region of the stomach to the pyloric opening of the stomach on the right. The greater curvature (not labelled) is longer and runs along the left side of the stomach. The body (C) forms the part of the stomach between the fundus and pyloric portions.

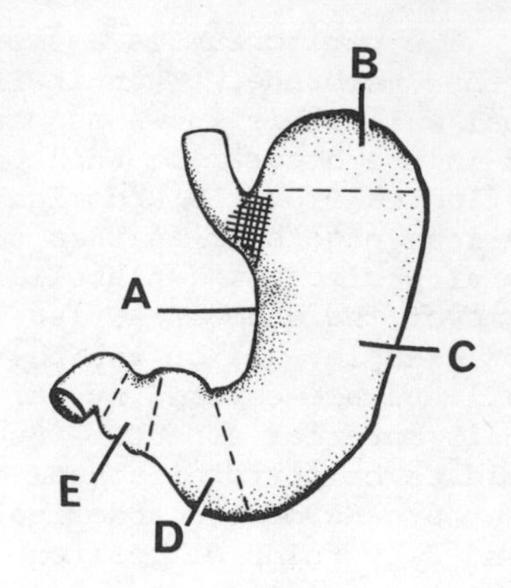

11. C. One of the anal columns is indicated by C in the diagram and its anal valve (D) is located at its inferior end. The anal valves mark the pectinate line. The internal (E) and external anal (A) sphincters are indicated. The internal rectal plexus of veins is not shown in this diagram but is located within lamina propria of the rectal and anal canals. The transverse fold(s) (B) project into the lumen of the rectum and may function to support the rectal contents.

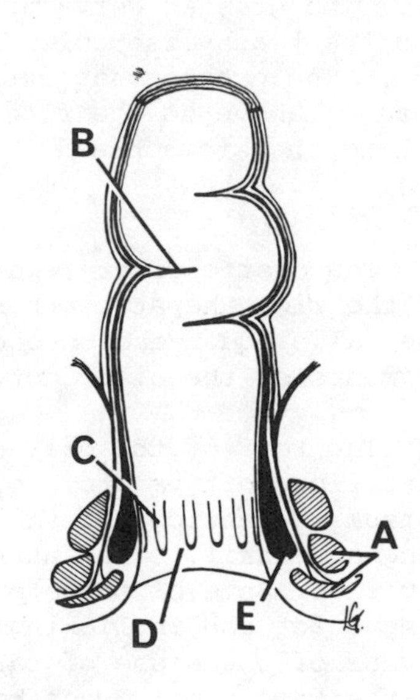

12. E. The arrow in the diagram indicates the hepatic artery. The hepatic, splenic (A) and left gastric (B) arteries are all branches of the celiac trunk. The left gastric, along with branches from the hepatic artery, carry blood to the stomach, pancreas, and the proximal part of the duodenum. The superior mesenteric artery (C) supplies the distal portion of the duodenum, jejunum, ileum, cecum, ascending colon, and the proximal half of the transverse colon. The inferior mesenteric artery (not illustrated in the diagram) supplies the rest of the transverse colon, descending colon, sigmoid colon, and rectum with arterial blood.

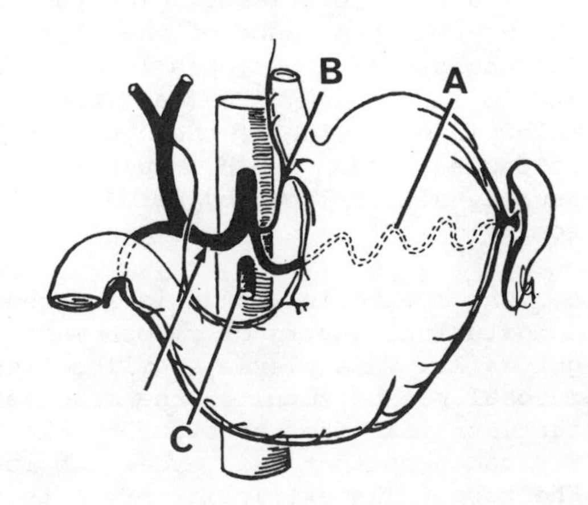

13. B. The pouch-like sacculations of the large intestine are called haustra. These haustra are not permanent structures but are involved in the formation, transport, and evacuation of feces. The haustra are formed either because of the shortness of the taeniae coli or the special arrangement of the circular muscle layer. The taeniae coli, of which there are three, are accumulations of the outer longitudinal (smooth) muscle layer which form single muscle bands. Another characteristic feature of the large intestine is the numerous appendices epiploicae, which are small masses of fat, enclosed in peritoneum extending from its surface. There are no villi in the mucous membrane of the large intestine.

MULTI-COMPLETION QUESTIONS

DIRECTIONS: In each of the following questions or incomplete statements, ONE OR MORE of the completions given is correct. At the lower right of each question, circle A if 1, 2, and 3 are correct; B if 1 and 3 are correct; C if 2 and 4 are correct; D if only 4 is correct; and E if all are correct.

1. FUNCTIONS OF THE TONGUE INCLUDE INVOLVEMENT IN:
   1. Speaking
   2. Swallowing
   3. Mastication
   4. Peristalsis

   A B C D E

2. THE GALLBLADDER:
   1. Lies on the inferior surface of the liver
   2. Absorbs water through its mucous membrane
   3. Is emptied when stimulated by a hormone
   4. Holds about 45 ml of bile

   A B C D E

3. THE MAJOR SALIVARY GLANDS:
   1. All empty their secretions into the mouth below the tongue
   2. Include the sublingual and submandibular glands
   3. Are ductless and secrete directly into the blood stream
   4. Produce both serous and mucus secretions

   A B C D E

4. THE MUCOUS MEMBRANE INCLUDES WHICH OF THE FOLLOWING?
   1. The muscularis mucosae
   2. An epithelial layer
   3. The lamina propria
   4. A serosal layer

   A B C D E

5. GENERAL FUNCTIONS OF THE STOMACH INCLUDE THE:
   1. Storage of consumed food
   2. Absorption of many of the digested nutrients
   3. Mixing of the gastric contents
   4. Secretion of alkaline enzymes

   A B C D E

6. THE PORTAL VEIN OF THE GUT:
   1. Receives some oxygenated blood from the aorta
   2. Is formed in part by the splenic vein
   3. Bypasses the inferior vena cava on its way to the heart
   4. Drains into the sinusoids of the liver

   A B C D E

7. WHICH OF THE FOLLOWING STATEMENTS IS(ARE) TRUE FOR THE LIVER?
   1. Receives both venous and arterial blood
   2. The bile ducts lie adjacent to the central vein
   3. Its cells form anatomical units called lobules
   4. The portal vein drains into the inferior vena cava

   A B C D E

A	B	C	D	E
1,2,3	1,3	2,4	Only 4	All correct

8. THE SMALL INTESTINE:
   1. Has a fixed portion called the duodenum
   2. Is the major site for digestion
   3. Has three separate segments
   4. Receives most of its blood supply from the superior mesenteric artery

   A B C D E

9. THE RECTUM:
   1. Has transverse folds
   2. Is continuous with the anal canal
   3. Lies within the pelvis
   4. Forms the largest segment of the large intestine

   A B C D E

10. KUPFFER CELLS ARE:
    1. Located in the liver
    2. Involved in the production of bile
    3. Active macrophages
    4. More numerous than the hepatic cells

    A B C D E

11. WHICH OF THE FOLLOWING IS(ARE) ASSOCIATED WITH THE SMALL INTESTINE?
    1. Haustra
    2. Appendices epiploicae
    3. Taeniae coli
    4. Villi

    A B C D E

12. WHICH OF THE FOLLOWING STATEMENTS IS(ARE) CORRECT FOR THE STOMACH?
    1. Has a sphincter formed by circularly arranged muscle fibers
    2. Its mucous membrane is thrown into rugae
    3. Has tubular glands that empty into pits
    4. Its chief cells secrete mucus

    A B C D E

13. THE APPENDIX:
    1. Attaches to the apex of the cecum
    2. Is supplied by a branch of the inferior mesenteric artery
    3. Contains great quantities of lymphoid tissue
    4. On cross section is a solid organ

    A B C D E

14. FUNCTIONS OF THE LARGE (BOWEL) COLON INCLUDE THE:
    1. Absorption of large quantities of water
    2. Release of enzymes for protein catabolism
    3. Secretion of great quantities of mucus
    4. Control over the secretion of pancreatic hormones

    A B C D E

15. WHICH OF THE FOLLOWING STATEMENTS IS(ARE) CORRECT FOR THE ANAL CANAL?
    1. Passes through the levator ani muscles
    2. Forms the terminal portion of the large intestine
    3. The external anal sphincter consists of skeletal muscle
    4. Is approximately 10 cm in length

    A B C D E

ANSWERS, NOTES AND EXPLANATIONS

1. A, 1, 2, and 3 are correct. The tongue, composed of vertical, horizontal, and oblique bundles of striated muscle, is involved in speaking, swallowing, and mastication. Motor fibers reach the tongue by the hypoglossal nerve. The lingual nerve carries sensory fibers from the anterior two-thirds of the

tongue, whereas the facial nerve carries taste sensation from the same region. The posterior third of the tongue is innervated by the glossopharyngeal nerve, carrying both general sensation and taste fibers.

2. E, All are correct. The gallbladder lies in a fossa on the inferior visceral surface of the liver, where it receives bile from the liver by the common bile duct. The gallbladder is covered by peritoneum and consists of a terminal fundic portion, a body, and a neck portion which is continuous with the cystic duct. Bile is continuously being produced by the liver and is stored in the gallbladder until it is needed by the duodenum. About 800-1000 ml of bile is produced per day by the liver and the gallbladder stores about 30-70 ml of bile. Bile is concentrated in the gallbladder by continual absorption of water, sodium, chloride, and other electrolytes by its mucosal lining. The gallbladder is emptied when the sphincter (Oddi) of the bile duct relaxes and the muscle of its wall contracts, expelling the concentrated bile into the duodenum. Fat and protein in the lumen of the duodenum stimulates the release of a hormone cholecystokinin which both relaxes the sphincter of Oddi and stimulates the contraction of the smooth muscle within the wall of the gallbladder.

3. C, 2 and 4 are correct. Besides the numerous small salivary glands of the oral mucosa or submucosa there are three large glands that secrete saliva into the oral cavity. They are the paired parotid, sublingual, and submandibular (submaxillary) glands. The smaller salivary glands secrete continuously, whereas the larger glands secrete when they are stimulated by either thermal, chemical, olfactory, or psychic stimuli. The saliva is a mixture of serous and mucus secretions plus amylase, an enzyme that splits starch. The sublingual and submandibular glands empty their saliva into the oral cavity through openings found below the tongue. The parotid duct opens into the oral cavity on the cheek opposite the second upper molar tooth.

4. A, 1, 2, and 3 are correct. The mucous membrane (mucosa) consists of an outer epithelial layer, a supporting layer of loose connective tissue (lamina propria), and an outer layer formed by a thin layer of smooth muscle (the muscularis mucosae). The layer of epithelium is regionally specialized for different digestive functions. The oral cavity, pharynx, esophagus, and anal canal are all lined by stratified squamous epithelium, whereas the remaining segments of the digestive tube are lined by specialized simple columnar epithelium. The cells forming the columnar epithelium are specialized to function in absorption, or the secretion of either mucus or digestive enzymes.

5. B, 1 and 3 are correct. The general functions of the stomach include the storage of the ingested food and its mixing with gastric secretions. The glands of the stomach secrete HCl, pepsinogen (the inactive form of pepsin which acts on proteins), and mucus which both protects the lining cells of the stomach and acts as a lubricant. Rhythmic contractions of the smooth muscle found within the wall of the stomach serve to mix the food with the gastric secretions.

6. C, 2 and 4 are correct. The portal vein of the gut is formed by the union of the splenic and superior mesenteric veins. The splenic vein receives the inferior mesenteric vein as a tributary. These veins drain the blood from most of the gastrointestinal tract and along with it all the products of digestion. The portal vein empties into the liver sinusoids. The hepatic cells utilize the products of digestion to perform the processes of anabolism and catabolism. The portal vein carries unoxygenated blood, whereas hepatic arteries carry oxygenated blood to the liver. The hepatic veins drain blood from the lobes of the liver and carry it to the inferior vena cava.

7. A, 1, 2, and 3 are correct. The hepatic artery, a branch of the celiac trunk, carries oxygenated blood to the liver and the portal vein carries unoxygenated blood to the liver. The blood carried to the liver by the portal vein also contains the products of digestion which have been absorbed mainly by the mucosa of the small intestine. Branches of the hepatic artery and portal vein along with the intrahepatic bile duct form the portal triad which lies at the periphery of the hepatic lobules. The lobules are formed by pillars of hepatic cells which are arranged so that they radiate from the central vein. The central vein of each lobule carries blood to the hepatic veins and they in turn drain into the inferior vena cava. The bile canaliculi carry bile, which is secreted by the hepatic cells, to the bile ducts located at the periphery of the hepatic lobules.

8. E, All are correct. The small intestine serves to pass luminal contents from the stomach to the large intestine. The small intestine is the site where most of the digestion and absorption takes place in the gastrointestinal tract. It consists of a short (25 cm), C-shaped portion called the duodenum, a middle portion (60 cm) called the jejunum, and a terminal portion (60 cm) called the ileum. The superior mesenteric artery supplies blood to the small intestine except for the upper half of the duodenum which receives its blood from branches of the celiac trunk.

9. A, 1, 2, and 3 are correct. The rectum (about 15 cm) is that portion of the large intestine lying between the sigmoid colon and the anal canal. It lies anterior to the sacrum within the pelvic cavity and has peritoneum on its upper anterior and lateral surfaces. The rectum has no peritoneum on its lower portion. Usually it has three lateral curvatures which coincide with transverse rectal folds of mucosa. These folds may be involved in supporting the rectal contents and in this way help protect the anal sphincters from the weight of the fecal mass.

10. B, 1 and 3 are correct. The Kupffer cells of the liver are active macrophages, lining the sinusoids of the liver. They frequently contain engulfed erythrocytes in various stages of disintegration, pigment deposits, and granules rich in iron. They are less numerous than the hepatic cells which produce bile. The Kupffer cells may be derived from circulating monocytes which represent the free mononuclear macrophages of other organs.

11. D, Only 4 is correct. The small intestine has special structural features that aid one of its major functions--absorption. The mucous membrane of the small intestine is thrown into inward projecting folds called plicae circularis and even smaller projections from the surface of these folds called villi. Both of these structures aid in increasing the surface area of the small intestine. To further increase the surface area of the small intestine the villi are lined by columnar cells which themselves have small microvilli that form a brush border. These microvilli greatly increase the total surface area of each cell. The haustra, appendices epiploicae and taeniae coli are all structural features of the large intestine.

12. A, 1, 2, and 3 are correct. The outer layer of the stomach wall consists of a serosal layer which covers the muscularis externa. In the stomach the muscularis externa consists of three separate muscle layers: an outer longitudinal, a middle circular, and an inner oblique layer. The middle circular layer for the pylorus forms a thick circular mass which forms the pyloric sphincter. The muscularis mucosae consists of a double layer of smooth muscle, whereas the lamina propria is rather sparse and lies between the gastric glands which empty into the pits. The lining of the empty stomach is thrown into long longitudinal folds (rugae) which allow for its expansion. The mucous membrane of the stomach consists of glands that empty into the

bottom of gastric pits. The pits and ridges between are lined by tall columnar epithelial cells. The gastric glands are slightly different in the cardiac, fundic, and pyloric regions. In general, they contain: chief (zymogenic) cells which secrete pepsinogen; parietal (oxyntic) cells which secrete HCl and an intrinsic factor; neck mucous cells which secrete mucus; and argentaffin cells which secrete the hormone gastrin. The pyloric glands consist mainly of mucus secreting cells.

13. B, 1 and 3 are correct. The appendix (veriform appendix), about 9 cm long, arises from the apex of the cecum at the junction of the three taeniae coli. It has a lumen which is continuous with the cecum and a mucosa which is largely infiltrated with lymphoid tissue. The appendix lacks a true mesentery and is supplied by the appendicular artery, which is a distal branch of the superior mesenteric artery.

14. B, 1 and 3 are correct. The large bowel functions in removing large quantities of water from its lumenal contents along with Na^+ and Cl^-. The contents of the large intestine are moved along its length by rather sluggish contractions and the aid of large amounts of mucus secreted by mucous secreting cells. No enzymes are secreted into the large bowel and it has no control over pancreatic secretion. Digestion of food and absorption of the products of digestion takes place in the small intestine.

15. A, 1, 2, and 3 are correct. The anal canal is the terminal 3 cm of the large intestine. It passes through the levator ani and below this muscle is surrounded by the external anal sphincter (skeletal muscle). The inner circular layer of the muscularis externus (smooth muscle) becomes thickened in the anal region and forms the internal anal sphincter. Both of these muscles play an important role in constricting the distal end of the anal canal and thus controlling defecation.

FIVE-CHOICE ASSOCIATION QUESTIONS

DIRECTIONS: Each group of questions below consists of a numbered list of descriptive words or phrases accompanied by a diagram with certain parts indicated by letters, or by a list of lettered headings. For each numbered word or phrase, SELECT THE LETTERED PART OR HEADING that matches it correctly. Then insert the letter in the space to the right of the appropriate number. Sometimes more than one numbered word may be correctly matched to the same lettered part or heading.

1. _____ Parietal cells
2. _____ Hepatic cells
3. _____ Chief cells
4. _____ Goblet cells
5. _____ Acinar cells

A. Pancreatic juice
B. Bile
C. Hydrochloric acid
D. Mucus
E. Pepsinogen

ASSOCIATION QUESTIONS

6. _____ The component of the colon suspended by a mesentery

7. _____ The hepatic flexure

8. _____ The ileocecal junction

9. _____ A portion of the colon which has two openings into it

10. _____ The splenic flexure

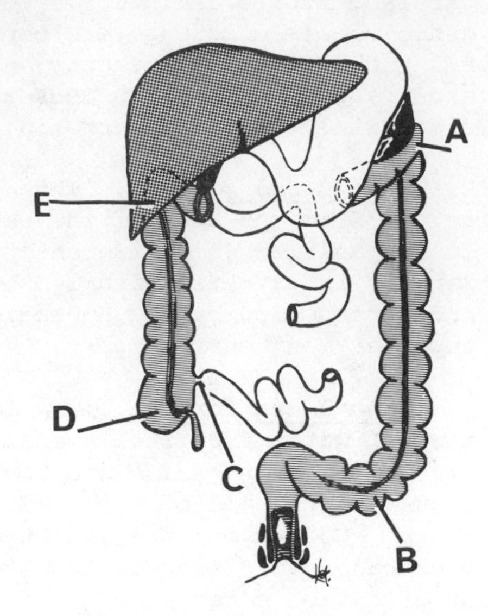

------------------------ ANSWERS, NOTES AND EXPLANATIONS ------------------------

1. C. The low pH of the gastric contents is because of the hydrochloric acid which is secreted by the parietal cells of the gastric glands. They are scattered among the chief (zymogenic) cells of the glands. The parietal cells also secrete the gastric intrinsic factor which binds to vitamin B_{12} of dietary origin and facilitates its absorption by the intestine.

2. B. The hepatic cells of the liver secrete bile into the bile canaliculi which eventually drain into the bile ducts. The liver cells detoxify toxic substances found in the blood, reaching the liver by way of the intestines or the general circulation. The products of this detoxification process or their harmless conjugates are excreted into the bile. Bile is a complex fluid that can be regarded as a secretion since it plays an important role in digestion. It also serves as a vehicle to eliminate detoxified waste and potentially harmful materials to the intestines where it is ultimately eliminated.

3. E. The chief (zymogenic) cells are located in the glands of the stomach and secrete pepsinogen, the precursor of the enzyme pepsin. This precursor is released by parasympathetic (cholinergic) stimulation into the lumen of the stomach. Here the low pH of the gastric contents, because of the presence of HCl, converts pepsinogen to pepsin. This enzyme, which is optimally active at a pH of 2, cleaves peptide bonds and is therefore important in the digestion of proteins.

4. D. Goblet cells are irregularly scattered among the absorptive cells lining the intestines. Their shape often resembles that of a wine glass (goblet). The secretory product of this cell is mucus, which serves both as a lubricant and as a protector of the surface epithelium. The mucus consists of mucoproteins and sulfated polysaccharides.

5. A. The pancreas has both endocrine and exocrine functions. The endocrine function of the gland is carried out by hormones produced by the cells of the islets of Langerhans and the exocrine portion is produced by the acinar cells which form a lobule. These cells secrete a pancreatic juice that contains proteases, nucleases, and lipases which are enzymes for the three major classes of nutrients (proteins, carbohydrates, and fats).

6. B. The sigmoid colon is suspended by a mesentery called the sigmoid mesocolon. The other portion of the colon suspended by a mesentery is the transverse colon which is that portion of the colon lying between the hepatic (right) and splenic (left) flexures.

7. E. The hepatic (right colic) flexure is the portion of the colon which bends (flexes) and marks the junction between the ascending and transverse colons. It lies behind the large right lobe of the liver and in front of part of the right kidney.

8. C. The ileocecal junction marks the entrance of the distal end of the ileum into the cecum of the large bowel. There is a valve at this junction but it is thought to serve little, if any, mechanical importance.

9. D. The cecum is a dilated portion of the colon and lies below the entrance of the ileum (ileocecal junction) into the large intestine. The cecum also has the lumen of the appendix opening into it. The appendix is attached to the apex of the cecum but it lies in a variety of positions.

10. A. The splenic (left colic) flexure lies inferior to the spleen and marks the junction between the transverse and descending colon.

DIGESTION, ABSORPTION, AND METABOLISM

F I V E - C H O I C E C O M P L E T I O N Q U E S T I O N S

DIRECTIONS: Each of the following questions or incomplete statements is followed by five suggested answers or completions. SELECT THE SINGLE BEST ANSWER in each case and then circle the appropriate letter at the lower right of each question.

1. SYMPATHETIC NERVE INNERVATION TO THE SMALL INTESTINE:
   A. Stimulates peristalsis
   B. Inhibits mucus secretion
   C. Stimulates the secretion of enzymes
   D. Inhibits the absorption of water
   E. None of the above

   A B C D E

2. THE SEGMENT OF THE GASTROINTESTINAL TRACT IN WHICH MOST ABSORPTION TAKES PLACE IS THE ______.
   A. Mouth
   B. Stomach
   C. Small intestine
   D. Large intestine
   E. None of the above

   A B C D E

3. THE PREFERRED SOURCE OF ENERGY FOR THE BODY IS FROM ______.
   A. Glucose
   B. Amino acids
   C. Fatty acids
   D. Vitamins
   E. None of the above

   A B C D E

4. THE CENTERS OF THE BRAIN CONTROLLING BODY TEMPERATURE ARE LOCATED IN THE ______.
   A. Midbrain
   B. Cerebellum
   C. Hypothalamus
   D. Medulla oblongata
   E. None of the above

   A B C D E

SELECT THE SINGLE BEST ANSWER

5. WHICH OF THE FOLLOWING IS NOT TRUE FOR THE LARGE INTESTINE?
   A. Secretes no enzymes
   B. Absorbs large quantities of water
   C. Secretes large quantities of mucus
   D. Houses bacteria that synthesize vitamin K
   E. Parasympathetic fibers inhibit its emptying

   A B C D E

6. WHICH OF THE FOLLOWING IS NOT INVOLVED IN THE PRODUCTION OF HEAT FOR THE BODY?
   A. Muscular activity
   B. High protein intake
   C. Hormonal stimulation
   D. Active secretion by organs
   E. Parasympathetic innervation of BMR

   A B C D E

7. AMINO ACIDS THAT CANNOT BE FORMED BY THE BODY ARE CALLED ______.
   A. Complete
   B. Conjugated
   C. Essential
   D. Non-essential
   E. None of the above

   A B C D E

8. THE MAJOR SOURCE OF HEAT LOSS IS THE _______.
   A. Skin
   B. Lungs
   C. Urine
   D. Feces
   E. None of the above

   A B C D E

9. AN ENZYME SECRETED BY THE STOMACH IS:
   A. Secretin
   B. Pancreozymin
   C. Cholecystokinin
   D. Enterogastrone
   E. None of the above

   A B C D E

10. THE INTESTINAL CRYPTS DO NOT SECRETE:
    A. Lactase
    B. Maltase
    C. Bile salts
    D. Proteases
    E. Enterokinase

    A B C D E

11. THE GALLBLADDER EMPTIES ONCE IT IS STIMULATED BY THE HORMONE ________.
    A. Cholecystokinin
    B. Enterogastrone
    C. Pancreozymin
    D. Secretin
    E. Gastrin

    A B C D E

12. WHEN ACETOACETIC ACID OR ITS METABOLIC DERIVATIVES ACCUMULATE IN THE BLOOD, WE HAVE A CONDITION CALLED ________.
    A. Obesity
    B. Ketosis
    C. Kyphosis
    D. Keratinization
    E. None of the above

    A B C D E

13. WHICH OF THE FOLLOWING HAS NO INFLUENCE ON THE RATE AT WHICH THE STOMACH EMPTIES?
    A. Motility of the fundic region of the stomach
    B. Type of food making up the duodenal contents
    C. Fluidity of the chyme in the duodenum
    D. Amount of chyme in the small intestine
    E. Degree of acidity of the duodenal contents

    A B C D E

1. B. Sympathetic or adrenergic innervation of the small intestine decreases its motility and tone of its smooth muscle, causes general contraction of sphincters, and generally inhibits the secretion of substances into its lumen. The parasympathetic (cholinergic) impulses tend to increase motility and tone, cause general relaxation of sphincters, and stimulate the release of secretions.

2. C. The greatest amount of food is absorbed through the mucous membrane of the small intestine. Water is absorbed through the wall of the large intestine in great quantities and through the stomach wall in smaller quantities. No absorption of food takes place through the wall of the oral cavity or the wall of the stomach. However, a few drugs are absorbed directly through the walls of the oral cavity and the stomach.

3. A. Glucose is the preferred source of fuel for energy production by the cells of the body. If there is not enough glucose available, then new glucose (gluconeogenesis) is formed first from the fat stores of the body and secondly from amino acids (proteins). Vitamins are not a source of energy at all.

4. C. The hypothalamus contains the main heat-regulating centers of the brain. These centers receive an input as to the temperature of the body and its external environment from the temperature of the blood passing near them and by reflexes from temperature receptors located in the skin.

5. E. The large bowel is primarily involved in absorbing large quantities of water from its luminal contents. The large number of glands lining its wall secrete copious quantities of mucus which aids in the movement of its contents. No enzymes are secreted into the contents of the large intestine. Bacteria that are abundant in the large intestine synthesize both vitamin K and certain components of the vitamin B complex. These products in turn are absorbed into the body for utilization. The anal sphincters of the large intestine are relaxed by parasympathetic fibers which aid (facilitate) the emptying of the large bowel. Sympathetic stimulation of these sphincters inhibits the opening of the anus.

6. E. The production of heat for the body is a by-product of the oxidation of carbohydrates, fats, and proteins for useful energy in cells of the body. The greatest amount of body heat is produced by skeletal muscle activities of the body. Active secretion by organs (glands) such as the liver also provides large quantities of heat. Hormonal and sympathetic stimulation also plays a role in heat production. Hormones from the thyroid gland increase cellular metabolism and sympathetic stimulation also plays a role in controlling the metabolic rate of the body. The intake of foods, such as proteins, yields great quantities of heat as do activities such as smooth muscle contraction in the gut.

7. C. Amino acids, derived from protein catabolism, are important sources for building new protoplasm, blood proteins, hormones, and enzymes. All amino acids are important for the body and they are either attained from the food we eat or are synthesized in the liver. Those that cannot be synthesized by the body are called essential, since it is essential that they be obtained from our diet. Those amino acids that can be synthesized by the body (liver) for its own needs are called non-essential.

8. A. The major source of heat loss from the body is from the skin. Radiation, conduction, convection, and vaporization all play a role in the amount of heat lost from the skin. Other less important sources of heat loss from the body

are from the air expelled from the lungs and the waste products (urine and feces) which are eliminated from the body.

9. E. The principal enzyme secreted by the chief cells of the stomach is pepsinogen. This inactive enzyme enters the stomach and comes in contact with previously formed pepsin and hydrochloric acid and is immediately activated to form the enzyme pepsin. Pepsin is a highly active proteolytic enzyme in a highly acid medium (pH of 2.0) but quickly becomes inactivated as the pH of the gastric contents rises. The acidic medium of the stomach is produced by the HCl secreted in the stomach by the parietal cells. The gastric glands of the pyloric region contain fewer chief and parietal cells but possess an increased number of mucous cells. In this region of the stomach these cells secrete an alkaline mucus which helps not only to protect the mucous lining of the stomach but also deactivates pepsin by raising the pH of the luminal contents.

10. C. The intestinal crypts do not secrete bile salts. These salts are secreted by the liver cells in the form of bile. They aid in emulsifying fats and the absorption of fatty acids, cholesterol, and other lipids from the small intestine. The two enzymes found in the intestinal contents are enterokinase, which activates chymotrypsin to trypsin, and a small amount of amylase. However, the epithelial cells of the mucosa of the small intestine produce enzymes that digest foods as they are being absorbed through the epithelium. These include: the peptidases which split proteins, four enzymes that split disaccharides into monosaccharides (sucrose, maltase, isomaltase, and lactase); and lipases for splitting neutral fats into glycerol and fatty acids. These enzymes are thought to be located in the brush border of the epithelial cells of the mucous membrane and presumably cause the hydrolysis of foods on the outer surface of the microvilli prior to absorption of the end products of digestion.

11. A. The gallbladder is stimulated to empty after a meal containing high concentrations of fat. The presence of large quantities of fats (and protein) entering the duodenum stimulates the release of a hormone called cholecystokinin from the intestinal mucosa. This hormone reaches the gallbladder by the blood stream and causes the muscle of its wall to contract. Contraction of the muscle in the wall of the gallbladder in some way inhibits the sphincter of Oddi which relaxes and permits the bile to enter the duodenum. It is thought that there is either a direct neurogenic or a myogenic reflex from the gallbladder to the sphincter of Oddi so its relaxation is coordinated with the contraction of the wall of the gallbladder.

12. B. The oxidation of fatty acids in the liver, by a complex set of reactions, produces a number of molecules of acetyl CoA. Acetyl CoA is utilized by the liver cells to produce energy, however, when there is an excess of acetyl CoA, the cells condense two of these molecules to form acetoacetic acid. This ketoacid diffuses into the blood stream and is utilized by other tissues to produce energy by breaking it once again down to acetyl CoA. Normally there are low levels of acetoacetic acid (ketones) in the blood (rarely exceeding 1 mg per cent). When excessive amounts of acetoacetic acid or its derivatives (betahydroxybutyric acid and acetone) accumulate in the blood (ketonemia) a state of ketosis exists. Ketosis may become severe when carbohydrate metabolism is depressed (untreated diabetes mellitus) and the body derives most of its energy from the metabolism of fatty acids. The amount of ketones builds up in the blood because the tissues of the body can metabolize only limited quantities. This build up in the blood causes large quantities of them to be excreted by the kidney (ketonuria). The accumulation of ketones (acids) in the blood produces metabolic acidosis by depleting the basic components of the buffer system of the body and by removing sodium along with the ketones excreted by the kidney.

13. A. Stomach emptying is promoted by peristaltic waves in the antrum (pyloric) of the stomach and is opposed by the resistance produced by the pyloric sphincter. The rate of emptying is determined primarily by the degree of antral peristaltic waves and is controlled mainly by feedback signals from the duodenum and secondly by factors produced in the stomach. Duodenal control of stomach emptying includes the enterogastric reflex and to a lesser extent hormonal feedback. These two systems work together to slow the rate of emptying when there is too much chyme present in the small intestine, or the chyme is too acidic, contains too much protein or fat, is hypotonic or hypertonic, or is irritating the mucous membrane. In this way the rate at which the stomach empties is limited to the amount of chyme that the small intestine can process. The stomach also controls its own emptying by the degree of its own peristalsis and the degree of filling.

MULTI-COMPLETION QUESTIONS

DIRECTIONS: In each of the following questions or incomplete statements, ONE OR MORE of the completions given is correct. At the lower right of each question, circle A if 1, 2, and 3 are correct; B if 1 and 3 are correct; C if 2 and 4 are correct; D if only 4 is correct; and E if all are correct.

1. GASTRIC JUICE:
   1. Is highly acidic because of its HCl content
   2. May be released by the mere thought of food
   3. Is controlled in part by a hormone
   4. Becomes neutralized only in the duodenum

   A B C D E

2. WHICH OF THE FOLLOWING STATEMENTS IS(ARE) TRUE FOR GLYCOLYSIS?
   1. Takes place in the mitochondria of a cell
   2. Requires large quantities of oxygen
   3. Converts pyruvic acid to glucose
   4. Most importantly produces energy

   A B C D E

3. THE TRICARBOXYLIC ACID (TCA) CYCLE:
   1. Takes place in mitochondria
   2. Utilizes oxygen
   3. Produces 34 ATP molecules
   4. Anabolizes amino acids

   A B C D E

4. ABSORBED LIPIDS ARE TRANSPORTED IN THE BLOOD AS:
   1. A unit attached to free hemoglobin molecules
   2. A complex with protein
   3. Particles loosely combined with carbohydrates
   4. Chylomicrons

   A B C D E

5. PROTEIN METABOLISM:
   1. Provides a source of energy for the body
   2. Leads to an increase in amino acids in the blood
   3. Takes place in the liver
   4. Produces ammonia as an end product

   A B C D E

6. PHYSICAL PROCESSES INVOLVED IN DIGESTION INCLUDE:
   1. Elimination of undigestable products
   2. Storing quantities of semidigested food
   3. Chewing of food in the oral cavity
   4. Rhythmic contractions that move the luminal contents along

   A B C D E

A	B	C	D	E
1,2,3	1,3	2,4	Only 4	All correct

7. FUNCTIONS OF VITAMINS INCLUDE WHICH OF THE FOLLOWING?
   1. Help regulate many metabolic processes
   2. Act as coenzymes in many reactions
   3. May act as important catalysts
   4. Produce substantial quantities of energy

   A B C D E

8. A POSITIVE NITROGEN BALANCE DOES NOT EXIST IN:
   1. Growing children
   2. Pregnant women
   3. Patients recovering from surgery
   4. Diseases which we classify as wasting

   A B C D E

9. WHICH OF THE FOLLOWING STATEMENTS IS(ARE) TRUE FOR SWALLOWING?
   1. Can be divided into two distinct stages
   2. Is controlled in part by voluntary and involuntary mechanisms
   3. A bolus takes 2-3 seconds to transverse the esophagus
   4. At one stage of swallowing the air passages are closed

   A B C D E

10. SALIVARY AMYLASE (PTYALIN):
    1. Is secreted by the parotid gland
    2. May continue to function in the stomach
    3. Acts specifically on starch
    4. Emulsifies fats in the oral cavity

    A B C D E

11. THE EXOCRINE SECRETION OF THE PANCREAS IS:
    1. In excess of 1,500 ml per day
    2. Controlled mainly by hormones
    3. Under control of two gastric hormones
    4. Continuous

    A B C D E

12. VITAMINS NEEDED BY THE BODY ARE OBTAINED FROM:
    1. The natural food stuffs we ingest daily
    2. Chemical reactions performed in the liver
    3. Bacterial reactions in the large intestine
    4. The synthesis of compounds in the pancreas

    A B C D E

13. THE PROCESSES INVOLVED IN ABSORPTION OF THE PRODUCTS OF DIGESTION INCLUDE:
    1. Osmosis
    2. Active transport
    3. Diffusion
    4. Filtration

    A B C D E

14. WHICH OF THE FOLLOWING IS(ARE) RESPONSIBLE, IN PART, FOR ABSORPTION TAKING PLACE IN THE SMALL INTESTINE?
    1. Increased length of this segment
    2. Products of digestion are present
    3. Excellent blood and lymphatic supply
    4. Increased absorptive surface of the mucosa

    A B C D E

------------------------ ANSWERS, NOTES AND EXPLANATIONS ------------------------

1. A, 1, 2, and 3 are correct. Gastric juice contains pepsin, a proteolytic enzyme, and HCl, an acid which makes it very acidic. Gastric secretions are regulated by both the nervous and hormonal mechanisms. The sight, smell, thought, or taste of food sends nerve signals to the dorsal nucleus of the vagus nerve which in turn stimulates the cells of the stomach to secrete

pepsinogen and HCl. In addition, a gastric hormone called gastrin is secreted into the blood stream, which in turn stimulates the cells of the stomach to secrete their enzymes. The mere presence of food in the stomach or the release of substances called secretagogues from certain foods causes gastrin to be released. Normally large amounts of viscid alkaline mucus are secreted by the mucous glands of the pyloric glands. This alkaline mucus helps neutralize the chyme before it enters the duodenum. Gastric secretions are inhibited by the enterogastric reflex. This reflex is initiated by distension of the small intestine, presence of acid in the duodenum, presence of protein breakdown products, or irritation of the duodenal mucosa.

2. D, Only 4 is correct. Glycolysis is an anaerobic (in the absence of oxygen) process which takes place in the cytoplasm of a cell. This anaerobic process converts glucose to pyruvic acid and in doing so produces a quantity of heat and energy in the form of two molecules of adenosine triphosphate (ATP). Glycolysis, an anaerobic process, is the first step in the catabolism of glucose. The pyruvic acid produced in this first step of glucose breakdown then enters the citric acid (Krebs') cycle and by this aerobic process produces more ATP and heat.

3. A, 1, 2, and 3 are correct. The citric acid (Krebs') cycle is the aerobic component of glucose metabolism and takes place in the mitochondria of all cells. This process breaks down and oxidizes pyruvic acid to yield carbon dioxide and water. This process, unlike glycolysis which functions in a non-oxygen atmosphere, will not function unless there is an adequate supply of oxygen. The most important function of Krebs' cycle is the production of energy in the form of ATP (adenosine triphosphate). One molecule of pyruvic acid produces 17 molecules of ATP. Therefore, since two molecules of pyruvic acid enter the cycle there is a total of 34 molecules of ATP produced. Since the anaerobic process (glycolysis) of glucose metabolism produces two ATP molecules then the process of glucose catabolism produces a total of 36 molecules of ATP for every single glucose molecule. Krebs' cycle may also utilize keto acid, a product of protein catabolism, as an alternate source of energy.

4. C, 2 and 4 are correct. Lipids are absorbed through the wall of the small intestine into the lacteals (lymphatic vessels) and small venules. In the blood stream the lipids are transported partly as chylomicrons and partly as lipoprotein complexes. Free fatty acids in the blood may also form a loose attachment with albumin.

5. E, All are correct. In protein metabolism, anabolism is primary and produces many substances such as protoplasm, blood proteins, hormones, and enzymes. Catabolism is secondary to anabolism. Proteins, in the form of amino acids, are broken down in two stages: the proteins are deaminated first in the liver cells with a resultant molecule of ammonia and a molecule of keto acid; and secondly, the keto acid may then enter the TCA cycle to be oxidized or it may be converted to glucose or lipids. So proteins, by entering the TCA cycle, may be a source of energy for the tissues of the body.

6. E, All are correct. All ingested food undergoes two basic types of activity to prepare it for absorption: mechanical (physical) and chemical. The physical activities are: 1) chewing the food, 2) storing quantities of food, 3) mixing the luminal contents, 4) moving the food along the tract, and 5) eliminating the undigestable foods or productions of digestion. The chemical activities involve the breaking down of complex compounds into smaller absorbable products. This process, called hydrolysis involves the combination of smaller products with water.

7. A, 1, 2, and 3 are correct. Vitamins are important components of many enzymatic or oxidative reactions. They may act as catalysts to speed up reactions which normally would not take place in the conditions dictated by the body. Some vitamins act as enzymes and others act as precursors to enzymes. Vitamins may also act as coenzymes which must be present for a given reaction to take place. Vitamins, however, do not provide a source of energy for the body.

8. D, Only 4 is correct. Protein anabolism and catabolism are normally in balance in the body. When this situation exists we talk of a nitrogen balance and in this case the amount of protein taken into the body equals that being eliminated. When the amount of protein catabolism exceeds the amount of protein anabolism, more nitrogen is lost from the body than is taken in and the condition is called negative nitrogen balance. This situation occurs in people on poor protein diets, starving, or suffering from tissue-wasting diseases. Positive nitrogen balance implies that more protein is taken into the body than is excreted. This situation exists in growing children, pregnant women, and patients recovering from surgery. In all these situations the proteins are either building new tissues or repairing old.

9. C, 2 and 4 are correct. Swallowing can be divided into three distinct stages. The first begins when the tongue detaches a portion of the contents of the mouth and thrusts it back into the pharynx. This first stage is a voluntary one. The movement of food into the pharynx marks the beginning of the second stage. Here the striated muscle of the pharynx closes the passage between the mouth and the nasopharynx to prevent regurgitation of food into the nose. Secondly, as the bolus passes the entrance to the larynx, the epiglottis folds over the glottis, and with the glottis closed respiration is briefly arrested. These events occur in less than a second. The third stage of swallowing is begun when the hypopharyngeal sphincter opens and allows the bolus to enter the esophagus. Sequential, ring-like contractions of the muscle of the esophagus form a peristaltic wave which pushes the bolus towards the stomach. The passage of food through the esophagus takes about 5-9 seconds. These last are mediated by impulses from the brain that are transmitted by fibers of the vagus nerve.

10. A, 1, 2, and 3 are correct. The presence of food in the mouth stimulates the secretion of saliva from three pairs of salivary glands and numerous small glands lying beneath the mucous lining of the mouth. The saliva from all of these glands contains ptyalin which initiates digestion by hydrolyzing starch. This enzyme is stable at pH values between four and 11 and continues to act upon starch in the portion of the gastric contents that has not become acidic. Pancreatic amylase also catalyses the break down of starch.

11. C, 2 and 4 are correct. Pancreatic juice secretion is controlled mainly by hormonal influences but nervous control also plays a part. When chyme enters the duodenum, it causes a release of secretin from the mucosa which stimulates the pancreas via the blood stream to secrete great quantities of water and bicarbonate. The duodenal mucosa also releases cholecystokinin which stimulates the release of the pancreatic enzymes. The water and especially the bicarbonate ions neutralize the acidic nature of the chyme entering the duodenum and also provides the optimal pH (8.0) for the function of the pancreatic enzymes. Parasympathetic innervation, via the vagal nerves, stimulate the pancreas to secrete moderate amounts of pancreatic enzymes. It should be remembered that the pancreas is continuously producing enzymes but only when it is stimulated does it put out large amounts of enzyme. The daily quantity of pancreatic juice secreted is about 1200 ml/day.

12. B, 1 and 3 are correct. Most vitamins needed for the body are obtained from the food we eat. However, a few are synthesized by the bacteria of the large

bowel and are absorbed into the blood stream. Vitamins are organic compounds which are required for normal growth, play a role in the transformation of energy, and the regulation of metabolism. They themselves produce no energy for the body.

13. A, 1, 2, and 3 are correct. Absorption through the gastrointestinal mucosa into the blood or lymph streams occurs by either osmosis or diffusion. These two processes are passive, in that they do not utilize energy produced by the cell. Absorption also occurs against an electrochemical gradient by the process of active transport, which utilizes energy produced by the cell.

14. E, All are correct. Most absorption of materials from the gastrointestinal tract takes place in the small intestine. The small intestine is structurally built to perform this important task. It is the longest segment of the digestive tract so its contents take a long time to completely pass through it. In addition to its length, its mucosa has many structures such as circular folds, villi, and microvilli which all increase the surface area of the small intestine for better absorption. Most of the gastrointestinal enzymes are secreted into the tubular system proximal to the jejunum, so that many products of digestion are available to absorption in the distal four-fifths of the small intestine. The small intestine also has a well developed venous (portal) and lymphatic system for carrying the products of digestion, either directly to the liver or to the venous system by the lymphatic vessels. All of these factors are important in normal absorption from the gastrointestinal tract.

FIVE-CHOICE ASSOCIATION QUESTIONS

DIRECTIONS: Each group of questions below consists of a numbered list of descriptive words or phrases accompanied by a diagram with certain parts indicated by letters, or by a list of lettered headings. For each numbered word or phrase, SELECT THE LETTERED PART OR HEADING that matches it correctly. Then insert the letter in the space to the right of the appropriate number. Sometimes more than one numbered word may be correctly matched to the same lettered part of heading.

1. _____ Involved in the synthesis of rhodopsin
2. _____ Synthesized by the bacteria of the large bowel
3. _____ Lack of which produces pellagra
4. _____ Involved in the maturation of red blood cells
5. _____ Promotes absorption of calcium through the intestinal wall

A. Vitamin K
B. Vitamin D
C. Vitamin A
D. Niacin
E. None of the above

6. _____ Produces two pyruvic acid and two ATP molecules
7. _____ The reversal of glycogenesis
8. _____ The process by which fats are converted to glucose
9. _____ The anabolism of glucose
10. _____ A hypoglycemic hormone

A. Glucagon
B. Gluconeogenesis
C. Glycolysis
D. Insulin
E. None of the above

------------------------ ANSWERS, NOTES AND EXPLANATIONS ------------------------

1. C. Carotene pigments are converted, on absorption, to vitamin A and it plays a role in the formation of rhodopsin or visual purple in the rods of the retina. Vitamin A is stored in the liver so that fresh liver, fish oils (cod liver) along with dark green and yellow vegetables, and fruits are excellent sources of this vitamin.

2. A. Enough vitamin K is produced by the bacteria of the large colon to provide the needs of the body. However, the needs of the body can be supplemented from foods such as liver, spinach, and cabbage. This vitamin acts as a coenzyme in the synthesis of prothrombin.

3. D. Severe niacin shortage in the diet produces the death of body tissues and is called pellagra. Niacin functions as important coenzymes (NAD, NADP) in the processes of utilizing foods for energy. Milder forms of this condition produce muscle weakness and poor glandular secretion.

4. E. Vitamin B_{12} is absorbed with the help of an intrinsic factor secreted by the mucosa of the stomach. It is actively involved in the maturation of red blood cells and when the intrinsic factor is missing, vitamin B_{12} is not absorbed. This condition leads to the decreased number of mature red blood cells and the clinical picture of pernicious anemia.

5. B. Vitamin D promotes the active transport of calcium through the intestinal wall, thus increasing calcium absorption. This vitamin is produced in the skin by the irradiation of 7-dehydrocholesterol by ultraviolet light.

6. C. Glucose catabolism consists of two successive sequences of events: glycolysis and the citric acid cycle (tricarboxylic acid cycle, Krebs' cycle or cellular respiration). Glycolysis consists of a number of intermediatory steps where glucose is converted, within the cytoplasm of a cell, to two pyruvic acid molecules and at the same time releases two molecules of ATP and a quantity of heat. The pyruvic acid molecules are combined with oxygen, in the citric acid cycle, to form carbon dioxide, water, heat, and most importantly, 34 molecules of ATP. These two processes together produce a total of 36 ATP molecules from a single mole of glucose. This large amount of energy is needed to carry out all the work of cells.

7. E. When an excess amount of glucose is being absorbed from the gut into the blood stream some of the glucose is converted to glycogen and stored mainly in liver and muscle cells. This process is called glycogenesis. The reverse of this process, glycogenolysis, converts glycogen back to glucose and takes place two or three hours after a meal when the blood glucose levels are below normal. Glycogenolysis takes place in the liver and adds the newly formed glucose molecules to the blood raising the glucose concentration back to its normal range. So these two processes are important components of maintaining blood glucose homeostasis. Glycogenolysis will only maintain the blood glucose at a fairly constant level for but a few hours, since the amount of glycogen stored in the body is minimal.

8. B. If, after high blood glucose levels have been reduced by converting some of the glucose to glycogen (glycogenesis) and to energy, there still remains an excess of glucose, this remaining portion will be converted to fat. This converted fat is stored in depots and the amount that can be stored is unlimited. Fats can be converted to glucose in the liver by a process called gluconeogenesis. Glucose can also be synthesized from proteins in the liver. These processes are also important factors in the homeostasis of blood glucose.

9. E. Glycogenesis, or the formation of glycogen from glucose, is a process of glucose anabolism. Cells take up glucose and by a series of reactions, requiring ATP, produce glycogen for storage. Glycogen molecules are not permanently stored in the cell, but with time are converted back to glucose-6-phosphate or back to glucose (glycogenolysis). Only liver, intestinal mucosal, and kidney tubule cells are capable of converting glycogen back to free glucose. Other cells convert glycogen to glucose-6-phosphate only and since they do not contain glucose phosphatase, are unable to complete the converting of glucose-6-phosphate to glucose.

10. D. Insulin is considered to be a hypoglycemic hormone because it tends to decrease the blood glucose levels. It is known that an insulin deficiency (diabetes mellitus) produces a decreased glycogenesis with a resultant low glycogen storage, decreased glucose breakdown and an increased level of blood glucose. Although the exact mechanism of insulin is not known, insulin is involved in glucose transport into the cell and in increasing the activity of glucokinase (an enzyme converting glucose to glucose-6-phosphate). So insulin aids the movement of glucose out of the blood stream to be utilized by cells of the body or to be stored as glycogen. Glucagon, epinephrine, growth hormone, ACTH and glucocorticoids are hyperglycemic hormones which tend to increase blood glucose.

15. THE URINARY SYSTEM

LEARNING OBJECTIVES

Be Able To:

- Describe the organs forming the urinary system and give the general functions of each.
- Describe the kidney with reference to the following: location, fascial layers, size, hilus and its contents, renal sinus, blood and nerve supply, and venous drainage.
- On a diagram of a frontal section of the kidney and ureter identify each of the following structures: capsule, cortex, renal column, medulla, renal pyramid, papilla, minor and major calyces, renal pelvis, and branches of the renal artery (interlobar, arcuate, and interlobular).
- Give the number of nephrons in the kidney and describe the microscopic structure of the nephron with reference to the following: renal corpuscle, Bowman's capsule, glomerular membrane, afferent and efferent arterioles, juxtaglomerular cells, macula densa, proximal convoluted tubule, loop of Henle, distal convoluted tubule, collecting tubule, and vasa recta.
- Describe the course, length, functions, and entrance of the ureters into the urinary bladder.
- Describe the anatomy of the urinary bladder and compare and contrast the anatomy of the female and male urethra.
- Discuss the innervation of the urinary bladder and urethra with reference to micturition.
- Describe the formation of urine under the following headings: glomerular filtration, active and passive transport in tubular reabsorption, tubular secretion, and tubular excretion.

  Describe the effects of antidiuretic hormone (ADH) and aldosterone on the formation of the urine and factors that produce diuresis.
- Describe the normal specific gravity, constituents, and daily output of urine.

- Describe the meaning of the following terms: cystitis, renal colic, hematuria, myelitis, renal calculi, pyuria, albuminuria (proteinuria), glycosuria, ketonuria, diuresis, oliguria, and polyuria.

RELEVANT READINGS

Anthony & Kolthoff - Chapter 16.
Chaffee & Greisheimer - Chapter 13.
Crouch & McClintic - Chapter 29.
Jacob, Francone & Lossow - Chapter 14.
Landau - Chapters 27 and 28.
Tortora - Chapter 21.

FIVE-CHOICE COMPLETION QUESTIONS

DIRECTIONS: Each of the following questions or incomplete statements is followed by five suggested answers or completions. SELECT THE SINGLE BEST ANSWER in each case and then circle the appropriate letter at the lower right of each question.

1. WHICH OF THE FOLLOWING IS NOT A FUNCTION OF THE KIDNEY?
   A. Activates vitamin C
   B. Is capable of synthesizing glucose
   C. Maintains the internal environment of the body
   D. Excretes the products of metabolic reactions
   E. Secretes a hormone involved with red blood cell maturation

   A B C D E

2. WHICH OF THE FOLLOWING STATEMENTS IS INCORRECT FOR THE MALE URETHRA?
   A. Lined by a mucous membrane
   B. Carries sperm to the external orifice
   C. The prostatic portion is about 7 cm long
   D. Its cavernous portion is the longest segment
   E. Its external sphincter can be closed voluntarily

   A B C D E

3. THE ARROW INDICATES THE _____.
   A. Renal pelvis
   B. Urethra
   C. Bladder
   D. Ureter
   E. None of the above

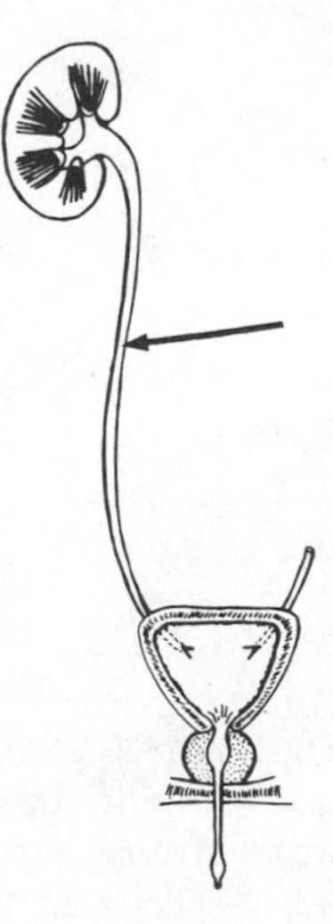

   A B C D E

SELECT THE SINGLE BEST ANSWER

4. THE NORMAL VOLUME OF URINE EXCRETED PER DAY IS ABOUT ______ LITERS.
   A. 0.75
   B. 1.50
   C. 2.25
   D. 3.00
   E. None of the above

   A B C D E

5. EPINEPHRINE FUNCTIONS AT WHICH OF THE FOLLOWING SITES?
   A. Podocytes
   B. Loop of Henle
   C. Proximal convoluted tubule
   D. Juxtaglomerular apparatus
   E. None of the above

   A B C D E

6. THE ARROW INDICATES THE _____.
   A. Glomerulus
   B. Proximal convuluted tubule
   C. Distal convoluted tubule
   D. Collecting duct
   E. Loop of Henle

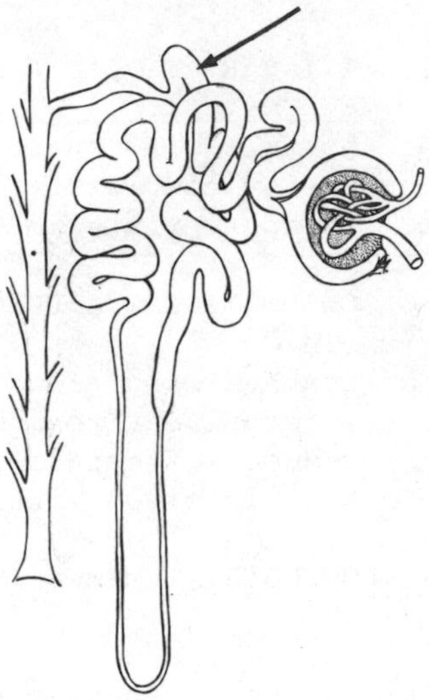

   A B C D E

7. THE KIDNEYS RECEIVE ABOUT ______ PER CENT OF THE TOTAL CARDIAC (BLOOD) OUTPUT PER MINUTE.
   A. 20
   B. 30
   C. 40
   D. 50
   E. None of the above

   A B C D E

8. WHICH OF THE FOLLOWING STRUCTURES IS NOT FOUND IN THE HILUM OF THE KIDNEY?
   A. Renal pelvis
   B. Renal arteries
   C. Renal nerves
   D. Renal veins
   E. Renal tubules

   A B C D E

9. THE PAPILLA PROJECTS INTO THE:
   A. Sinus
   B. Minor calyx
   C. Major calyx
   D. Renal pelvis
   E. None of the above

   A B C D E

10. WHICH OF THE FOLLOWING STATEMENTS IS NOT TRUE FOR THE URETER?
   A. Is a retroperitoneal structure
   B. Traverses the muscle layer of the bladder
   C. Connects the kidney with the urinary bladder
   D. Is approximately 25 cm in length
   E. Contains skeletal muscle in its wall

   A B C D E

SELECT THE SINGLE BEST ANSWER

11. THE RENAL COLUMN IS INDICATED BY _________.

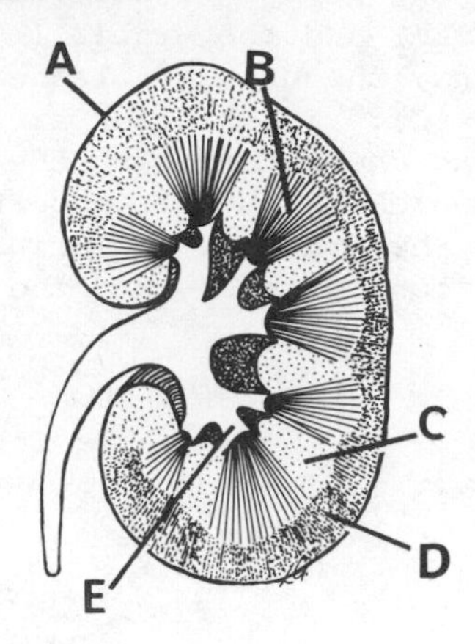

A B C D E

12. THE HORMONE INVOLVED WITH KIDNEY TUBULAR REABSORPTION OF SODIUM IS:
A. Renin
B. Thyroxine
C. Calcitonin
D. Testosterone
E. None of the above

A B C D E

------------------------ ANSWERS, NOTES AND EXPLANATIONS ------------------------

1. A. The kidneys have a number of functions, one of which is regulating the internal environment of the body. For example, the control of urinary water loss by the kidneys is the major automatic mechanism by which body water is regulated. The ions that determine the properties of the extracellular fluid such as sodium, potassium, calcium, magnesium, sulfate, phosphate, and hydrogen are all regulated mainly by the kidney. The kidney also acts in an excretory role, eliminating waste products of metabolism such as urea (protein catabolism), uric acids (nucleic acid catabolism), and creatine (from muscle creatine). The kidney also has excretory functions which eliminate foreign chemicals, such as drugs, pesticides, and food additives along with their metabolites. The kidney secretes two hormones: renin which is involved in raising blood pressure, and erythropoietin which is involved in the maturation process of red blood cells. The kidneys also activate vitamin D which is important in calcium absorption, normal calcification of bone, and development and maturation of teeth. The kidney is also a gluconeogenic organ and during prolonged fasting synthesizes glucose and releases it into the blood.

2. C. The male urethra unlike that of the female is also involved with the reproductive system by carrying spermatozoa to the external orifice of the penis. The male urethra consists of three segments: the prostatic, the membranous, and the cavernous (penile) portions. The prostatic portion (A in the diagram) passes from the bladder through the prostate (B) to the pelvic floor. The membranous portion (C) passes through the skeletal muscles making up the floor of the pelvis. A striated muscle sphincter surrounds this portion of the urethra. The last segment of the male urethra, the cavernous (penile) urethra (D), extends from the bulb of the urethra to the external orifice. The prostatic, membranous, and cavernous segments of the urethra, are about three, two, and 14 cm long, respectively.

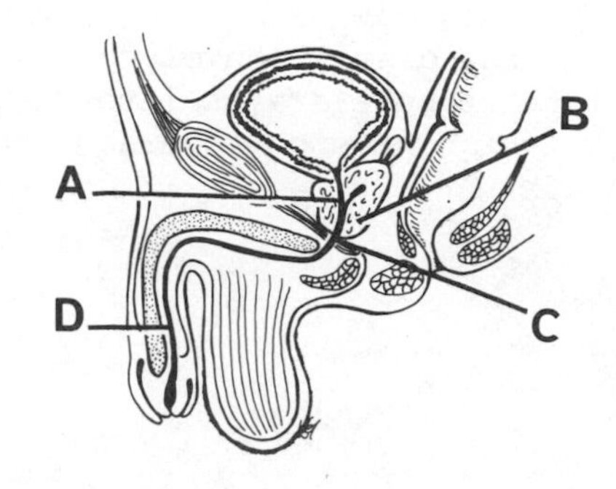

3. D. The urinary system consists of four basic components: two paired organs, the kidney (A in the diagram) and the ureter (D), and two single organs, the urinary bladder (B) and the urethra (C). The kidneys are responsible for producing and determining the nature of the urine, which is then transported out of the body by the ureters, the urinary bladder (stored for convenience), and the urethra.

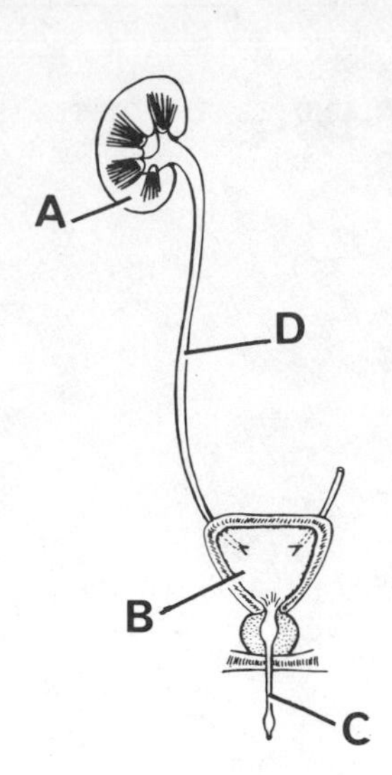

4. B. The normal volume of urine excreted per day by the kidneys is about 1.5 liters. A persistent increase in the daily volume of urine above this level is called polyuria and a temporary increase in the daily volume of urine is called diuresis. A decrease in the total volume of urine produced per day is called oliguria and if no urine is produced at all the condition is called anuria. Anuria, understandably, is incompatible to life.

5. D. Epinephrine functions at two sites in the kidney, the arterioles and the juxtaglomerular (JG) apparatus. If there is a decrease in the systemic arterial pressure, there is an increased sympathetic outflow to the kidneys as well as an increased level of circulating epinephrine from the suprarenal medulla. The sympathetic stimulation constricts the afferent arteriole which in turn stimulates renin secretion and activates the angiotensin system. The sympathetic neurons end in the immediate vicinity of the granular cells and the macula densa and they are thought to have an effect upon the JG apparatus and the secretion of renin.

6. C. This diagram illustrates the basic components of the nephron. The glomerulus (A in the diagram), formed from the afferent arterioles, lies within Bowman's capsule (F). Bowman's capsule opens into the first portion of the tubular system, the proximal convoluted tubule (B). The proximal tubule then enters into the loop of Henle (E) which is continuous with the distal convoluted tubule (C). The distal tubule drains into the collecting duct (D) which goes on to drain into the minor calyx.

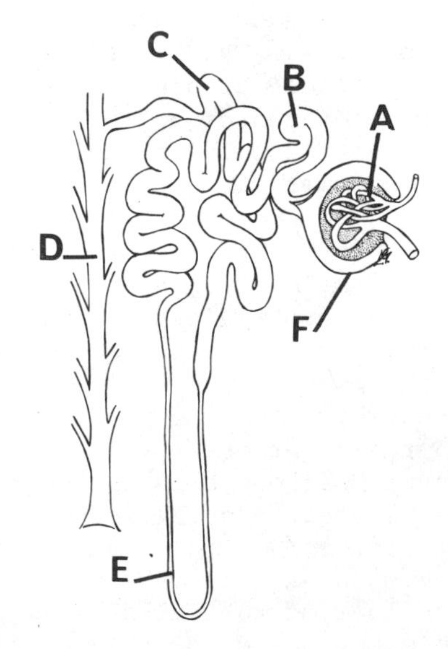

7. A. The total volume of blood flowing to the kidneys is about 1.1 liters per minute. The total cardiac output is approximately 5 liters per minute so the kidneys receive about 20-25 per cent of the total cardiac output. About 20 per cent of the blood reaching the kidney is filtered (filtration fraction)

through Bowman's capsule and the other 80 per cent passes via the efferent arterioles into the peritubular capillaries.

8. E. The medial border of the kidney contains a vertical fissure called the hilum. This fissure transmits the renal arteries and veins, renal nerves and the upper end of the ureter or renal pelvis. These structures entering or leaving the hilum constitute the renal pedicle.

9. B. The renal pyramids are cone-shaped masses making up the medullary portions of the kidney. The base of the pyramid lies adjacent to the renal cortex and its apex is directed towards the renal pelvis. The apex of the pyramid, the papilla, projects into a minor calyx. The papilla is studded with many openings of the collecting tubules which allows urine to enter the minor calyx. From here the urine passes into the major calyx and the renal pelvis, the upper portion of the ureter.

10. E. The ureter, a retroperitoneal structure, carries urine from the calyces of the kidney to the urinary bladder. It is a smooth muscle-containing tube about 25 cm long which at its distal end traverses the muscular (detrusor) wall of the bladder in an oblique fashion to enter the bladder. The ureter has both an abdominal and pelvic portion. It becomes constricted at three sites along its course: 1) where the renal pelvis narrows to enter the ureter proper, 2) where it crosses the pelvic brim, and 3) where it enters the wall of the bladder. Renal stones may lodge at these sites and produce urine stasis.

11. C. The substance of the kidney is composed of a centrally placed medulla and a peripherally placed cortex (D in the diagram). The medulla is made up of a number (8-18) of conical structures called renal pyramids (B). The pyramids have their base towards the cortex and their apices (papillae) project into the lumen of a minor calyx (E). Each pyramid is related laterally to downward extensions of cortical tissue called the renal columns (C). Each kidney is covered by a thin connective tissue capsule (A).

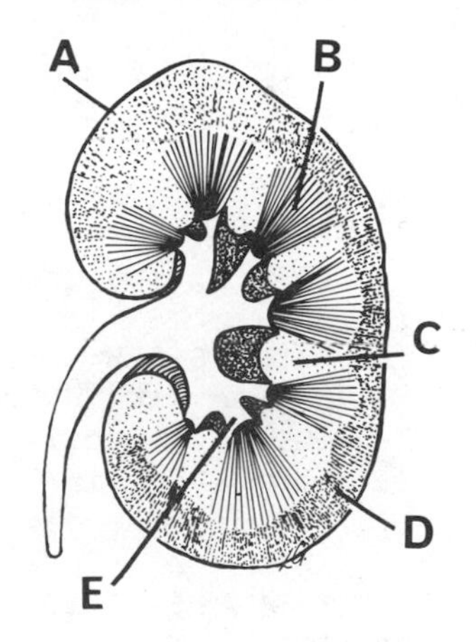

12. E. The hormone aldosterone and the other mineral corticoids secreted by the adrenal cortex, help maintain a proper balance of sodium in the body, controlling the tubular reabsorption of sodium. When an excessive amount of aldosterone is secreted, the active transport of sodium across the tubular membrane is high. Under these conditions potassium is excreted in large amounts in the urine. When aldosterone is secreted in decreasing amounts (Addison's disease) the tubular reabsorption of sodium declines and it is excreted. Along with the loss of sodium there is an important subsequent loss of body water. Under these conditions potassium is reabsorbed at a greater rate.

M U L T I - C O M P L E T I O N   Q U E S T I O N S

DIRECTIONS: In each of the following questions or incomplete statements, ONE OR MORE of the completions given is correct. At the lower right of each question, circle A if 1, 2, and 3 are correct; B if 1 and 3 are correct; C if 2 and 4 are correct; D if only 4 is correct; and E if all are correct.

1. THE CORTEX OF THE KIDNEY CONTAINS THE:
   1. Renal corpuscles (glomeruli)
   2. Distal convoluted tubules
   3. Proximal convoluted tubules
   4. First portions of the collecting tubules

   A B C D E

2. THE URINARY BLADDER:
   1. Lies anterior to the symphysis pubis
   2. Is lined by transitional epithelium
   3. Has visceral peritoneum on all of its surfaces
   4. Contains a smooth muscle layer called the detrusor

   A B C D E

3. WHICH OF THE FOLLOWING STATEMENTS IS(ARE) TRUE FOR THE JUXTA-GLOMERULAR APPARATUS?
   1. Is involved in controlling sodium reabsorption
   2. Includes the granular cells
   3. Consists of a specialized portion of the distal tubule
   4. Regulates the secretion of erythropoietin

   A B C D E

4. WHICH OF THE FOLLOWING STATEMENTS CONCERNING THE RENAL CORPUSCLE IS(ARE) CORRECT?
   1. Contains capillary loops which anastomose
   2. Encloses a space containing provisional urine
   3. Has a visceral layer formed by podocytes
   4. Includes the juxtaglomerular cells

   A B C D E

5. THE KIDNEYS ARE:
   1. Located within the peritoneal cavity
   2. Drained directly by the urethra
   3. About 18 cm in length
   4. Protected in part by the lower ribs

   A B C D E

6. WHICH OF THE FOLLOWING STATEMENTS IS(ARE) CORRECT FOR THE URINARY BLADDER?
   1. Aids in the expulsion of urine
   2. Becomes slightly distended when its volume reaches 500 ml
   3. Has three openings which pass through its wall
   4. Absorbs water through its epithelial lining

   A B C D E

7. THE FEMALE URETHRA HAS:
   1. Three unequal portions
   2. A length of about 8 cm
   3. An external opening posterior to the vaginal orifice
   4. A striated muscle sphincter surrounding its orifice

   A B C D E

8. WHICH OF THE FOLLOWING STATEMENTS IS(ARE) CORRECT FOR TUBULAR REABSORPTION?
   1. Glucose reabsorption has a well defined renal threshold
   2. Urea is reabsorbed by passive transport
   3. Most water is reabsorbed back into the circulating blood
   4. ADH accelerates the rate of water reabsorption

   A B C D E

A	B	C	D	E
1,2,3	1,3	2,4	Only 4	All correct

9. IN MICTURITION (URINATION) THE:
   1. External sphincter of the bladder is relaxed
   2. Reflex component causes the detrusor muscle to contract
   3. Micturition reflex may be modified by impulses from the brain
   4. Trigonal smooth muscle tone is overcome

   A B C D E

10. WHICH OF THE FOLLOWING CHARACTERISTICS WOULD FALL WITHIN THE NORMAL RANGE FOR A FRESH URINE SPECIMEN?
    1. Becomes cloudy on standing
    2. Specific gravity of 1.005
    3. An acidic pH
    4. A persistent amount of glucose

    A B C D E

11. THE AMOUNT OF ANTIDIURETIC HORMONE (ADH) SECRETED DEPENDS UPON WHICH OF THE FOLLOWING?
    1. Concentration of solute in the circulating blood
    2. Volume of water in the tubular filtrate
    3. Atrial blood pressure
    4. Permeability of distal convoluted tubules

    A B C D E

12. THE INTERNAL SPHINCTER OF THE BLADDER IS:
    1. Incorporated within the trigone region
    2. Composed of skeletal muscle
    3. Controlled by parasympathetic fibers
    4. Responsible for preventing constant dribbling of urine

    A B C D E

13. WHICH OF THE FOLLOWING IS(ARE) TRUE FOR GLOMERULAR FILTRATION?
    1. Large molecules (colloids) do not enter the urinary space
    2. Glomerular capillary hydrostatic pressure opposes filtration
    3. Takes place across the glomerular membrane
    4. About 100 liters of glomerular filtrate is produced per day

    A B C D E

ANSWERS, NOTES AND EXPLANATIONS

1. E, All are correct. The kidney can be divided into an outer cortex and an inner medulla. The following components of the nephron are located in the cortex: glomerulus, distal and proximal convoluted tubules, and the first or outer segments of the collecting tubules. The inner segments of the collecting tubules and the loop of Henle are located in the medulla.

2. C, 2 and 4 are correct. The urinary bladder lies behind the symphysis pubis (A in the diagram) in both males and females and in front of the uterus (C) in the female and in front of the rectum in the male. The bladder is covered superiorly by the visceral peritoneum (B) which attaches rather loosely to the bladder so that it will allow the urinary bladder to enlarge as it fills with urine. The wall of the bladder consists of smooth muscle called the detrusor and is lined by transitional epithelium. This epithelium is capable of flattening when the bladder is distended thus allowing the inner surface of the expanded bladder to be covered.

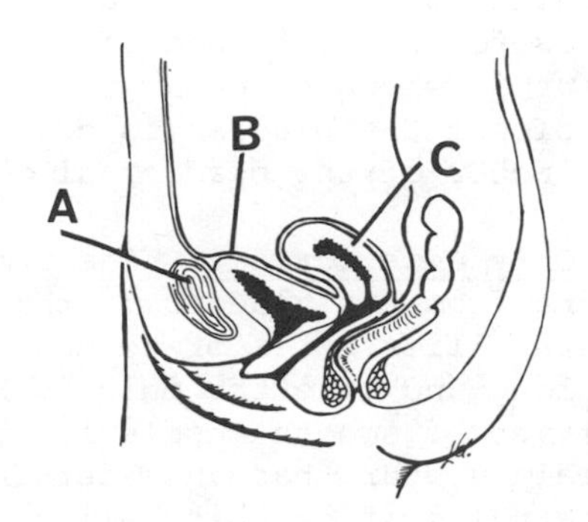

3. A, 1, 2, and 3 are correct. The juxtaglomerular apparatus is formed by the cells of the macula densa and the juxtaglomerular (granular) cells. The macula densa is a specialized portion of tubule that marks the transition from the ascending loop of Henle to the distal convoluted tubule. This segment lies between the afferent and efferent arterioles of its own glomerulus. The cells of the macula densa are always touching the granular cells of the afferent arteriole and are thought to contribute to the control of renin secretion. The granular cells are modified smooth muscle cells that secrete renin, a proteolytic enzyme. Angiotensinogen, which is synthesized and secreted by the liver and released into the blood, is activated to angiotensin I by renin. A converting enzyme activates angiotensin I to angiotensin II which is involved in vasoconstriction (increasing blood pressure), stimulating the outflow of the sympathetic nervous system, and acts upon the brain to stimulate thirst.

4. A, 1, 2, and 3 are correct. The renal corpuscle forms the first component of the nephron. The renal corpuscle is a double-walled cup consisting of an inner (visceral) layer of glomerular epithelium or podocytes, and an outer (parietal) layer of capsular epithelium, and a narrow space between the two called the capsular (Bowman's) space. The capsular space contains provisional urine which then flows into the next component of the nephron, the proximal convoluted tubule. In the center of this double-walled cup lies a capillary network which branches and anastomoses freely. Functionally, the blood enters this capillary network from the afferent arteriole and is separated from the fluid within Bowman's space by only the glomerular membrane. This membrane consists of capillary endothelial cells, a basement membrane, and podocytes of the visceral (inner) layer of Bowman's capsule. The juxtaglomerular cells do not form a part of the renal corpuscle but are both anatomically and functionally closely related to it.

5. D, Only 4 is correct. The kidneys are located outside the peritoneal cavity (retroperitoneal) of the abdomen, deep to the lower ribs which offer a measure of protection. The kidneys are about 12 to 13 cm long and are covered anteriorly by visceral peritoneum. The left kidney is usually slightly larger and higher than the right kidney. Each kidney is drained by a ureter (also a retroperitoneal structure) to the urinary bladder.

6. B, 1 and 3 are correct. The urinary bladder is a smooth muscle pouch whose mucous membrane is thrown into folds (rugae) when it is empty. It begins to become distended when about 250 ml of urine has accumulated. The bladder is lined by transitional epithelium that is capable of stretching to accommodate a large quantity of urine. The transitional epithelium does not reabsorb water from the urine. The floor area of the bladder is a smooth triangular area called the trigone. The anterior angle of the trigone is marked by an opening for the urethra and the two posterior angles of the trigone mark the sites for the entrance of the ureters. Distension of the bladder, beyond the point in which voluntary control prevents micturition, causes the detrusor muscle of the bladder to contract and a subsequent relaxation of the internal sphincter of the bladder allows the urine to enter the urethra.

7. D, Only 4 is correct. The female urethra is solely involved with carrying urine from the bladder to the external orifice. The external orifice of the urethra lies anterior to the vaginal opening. The female urethra is about 13 cm long and like the male is lined by a mucous membrane. The urethra also has a layer of smooth muscle in its wall which aids in moving the urine. The female urethra has an external sphincter of skeletal muscle surrounding its external orifice which allows one to voluntarily control the release of urine.

8. E, All are correct. Reabsorption removes many of the filterable plasma components from the tubule which were originally filtered at the glomerulus.

This reabsorption returns many of these components to the plasma so that they are either completely absent from the urine or are present in reduced quantities. Tubular reabsorption takes place either by passive mechanisms or by active transport as in the case of phosphate, sodium, and glucose. Reabsorption of chloride, urea, and water are carried out by passive mechanisms, although the reabsorption of water is dependent on the active reabsorption of sodium. About half of the urea is reabsorbed from the tubular filtrate because of a concentration gradient in the interstitial fluid. About 99 per cent of the water and sodium found in the tubular filtrate is reabsorbed. The antidiuretic hormone (ADH), secreted from the anterior pituitary, renders the cells lining the distal tubule and the collecting duct more permeable to water and therefore accelerates the rate of water reabsorption. In normal individuals, all glucose filtered across the glomerular membrane is reabsorbed by active transport. However, if there is an ever increasing amount of glucose filtered across the glomerular membrane, there is a point where glucose will appear in the urine. This is because the renal tubule can transport only limited amounts of glucose per unit time, primarily because the membrane carrier responsible for the transport becomes saturated.

9. E, All are correct. When the bladder contains enough urine (250 ml) to be distended, it initiates the stretch reflex by stimulating receptors in the bladder wall and proximal urethra. Sensory nerves conduct the impulses to the sacral cord and then reach the bladder again by way of the parasympathetic fibers. The reflex causes the detrusor muscle to contract by overcoming the internal sphincter (trigonal muscle) tone and relaxing the external sphincter of the bladder. Contraction of the bladder wall forces the urine out of the bladder through the relaxed internal and external sphincters. The micturition reflex is completely an automatic cord reflex, but it can be completely inhibited or facilitated by centers in the brain stem and the cerebral cortex.

10. A, 1, 2, and 3 are correct. Freshly voided urine is amber in color but usually becomes cloudy on standing. The normal kidneys are capable of varying the specific gravity of urine from 1.002-1.040. The pH of urine is usually acidic on normal diets but becomes alkaline on a pure vegetable diet. Glucose is a substance that has a well defined limit or renal threshold, which means that if a large amount of sugar (glucose) is consumed some will appear in the glomerular filtrate and will persist to show up in the urine. This is an example of alimentary glycosuria. Lactosuria may be found in pregnant or lactating women. So when glucose is detected in the urine these possible causes must be ruled out before diabetes mellitus is suspected.

11. B, 1 and 3 are correct. Antidiuretic hormone (ADH) secretion depends upon two factors: the atrial blood pressure and changes in body fluid osmolarity. An increase in atrial blood pressure triggers baroreceptors located in the right atrium and impulses resulting from this stimulation are transmitted via afferent nerves and ascending pathways to the hypothalamus, where they inhibit the ADH-producing cells of the pituitary. Osmoreceptors of the hypothalamus in some way, as yet unknown, detect changes in the osmolarity. An increase in osmolarity stimulates these receptors which then cause an increased rate of ADH production and secretion.

12. B, 1 and 3 are correct. The trigonal muscle composed of smooth muscle fibers surrounds the opening of the urethra and is called the internal sphincter of the bladder. This sphincter maintains tonic closure of the urethral opening until the pressure in the bladder rises high enough to overcome its tone and urine is allowed to escape into the urethra. The external sphincter of the bladder lies a few cm below the bladder and is formed by the skeletal muscle of the urogenital diaphragm. This external sphincter remains tonically contracted but it can be reflexively or voluntarily relaxed for urinating. Its

tonic contraction prevents constant dribbling of urine from the bladder. Parasympathetic fibers both excite the detrusor muscle of the bladder to contract and relax the internal sphincter to allow urine to leave the bladder.

13. B, 1 and 3 are correct. The glomerulus of the nephron behaves qualitatively like any other capillary of the body even though the glomerular membrane consists of a capillary endothelium, a basement membrane and a single layer of epithelial cells (podocytes). Filtration of the plasma takes place across this glomerular membrane and the major difference between the plasma of the glomerulus and the filtrate is that the filtrate contains little or no large molecules (colloids). The most important colloids of the plasma are the plasma proteins and they fail to cross the membrane. The force promoting filtration is the glomerular capillary hydrostatic pressure in Bowman's capsule and the colloid pressure in the glomerular capillary. The hydrostatic pressure of the glomerular capillary is higher than those forces opposing it, so there is a net filtration pressure forcing the plasma of the glomerulus through the membrane and into Bowman's space. Of course the plasma proteins remain within the glomerulus. The average volume of fluid filtered from the plasma into Bowman's capsule is about 170-180 liters or about 45 gallons per day. Remember that 1-2 liters of urine is produced per day so that about 99 per cent of the filtered water or about 168-178 liters are reabsorbed into the peritubular capillaries.

FIVE-CHOICE ASSOCIATION QUESTIONS

DIRECTIONS: Each group of questions below consists of a numbered list of descriptive words or phrases accompanied by a diagram with certain parts indicated by letters, or by a list of lettered headings. For each numbered word or phrase, SELECT THE LETTERED PART OR HEADING that matches it correctly. Then insert the letter in the space to the right of the appropriate number. Sometimes more than one numbered word may be correctly matched to the same lettered part or heading.

1. _____ A temporary increase in urine output
2. _____ Presence of pus in the urine
3. _____ A decrease in urine output
4. _____ Infection of the mucosa of the bladder
5. _____ Infection of the renal pelvis

A. Cystitis
B. Diuresis
C. Pyelitis
D. Oliguria
E. None of the above

6. _____ Site of greatest sodium reabsorption
7. _____ Becomes highly permeable to water in the presence of ADH
8. _____ Contains the macula densa
9. _____ Site of the countercurrent multiplier system
10. _____ Location of the granular cells

A. Distal convoluted tubule
B. Afferent arteriole
C. Proximal convoluted tubule
D. Loop of Henle
E. None of the above

1. B. Diuresis is a temporary increase in the amount of urine excreted. This condition may be due to a decreased amount of antidiuretic hormone (ADH) secreted by the anterior pituitary which makes the cells lining the distal convoluted tubule and collecting duct less permeable to water. Therefore there is more water in the urine, with a greater volume having a lower specific gravity or osmolarity. These characteristics are indicative of diabetes insipidus. Polyuria is the persistence of an increase in daily urine output.

2. E. The presence of pus in the urine is called pyuria. The presence of pus in the urine usually indicates an infection of the bladder or the ureter.

3. D. Oliguria is a term used to indicate a decrease in the daily output of urine as related to the fluid intake. Anuria is a total lack of urine formation.

4. A. A urinary tract infection may spread to the bladder, producing either an acute or chronic cystitis. Pyelonephritis is an infection of one or both kidneys and is usually brought on by a bacterial invasion. Acute pyelonephritis is extremely common and is usually benign.

5. C. Pyelitis is an inflammation of the pelvis of the kidney. It often produces tenderness or pain in the lower back and often bloody or purulent urine (pyuria).

6. C. The proximal convoluted tubule is the site of the greatest amount of sodium and water reabsorption from the nephron. About 65 per cent of the filtered sodium is reabsorbed from the filtrate by the time it has reached the end of the proximal tubule.

7. A. The distal convoluted tubule and the collecting duct both become highly permeable to water in the presence of antidiuretic hormone (ADH or vasopressin). The final urine volume is usually less than one per cent of the total filtered water. In the absence of ADH the tubular water permeability is very low and the normal sodium reabsorption takes place out of the tubule and duct. With the water not able to follow the sodium it remains to be excreted as a large volume of diluted urine.

8. A. The macula densa, a component of the juxtaglomerular apparatus, is found within the wall of the distal convoluted tubule. This segment of the tube lies in close proximity to the vascular pole of its own glomerulus and touches on the juxtaglomerular (granular) cells of the afferent arteriole. The macula densa is thought to be involved in controlling granular cell secretion of renin by sampling the sodium concentration of the early distal-tubular fluid. Renin is important in activating angiotensin, a powerful vasoconstrictor which tends to increase blood pressure.

9. D. The descending and ascending loop of Henle is involved in the countercurrent multiplier system which is involved in producing a highly concentrated urine. This system, by controlling the movements of NaCl and water in and out of the ascending and descending loop, concentrates the intercellular fluid of the interstitium of the medulla. The farther away one goes from the cortex, the higher the osmolarity of the interstitium becomes. The vasa recta of the loop helps to maintain this gradient. Simply put, the gradient allows the water to be reabsorbed from the distal convoluted tubule and the collecting duct into the interstitial spaces, provided ADH is present in adequate amounts. The removal of water from the urine both concentrates it and decreases the volume of the original glomerular filtrate.

10. B. The granular or juxtaglomerular cells are thought to be modified smooth muscle cells found in the wall of the afferent arteriole. These cells synthesize renin in the form of granules, which when secreted into the blood stream, activate angiotensin, a powerful vasoconstrictor that serves to increase blood pressure.

16. BODY FLUIDS AND ACID-BASE BALANCE

LEARNING OBJECTIVES

Be Able To:

- Describe the functions of body fluids and the balance between its intake and output.
- Compare and contrast the intracellular and extracellular body fluid compartments as to their: location, electrolyte composition, and volume.
- List the general functions of electrolytes found in body fluids.
- Describe the importance of sodium and potassium with reference to their maintenance of balance, depletion, and excesses.
- Describe the maintenance of calcium and phosphate balance, the abnormalities in their balance, and the functions of these two electrolytes.
- Describe the significance of magnesium, iron, and trace elements (minerals) in the body fluids.
- Define the term acid-base balance and describe the significance of buffer pairs.
- Discuss the importance of the carbonate buffer system in blood plasma.
- Describe the functions of the buffer systems and their relationships with respiration and kidney excretions.
- Describe mechanisms for producing metabolic acidosis and alkalosis and their consequences.

RELEVANT READINGS

Anthony & Kolthoff - Chapters 20 and 21.
Chaffee & Greisheimer - Chapter 19.
Crouch & McClintic - Chapter 21.
Jacob, Francone & Lossow - Chapter 16.
Landau - Chapter 28.
Tortora - None.

DIRECTIONS: Each of the following questions or incomplete statements is followed by five suggested answers or completions. SELECT THE SINGLE BEST ANSWER in each case and then circle the appropriate letter at the lower right of each question.

1. REGARDING THE BODY FLUID COMPARTMENTS WHICH OF THE FOLLOWING STATEMENTS IS CORRECT?
   A. The intracellular fluid volume equals 18 liters
   B. Plasma makes up a component of the extracellular compartment
   C. Interstitial fluid forms the smaller component of the extracellular volume
   D. Intracellular fluid represents about 30 per cent of body weight
   E. None of the above

   A B C D E

2. HYPERNATREMIA CAN BE DEFINED AS A(AN):
   A. Excessive retention of plasma sodium
   B. Reduction of concentration of plasma calcium
   C. Decrease in plasma potassium
   D. Increase in plasma phosphate levels
   E. None of the above

   A B C D E

3. A NORMAL PLASMA pH WOULD BE:
   A. 7.08
   B. 7.28
   C. 7.48
   D. 7.58
   E. None of the above

   A B C D E

4. WHICH OF THE FOLLOWING IS NOT A SOURCE OF WATER LOSS FROM THE BODY?
   A. Skin
   B. Lungs
   C. Stomach
   D. Kidneys
   E. Intestines

   A B C D E

5. WHICH OF THE FOLLOWING STATEMENTS DESCRIBES A BUFFER:
   A. Indicates that the pH of plasma is neutral
   B. Prevents marked changes of plasma pH
   C. Located only within the cytoplasm of cells
   D. Effective in neutralizing only strong acids
   E. None of the above

   A B C D E

6. THE LARGEST CONSTITUENT OF THE EXTRACELLULAR FLUID COMPARTMENT IS:
   A. Protein
   B. HCO_3^-
   C. Ca^{++}
   D. Cl^-
   E. Na^+

   A B C D E

7. HYPOVOLEMIA IS A(AN):
   A. Decreased blood volume below normal levels
   B. Increased interstitial volume above normal levels
   C. Decreased interstitial volume below normal levels
   D. Increased intracellular volume above normal levels
   E. None of the above

   A B C D E

ANSWERS, NOTES AND EXPLANATIONS

1. B. Body fluids can be divided into two compartments, that found within the cells or intracellular fluid, and that found outside the cells or extracellular fluid. The extracellular fluid compartment can be further divided into two compartments, that within the blood vessels or blood plasma and that outside the blood vessels or interstitial fluid. All three compartments differ as to

their composition and volume. The intracellular fluid compartment represents about 40 per cent body weight and has a volume of 28 liters. The interstitial fluid and blood plasma compartments represent 16 and four per cent body weight, or 11.2 and 2.8 liters, respectively.

2. A. Hypernatremia or excessive retention of plasma sodium occurs most frequently when the renal mechanisms for sodium excretion are disturbed. This condition causes many responses in the central nervous system such as irritability, increased reflex activity, or coma.

3. E. The normal plasma pH falls within the narrow range of 7.35 to 7.45. At this pH the number of hydroxyl ions slightly exceeds the number of hydrogen ions so the plasma is slightly basic. The narrow range of plasma pH is maintained by chemical buffers, and the acid-base functions of the lungs and the kidneys.

4. C. The body loses water from four sources: skin, lungs, kidneys, and intestines. Water is lost from the skin in the form of perspiration which functions to cool the body. The lungs lose a small amount of water by the moisture content in expired air. The intestine is also a normal source of water loss through the small amount of water which remains in the eliminated feces. The kidney tubules are by far the major source of water loss from the body. However, the amount of water loss in the urine is closely regulated to the intake of water into the body.

5. B. A buffer is a substance that prevents marked changes in the pH of a solution when an acid or a base is added to it. Buffers consist of two kinds of chemical substances and are often referred to as "buffer pairs". Most buffers of the body fluids consist of a weak acid and a salt of that acid. They function by replacing a strong acid (or base) with a weak acid (or base). In this way instead of a strong acid remaining in the solution and contributing many H+ to the solution to drastically lower the pH of the solution, the weaker acid replacing it will contribute less H+ to the solution and so the pH will fall only slightly. The same mechanism takes place when a weak base buffers a strong base and prevents marked changes in the pH of the solution.

6. E. The largest constituent of the extracellular (interstitial and plasma) fluid compartment is the sodium cation. Very little of it, if any, is found within the intracellular compartment of the body. The most abundant cation of the intracellular fluid compartment is potassium. In comparing the two compartments of the extracellular fluid, the blood plasma has more total electrolytes and protein than does the interstitial compartment. The most abundant anion of the extracellular fluid is chlorine and it is found in neglible amounts in the intracellular fluid.

7. A. Hypovolemia means a decreased blood volume below the normal level. The most common cause of hypovolemia is the hemorrhage of blood out of the circulatory system. The hemorrhaging of great quantities of blood decreases the systemic filling pressure and also produces a decrease in the venous return. As a result, the cardiac output falls below normal and hypovolemic shock occurs. Two other causes of hypovolemic shock are plasma loss from severe skin burns and dehydration from excessive sweating, diarrhea, vomiting, and kidney tubule damage.

MULTI-COMPLETION QUESTIONS

DIRECTIONS: In each of the following questions or incomplete statements, ONE OR MORE of the completions given is correct. At the lower right of each question, circle A if 1, 2, and 3 are correct; B if 1 and 3 are correct; C if 2 and 4 are correct; D if only 4 is correct; and E if all are correct.

1. WHICH OF THE FOLLOWING SOURCES OF WATER LOSS IS(ARE) ADJUSTED TO THE INTAKE VOLUME?
   1. Intestine
   2. Lungs
   3. Skin
   4. Kidneys

   A B C D E

2. ELECTROLYTES:
   1. Help maintain electroneutrality of body fluids
   2. Maintain osmotic equilibrium between fluid compartments
   3. Regulate neuromuscular activities
   4. Provide a milieu for chemical reactions to take place

   A B C D E

3. WHICH OF THE FOLLOWING STATEMENTS IS(ARE) CORRECT FOR THE TERM pH?
   1. Indicates the degree of acidity or alkalinity
   2. Concentration of hydrogen ions in a solution
   3. A pH of 7 represents neutrality
   4. A high pH indicates more OH^- than H^+

   A B C D E

4. MECHANISMS INVOLVED IN ACID-BASE BALANCE IS(ARE):
   1. Elimination by the intestines
   2. Renal secretions of acids and bases
   3. Secretion through the skin
   4. The actions of buffer pairs

   A B C D E

5. WHICH OF THE FOLLOWING GENERAL STATEMENTS IS(ARE) TRUE FOR FLUID BALANCE?
   1. Blood volume is maintained over interstitial fluid volume
   2. Input control is more important than output control
   3. Ideally intake equals output
   4. Fluid distribution in the body is of little importance

   A B C D E

6. CORRECT STATEMENTS CONCERNING BUFFERS INCLUDE:
   1. Are capable of maintaining pH homeostasis by themselves
   2. Usually work in pairs
   3. Function only in extracellular fluids
   4. A strong base is replaced by a weak base

   A B C D E

7. EDEMA:
   1. Tends to be a self-limiting process
   2. May be caused by an increase in effective filtration pressure
   3. Can be produced by the retention of large amounts of salts
   4. Is defined as water retention by the cells of the body

   A B C D E

8. WHICH OF THE FOLLOWING STATEMENTS IS(ARE) CORRECT FOR THE INTRA-CELLULAR VOLUME?
   1. Equals about 28 liters
   2. Represents about 20 per cent of body weight
   3. Is larger than the interstitial fluid volume
   4. Is smaller than the plasma volume

   A B C D E

A	B	C	D	E
1,2,3	1,3	2,4	Only 4	All correct

9. WHICH OF THE FOLLOWING STATEMENTS IS(ARE) CORRECT FOR THE MECHANISMS OF RESPIRATION INVOLVED IN pH CONTROL?
   1. A blood pH above 7.45 causes hypoventilation
   2. Decreased blood pH below normal stimulates respiration
   3. Involves chemoreflexes of the carotid body
   4. Prolonged hyperventilation will produce alkalosis

A B C D E

------------------------ ANSWERS, NOTES AND EXPLANATIONS ------------------------

1. D, Only 4 is correct. The kidney is the only organ which adjusts the amount of water it loses, in the urine, to the volume of water taken into the body. Since the glomerular filtration rate is fairly constant, the water loss is controlled by the rate of tubular reabsorption. Two hormones, ADH and aldosterone, are involved in determining the rate of water reabsorbed by the kidney tubules. For example, when one drinks a large volume of fluid, there is a rapid increase in the volume of urine excreted following a short delay of about one-half hour. This delay is the time that it takes to destroy the ADH present at the time the fluid is taken in. This phenomenon is called water diuresis. The decreased ADH level inhibits tubular reabsorption of water and so the large amount of fluid taken in is lost to the urine.

2. E, All are correct. Electrocytes, both anions and cations, of the body fluid compartments are essential for normal functioning of the body. In general, they are directly involved in maintaining the electroneutrality of the body fluids and maintaining the osmotic equilibrium between the extracellular and intracellular fluids. These electrolytes also maintain proper conditions for chemical reactions to take place within the body and also help regulate neuromuscular activities. Many of the electrolytes, besides having these general functions, also perform many specialized tasks.

3. E, All are correct. The term pH is an indication of the hydrogen ion (H+) concentration in a solution. The term pOH indicates the hydroxyl ion (OH^-) concentration of a solution. The sum of pH and pOH equals 14. The smaller the pH number, the greater the number of hydrogen ions or the more acidic the solution. A pH of 7 indicates a neutral solution or one containing equal numbers of hydrogen and hydroxyl ions. The higher the pH the smaller the number of (H+) found in a solution and the more basic it is.

4. C, 2 and 4 are correct. The mechanisms involved in maintaining the acid-base balance of the extracellular and intracellular compartments are the chemical buffers, the respiratory system, and the kidneys. The chemical buffer pairs exchange weak acids for strong acids (reducing the H+ concentration) and exchange weak bases for strong bases (reducing the OH^- concentration). Both of these exchanges tend to push the pH of a solution towards neutrality. The respiratory system regulates the acid half of the carbonate pair (H_2CO_3) by forming CO_2 and H_2O and blowing off the CO_2. The kidney regulates the $NaHCO_3$ (base half of the carbonate pair) level by exchanging (H+) for the Na+ so that (H+) is secreted and the Na+ is returned to the blood.

5. B, 1 and 3 are correct. Fluid balance or homeostasis can only be maintained if the amount of fluid entering (input) the body equals the amount of fluid being lost (output) from the body. So ideally intake equals output. The mechanisms that are used to vary the output of body fluids are crucial in maintaining homeostasis. However, those mechanisms adjusting intake are not as crucial as those adjusting output. The fluid balancing devices functioning between the

three fluid compartments take place very rapidly. The blood (plasma) volume is very important and is maintained at the expense of the interstitial fluid volume.

6. C, 2 and 4 are correct. Most buffer pairs consist of a weak acid and a salt of that acid and help to maintain the pH of all body fluids within a narrow range. They function by replacing a strong base (or acid) with a weak base (or acid) so that effect of the strong base (or acid) being added to a solution will be minimized in its effect on the pH of the resultant solution. Buffering alone cannot maintain homeostasis of pH since (H+) are continually being added to the blood as a result of metabolic processes of the body. The body must have a method of disposing of excess hydroxyl ions. The buffer pairs function concurrently with the lungs and kidneys to remove them from the blood. Only in this way can the body maintain its blood pH within narrow limits (7.35-7.45).

7. A, 1, 2, and 3 are correct. Edema is the accumulation of fluid within the interstitial spaces of the body. A change in either effective filtration pressure or oncotic pressure can upset the dynamic equilibrium between the blood and the interstitial fluid compartments. If the effective filtrate pressure exceeds the oncotic pressure, edema will occur. The effective filtration pressure tends to rise in certain cardiac conditions or venous obstructions and often produces edema. On the other hand oncotic pressure may fall in some kidney diseases and edema may ensue. The retention of large amounts of salt (Na+) within the body is usually accompanied by the retention of large amounts of water, thus producing edema. With the influx of fluid into the interstitial spaces and the resultant rise in pressure, an equilibrium soon is reached so that the edema fluid will remain stable, thus edema is a self-limiting process.

8. B, 1 and 3 are correct. The intracellular volume at 28 liters is larger than the combined interstitial fluid (11.2 liters) and plasma volumes (2.8 liters) making up the extracellular volume. The intracellular fluid represents about 40 per cent body weight, whereas the interstitial fluid and plasma represent about 16 and four per cent, respectively. Each of these fluid compartments are separated from each other by cellular membranes and any changes taking place in the volumes of these fluids obey the laws governing membrane transport.

9. E, All are correct. The neurons of the respiratory centers of the brain and the chemoreflexes of the carotid body are two devices by which respirations adjust to blood pH and in turn adjust pH. For example, neurons of the respiratory center detect changes in the CO_2 and pH levels of the blood and respond accordingly. Respirations are increased in both number (hyperventilation) and depth, when the CO_2 level of the blood rises beyond certain levels or arterial blood pH falls below 7.38. These respiratory changes will eliminate more CO_2, reduce carbonic acid and hydrogen ions, and increase the pH of the blood back to normal levels. Prolonged hyperventilation will cause the pH of the blood to rise above 7.45, thereby producing alkalosis. Normally, a blood pH about 7.45 will have a negative effect upon the respiratory centers and will produce a decrease in respiration or hypoventilation. Hypoventilation produces a fall in the pH of the blood and if it carries on too long will produce acidosis.

DIRECTIONS: Each group of questions below consists of a numbered list of descriptive words or phrases accompanied by a diagram with certain parts indicated by letters, or by a list of lettered headings. For each numbered word or phrase, SELECT THE LETTERED PART OR HEADING that matches it correctly. Then insert the letter in the space to the right of the appropriate number. Sometimes more than one numbered word may be correctly matched to the same lettered part or heading.

1. _____ Contains the largest amount of protein
2. _____ Due to increased amounts of parathyroid hormone
3. _____ Site of highest K+ concentration
4. _____ Excessive amounts of potassium produces
5. _____ Contains a minimal amount of protein
6. _____ Excess sodium in the plasma is called

A. Cell fluid
B. Hyperkalemia
C. Interstitial fluid
D. Hypercalcemia
E. None of the above

------------------------ ANSWERS, NOTES AND EXPLANATIONS ------------------------

1. A. The intracellular fluid compartment of the body contains the largest amount of protein found in the three fluid compartments. The plasma compartment also contains an appreciable amount of protein anions, whereas the interstitial fluid contains hardly any protein anions. This is the major difference between the composition of the two compartments making up the extracellular fluid compartment.

2. D. An increased secretion of parathyroid hormone causes an elevation in the amount of calcium in the plasma or hypercalcemia. The most common cause for hypercalcemia is a parathyroid hormone secreting tumor. Hypercalcemia produces a general depression of the nervous system with reduced reflex activities, and the depression of muscle contractility which produces skeletal muscle weakness. It also produces a loss of appetite, constipation, and may be responsible for the induction of kidney stones. Hypocalcemia is a reduced plasma calcium level. Low calcium levels increase the excitability of neurons which then produce muscle tetany. Muscle tetany is often fatal.

3. A. The major cation of the intracellular fluid is potassium, whereas the major cation of the two extracellular compartments is sodium.

4. B. Potassium excess in the body is called hyperkalemia and develops when the kidney tubules are unable to excrete adequate amounts. This may occur when the aldosterone levels fall or the kidneys are severely damaged as in acute renal failure. Hypokalemia, a decrease in the plasma concentration level of potassium, may occur because of a shift in potassium from the extracellular fluid compartment to the intracellular fluid compartment.

5. C. The interstitial fluid compartment contains little, if any, protein. Most protein (anions) are found in the intracellular compartment although there are appreciable amounts in the plasma component of the extracellular fluid compartment.

6. E. Hypernatremia or excess levels of sodium in the plasma occurs when the mechanisms of sodium excretion in the kidney are disturbed. Hypernatremia produces central nervous system responses such as: increased reflex activities,

increased body temperatures, mental impairments, irritability and finally coma. Hyponatremia or loss of body sodium is usually caused by massive losses of sodium from the gastrointestinal tract.

17. THE REPRODUCTIVE SYSTEM

LEARNING OBJECTIVES

Be Able To:

- Give the general functions of the reproductive system and list the major (primary) and accessory (secondary) organs of this system in the female and male.
- Describe the anatomy of the uterus with reference to the following: size, shape, named parts, cavities, layers forming the wall, blood supply, location, and ligaments which support it.
- Describe the parts of the uterine tube and give its functions.
- For the ovary describe its size, location, structure, support, and the stages of development of an ovarian follicle.
- Describe the location, structure and functions of the vagina and female external genitalia, including the perineum.
- Describe the female breast as to its location, structure, function, and the mechanisms involved in controlling lactation.
- Compare and contrast oogenesis and spermatogenesis with special emphasis on the importance of meiosis.
- For each of the following female sexual cycles discuss the functions, the structures involved, and the mechanisms involved in their control: ovarian, endometrial (menstrual), myometrial, and gonadotropic.
- Define the terms menarche and menopause.
- Describe the testes under the following headings: size, location, structure, function, and the effect of anterior pituitary hormones.
- Describe the location, structure, and functions of the epididymis, efferent ductules, vas deferens (ductus deferens), ejaculatory duct, seminal vesicles, prostate gland, and bulbourethral glands.
- Describe the structure and function of the following: scrotum, penis, and spermatic cord.

- Describe the composition of semen and the factors involved in male fertility.

- Define the following terms: phimosis, epispadias, hypospadias, circumcision, cryptorchism, episiotomy, hermaphrodite, and pseudohermaphrodite.

RELEVANT READINGS

Anthony & Kolthoff - Chapters 17, 18, and 19.
Chaffee & Greisheimer - Chapter 21.
Crouch & McClintic - Chapters 30 and 33.
Jacob, Francone & Lossow - Chapter 17.
Landau - Chapters 30 and 31.
Tortora - Chapter 18.

FIVE-CHOICE COMPLETION QUESTIONS

DIRECTIONS: Each of the following questions or incomplete statements is followed by five suggested answers or completions. SELECT THE SINGLE BEST ANSWER in each case and then circle the appropriate letter at the lower right of each question.

1. SPERMATOZOA ENTER THE HEAD OF THE EPIDIDYMIS FROM THE:
   A. Rete testis
   B. Efferent ductules
   C. Ductus deferens
   D. Tunica vaginalis
   E. Seminiferous tubules

   A B C D E

2. THE STRUCTURE REMAINING IN THE OVARY FOLLOWING THE RELEASE OF THE OVUM IS THE ______.
   A. Corpus luteum
   B. Primary follicle
   C. Corpus albicans
   D. Mature follicle
   E. None of the above

   A B C D E

3. THE CERVIX OF THE UTERUS IS INDICATED BY ______.

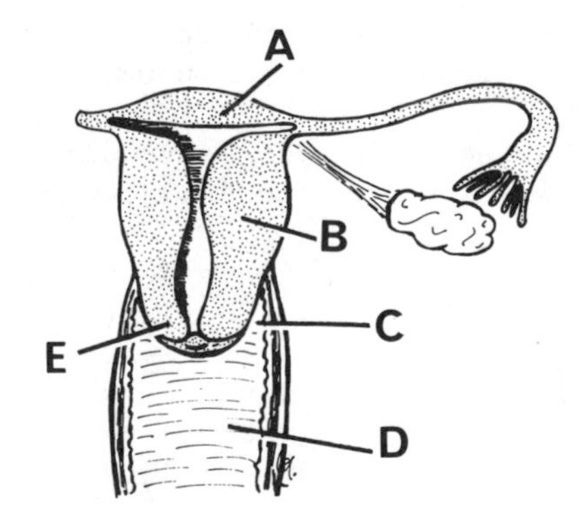

   A B C D E

4. WHICH OF THE FOLLOWING WOULD BE INDICATIVE OF TESTOSTERONE HYPOSECRETION?
   A. Increased muscle development in the shoulder
   B. A thickening of the long bone
   C. The development of a beard
   D. A deepening of the voice
   E. A small penis

   A B C D E

SELECT THE SINGLE BEST ANSWER

5. THE ARROW INDICATES THE ________.
   A. Seminiferous
   B. Rete testis
   C. Duct of epididymis
   D. Ductus deferens
   E. Efferent ductules

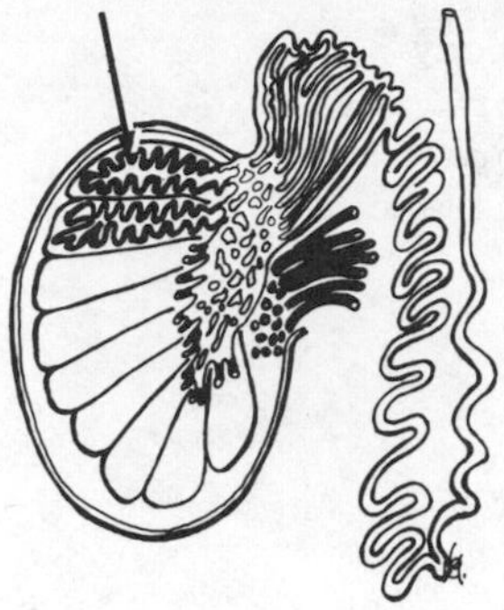

A B C D E

6. THE UTERINE ARTERY IS A BRANCH OF THE ______ ARTERY.
   A. Lumbar
   B. Ovarian
   C. External iliac
   D. Internal iliac
   E. None of the above

A B C D E

7. WHICH OF THE FOLLOWING IS NOT CONSIDERED TO BE PART OF THE EXTERNAL GENITALIA?
   A. Cervix
   B. Clitoris
   C. Mons pubis
   D. Labia minora
   E. Greater vestibular glands

A B C D E

8. THE COMPONENT OF THE DUCT SYSTEM EMPTYING INTO THE URETHRA IS THE ________.
   A. Rete testis
   B. Vas deferens
   C. Seminiferous tubules
   D. Ductus epididymis
   E. Ejaculatory duct

A B C D E

9. IN THIS CROSS-SECTIONAL DRAWING OF THE PENIS THE CORPUS CAVERNOSUM IS INDICATED BY ______.

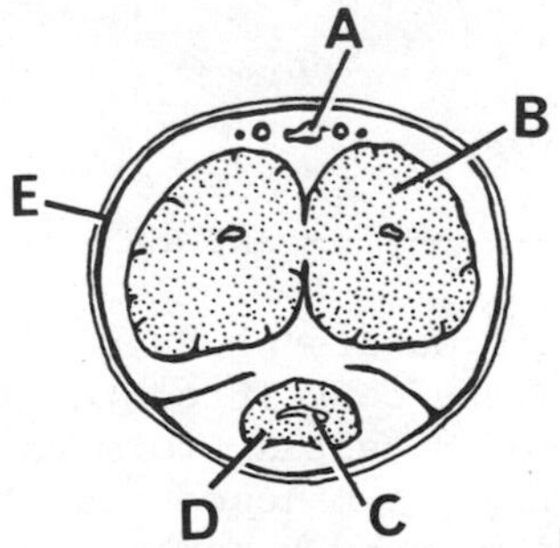

A
B
E
D
C

A B C D E

10. SUCKLING STIMULATES THE SECRETION OF:
   A. Follicular stimulating hormone
   B. Estrogen
   C. Progesterone
   D. Luteinizing hormone
   E. None of the above

A B C D E

SELECT THE SINGLE BEST ANSWER

11. THE ARROW INDICATES THE _____.
    A. Clitoris
    B. Perineal body
    C. Labia majora
    D. Labia minora
    E. None of the above

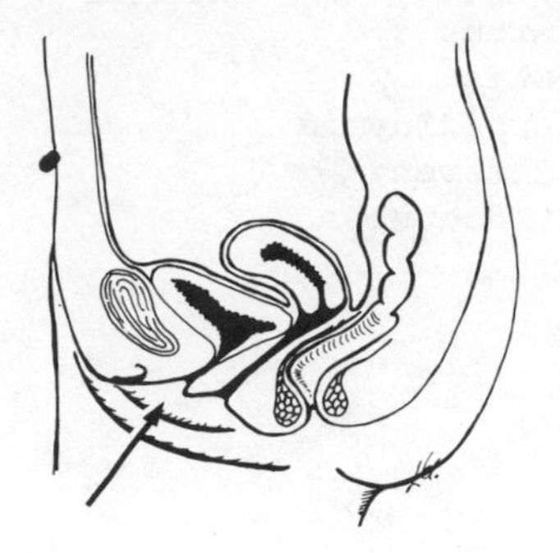

A B C D E

12. THE GLANS PENIS IS INDICATED BY _____.

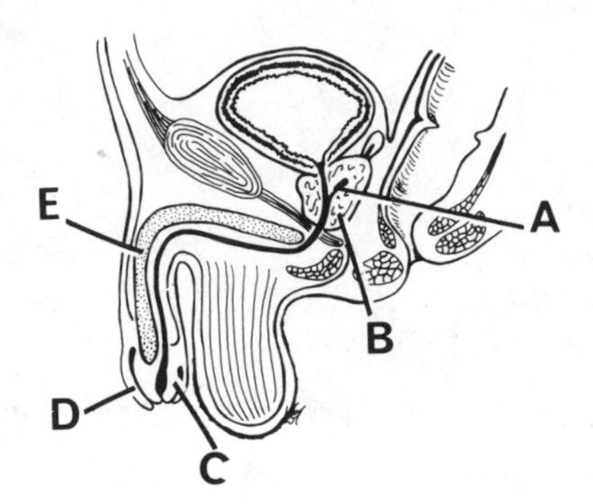

A B C D E

13. A SECONDARY SEX ORGAN IS THE _______.
    A. Testis
    B. Uterus
    C. Epididymis
    D. Prostate
    E. Vagina

A B C D E

------------------------ ANSWERS, NOTES AND EXPLANATIONS ------------------------

1. B. Spermatozoa are produced in the convoluted seminiferous tubules of the testes. These convoluted tubules become straight tubules and in the mediastinum of each testis, form an elaborate network of canals called the rete testis. The spermatozoa pass from the seminiferous tubules into the rete testis and then by a number of efferent ductules, enter the head of the epididymis. The duct of the epididymis, lying on the posterior margin of the testis, conveys the spermatozoa to the ductus deferens which makes up part of the spermatic cord. The tunica vaginalis is not a tubular structure of the testis but forms its outer layer.

2. A. The primary follicles, consisting of an ovum and a ring of follicular cells, are found in great numbers in the ovary of a child. Most of these primary follicles will degenerate (atresia) with time, but under the influence of FSH, some of these primary follicles will mature to form secondary follicles. The ovum at primary follicle stage is called a primary oocyte and it undergoes a meiotic (reduction) division to form a haploid secondary

oocyte. The secondary oocyte begins to undergo a normal mitotic division which is completed later at ovulation. The ovum is released from the follicular cells and may either become fertilized or it simply may be lost. The corpus luteum consists of the follicular cells that remain at the surface of the ovary after the ovum has been extruded. The future of the corpus luteum depends on whether the ovum is fertilized or not. If the ovum does not become fertilized the corpus luteum grows for about 12 days and then regresses to eventually form the corpus albicans, a small amount of scar tissue. If the ovum becomes fertilized, the corpus luteum forms the corpus luteum of pregnancy and secretes large amounts of progesterone to prepare the endometrium of the uterus for the implantation of the zygote. Progesterone secreted by the corpus luteum of pregnancy is also responsible for inhibiting muscular contractions of the uterine wall, prevents menstruation and further ovulation, and later helps to promote the development of secretory tissue in the breast. The luteinizing hormone (LH) of the pituitary is responsible for the development of the corpus luteum of pregnancy.

3. E. The uterus is a pear-shaped structure which measures about 7.5 cm long and 5 cm wide. The uterus can be divided into an upper part called the body (B in the diagram) and a lower part called the cervix (E). That portion of the body lying above the entry of the uterine tubes into the uterus is called the fundus (A). At the other end of the uterus the cervix projects downward into the upper portion of the vagina (D) and the opening of the cervix is called the external os. On either side of that portion of the cervix projecting into the vagina there is a space called the fornix (C).

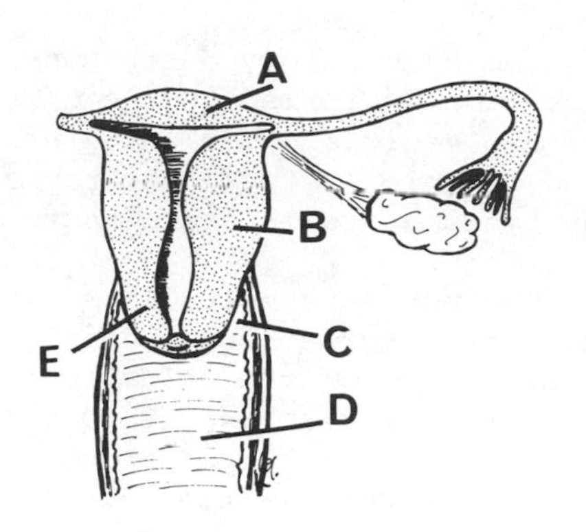

4. E. Testosterone secretion, after puberty, causes the penis, testis, and scrotum to increase in size until about the age of 20. Testosterone is also responsible for the development of the secondary sexual characteristics in the male. The secondary sex characteristics stimulated by normal testosterone levels include: the masculine distribution of body hair (pubic, axillary, trunk, and limbs), the involvement in baldness, the deepening of the voice due to laryngeal hypertrophy and enlargement, the increase in skeletal muscle mass, the increase in bone growth and calcium retention, and the closing of the epiphyses of the shafts of long bones. All of these characteristics of "maleness" would fail to develop with a hyposecretion of testosterone.

5. A. The lobes of the testis contain narrow coiled tubes called seminiferous tubules (A). These tubes are the site of spermatogenesis and in cross-section the mature spermatozoa would be found in the center of the lumen and the spermatocytes (germ cells) are located at the periphery of the tube. The seminiferous tubules pass into the mediastinum of the testis where they enter an elaborate network of canals called the rete testis (B). These canals flow into the straight efferent ductules (E) that enter the head of the epididymis and eventually form the duct of the epididymis (C). The epididymis lies at the

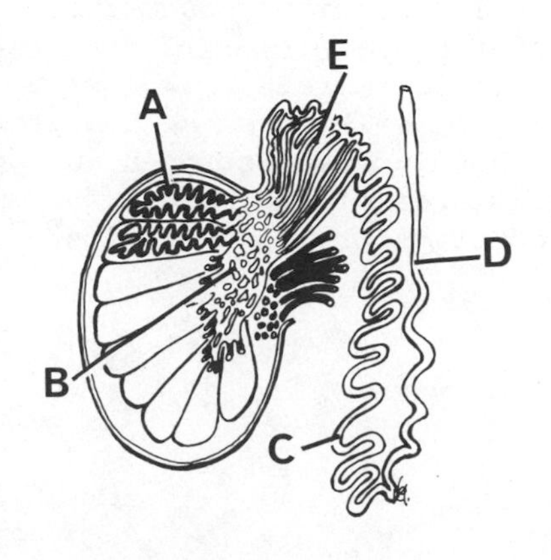

posterior margin of the testis and passes inferiorly where it joins with the ductus deferens (D) of the spermatic cord.

6. D. The uterine arteries are usually separate branches of the internal iliac arteries. It is homologous to the artery of the ductus deferens in the male. Each artery supplies the great needs of the uterus for blood and has branches which also help to supply the vagina, the uterine tube, and the ovary. These three structures have their own arterial supply but receive additional blood supply from the uterine artery. These structures all have a very good blood supply.

7. A. The external genitalia (vulva) of the female consist of: the mons (veneris) pubis, the labia minora and majora, the clitoris, and the greater vestibular glands. The external genitalia, along with the mammary glands, are considered to be accessory structures of the female reproductive system. The primary sex organs of the female are considered to be the ovaries, the uterine tubes, the uterus, and the vagina.

8. E. The ejaculatory duct opens into the prostatic segment of the urethra. This duct is formed by the union of the ductus deferens and the duct of the seminal vesicle. The entry of the ejaculatory duct into the urethra marks the site of union between the reproductive and urinary systems in the male. These two systems are separate in the female.

9. B. The penis is normally a flaccid structure, consisting of three longitudinal columns of erectile tissue. Two of the columns, the **Corpora cavernosa** (B in the diagram), make up the dorsal portion of the penis. The corpus spongiosum (D), the other column of erectile tissue, is found on the ventral aspect of the penis surrounding the penile urethra (C). When sexual stimulation occurs, the arteries, some of which are seen lateral to the deep dorsal vein (A), supplying the penis become dilated and blood enters and fills the spongy erectile tissue. The engorgement of the erectile tissue by blood makes the penis erect and at the same time compresses the veins draining the penis, and thus maintains blood within the penis. At the end of an ejaculation, the sympathetic nerves are thought to constrict the arteries, allowing the blood to enter the veins and as a consequence, the penis becomes flaccid. The fascia of the penis is indicated by E in the diagram.

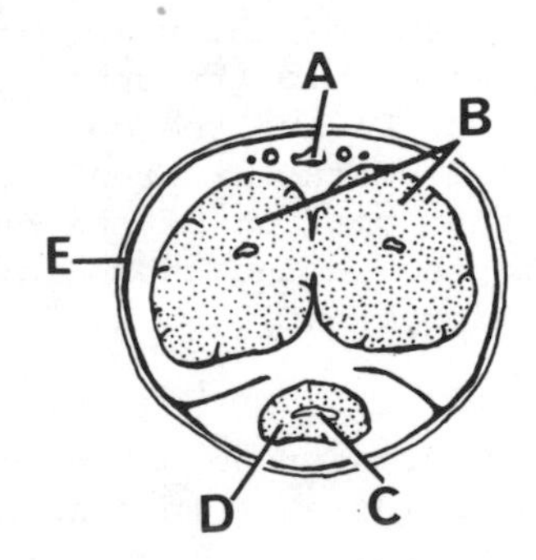

10. E. The physical activity of suckling the breast is thought to stimulate the anterior pituitary secretion of lactogenic hormone and the posterior pituitary secretion of oxytocin. The lactogenic hormone stimulates the alveoli of the breast to secrete milk. Oxytocin, on the other hand, stimulates the alveoli to eject milk into the ductal system, so that it can be removed from the tip of the nipple. Estrogen and progesterone are both involved in developing the structures of the breast for milk secretion, but they are not involved in milk production or its secretion.

11. D. The labia minora (D in the diagram) and labia majora (C) are two pairs of lips that serve to protect the clitoris (A), the orifice of the urethra and the orifice of the vagina. The labia majora consists of glands and fat and is covered by pigmented skin. The outer surface of the labia majora has hair, while the inner surface has none. The labia minora are free of hair and mark the lateral boundaries of the vestibule. The clitoris is a small structure consisting of erectile tissue which is homologous to the corpora cavernosa and glans of the penis. **The perineal body (B) is a fibronuscular mass of tissue lying between the vaginal and anal orifices.**

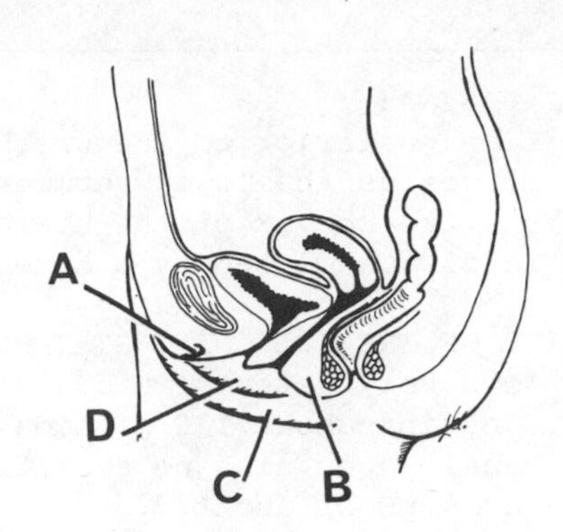

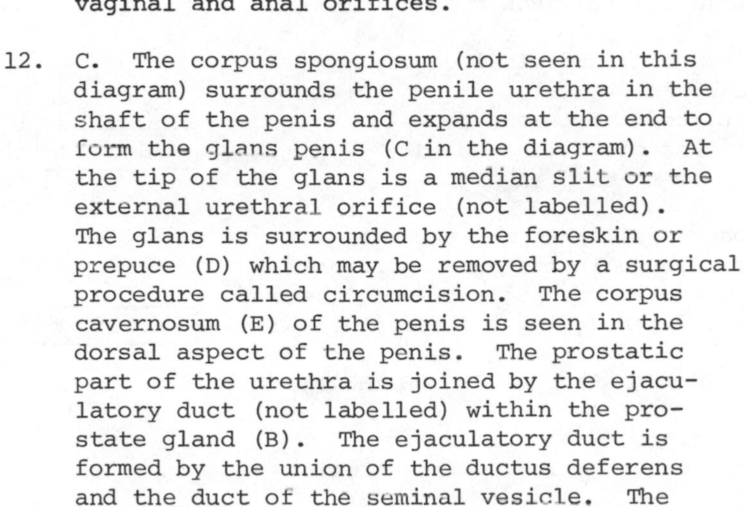

12. C. The corpus spongiosum (not seen in this diagram) surrounds the penile urethra in the shaft of the penis and expands at the end to form the glans penis (C in the diagram). At the tip of the glans is a median slit or the external urethral orifice (not labelled). The glans is surrounded by the foreskin or prepuce (D) which may be removed by a surgical procedure called circumcision. The corpus cavernosum (E) of the penis is seen in the dorsal aspect of the penis. The prostatic part of the urethra is joined by the ejaculatory duct (not labelled) within the prostate gland (B). The ejaculatory duct is formed by the union of the ductus deferens and the duct of the seminal vesicle. The seminal vesicle (A) lies behind and above the prostate gland and functions to provide a secretion for the semen.

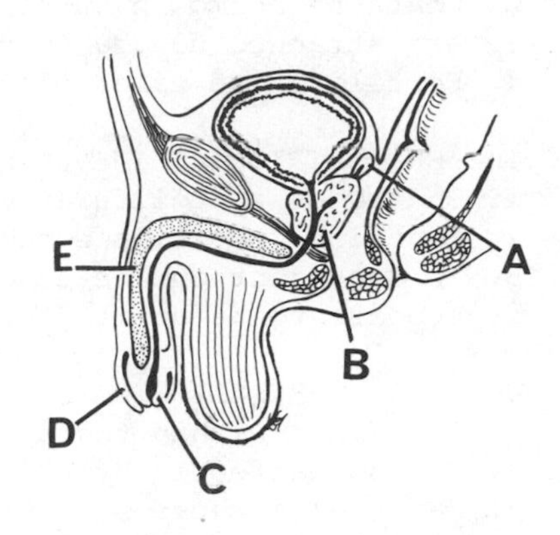

13. D. The male and female reproductive systems both have primary and accessory sexual structures. The primary sexual structures in the female are: paired gonads (ovaries), paired uterine tubes, and a single uterus and vagina. The accessory structures in the female include the mammary glands, mons pubis, labia majora and minora, clitoris, hymen, and vestibule. The primary sexual structures in the male are: paired gonads (testes) and a duct system which includes the epididymis, the ductus deferens, and ejaculatory ducts. The accessory structures in the male include seminal vesicles, the prostate gland, the bulbourethral glands, and the penis.

M U L T I - C O M P L E T I O N   Q U E S T I O N S

DIRECTIONS: In each of the following questions or incomplete statements, ONE OR MORE of the completions given is correct. At the lower right of each question, circle A if 1, 2, and 3 are correct; B if 1 and 3 are correct; C if 2 and 4 are correct; D if only 4 is correct; and E if all are correct.

1. THE BULBOURETHRAL GLANDS ARE:
   1. Located above the prostate gland
   2. Involved in the production of spermatozoa
   3. Responsible for secreting a fluid into the seminal vesicle
   4. Involved in secreting mucus providing lubrication during colitus.

A B C D E

A	B	C	D	E
1,2,3	1,3	2,4	Only 4	All correct

2. THE VAGINA:
   1. Lies posterior to the urethra
   2. Serves as the lower segment of the birth canal
   3. Has the cervix of the uterus projecting into it
   4. Is always closed by a hymen

   A B C D E

3. WHICH OF THE FOLLOWING STATEMENTS IS(ARE) TRUE FOR SEMEN?
   1. Is a very watery fluid
   2. Contains about 120 million sperm per ml
   3. Usually contains no secretions from the bulbourethral glands
   4. Has a pH of about 7.5

   A B C D E

4. WHICH OF THE FOLLOWING STATEMENTS IS(ARE) TRUE FOR THE UTERINE TUBE?
   1. Prevents bacteria from entering the abdominal cavity
   2. Contains a smooth muscle layer in its wall
   3. Are attached to the ovary at its distal end
   4. Fertilization takes place in its distal one-third

   A B C D E

5. THE ENDOMETRIUM OF THE UTERUS:
   1. Undergoes a proliferative phase following menstruation
   2. Contains many simple tubular glands in its outer portion
   3. Is dependent on the endocrine secretory activities of the ovary
   4. Appears to be at its thinnest depth during the secretory phase

   A B C D E

6. WHICH OF THE FOLLOWING STATEMENTS IS(ARE) CORRECT FOR THE TESTES?
   1. Contained within the scrotum
   2. Are not affected by FSH secretion
   3. Have cells whose secretions are controlled by the anterior pituitary
   4. Can function effectively within the abdominal cavity

   A B C D E

7. WHICH OF THE FOLLOWING STATEMENTS IS(ARE) TRUE FOR THE FEMALE BREAST?
   1. Their sole function is to produce milk for the newborn child
   2. The ampulla is a pigmented area surrounding the nipple
   3. Are separated from the pectoral muscles by a fascial layer
   4. Estrogen stimulates the development of the alveoli, the secreting cells

   A B C D E

8. WHICH OF THE FOLLOWING STATEMENTS IS(ARE) TRUE FOR THE WALL OF THE UTERUS?
   1. Its outer layer blends with the broad ligament
   2. Has an inner layer that is partly lost every 28 days
   3. Smooth muscle forms its thickest component
   4. The middle layer undergoes hyperplasia during pregnancy

   A B C D E

9. MEIOSIS:
   1. Plays a part in the maturation process of sex cells
   2. Consists of one nuclear division
   3. Produces haploid gametes
   4. Can be found in cells other than sex cells

   A B C D E

A	B	C	D	E
1,2,3	1,3	2,4	Only 4	All correct

10. AT OVULATION THE:
    1. Endometrium is continuing to grow
    2. Production of estrogen is at its highest level
    3. Follicular wall is ruptured
    4. LH level in the blood stream is low

    A B C D E

11. TESTOSTERONE IS:
    1. Responsible for the secondary male sexual characteristics
    2. Under the control of ICSH from the anterior pituitary
    3. Produced by the interstitial cells
    4. Necessary for the development of the male reproductive organs

    A B C D E

12. THE HORMONE(S) RESPONSIBLE FOR CONTROLLING ESTROGEN SECRETION IS(ARE) THE:
    1. Follicle-stimulating
    2. Progesterone
    3. Luteinizing
    4. Growth

    A B C D E

13. WHICH OF THE FOLLOWING STATEMENTS IS(ARE) TRUE FOR A SPERM?
    1. Consists of a head, middle piece, and a tail
    2. Contains 22 autosomes
    3. Is solely responsible for the sex of a zygote
    4. Has a great deal of cytoplasm

    A B C D E

14. MECHANISMS INVOLVED IN CONTROLLING LACTATION INCLUDE:
    1. Preparation of the breast by ovarian hormones
    2. Placental loss stimulates the secretion of lactogenic hormone
    3. An anterior pituitary hormone stimulates alveoli to secrete milk
    4. The alveoli eject milk into the ducts

    A B C D E

15. WHICH OF THE FOLLOWING STATEMENTS IS(ARE) TRUE FOR THE OVARY?
    1. The outer border of the medulla is marked by germinal epithelium
    2. Follicles are found only in the cortex
    3. Blood vessels are located within the cortex
    4. Is supported by a number of ligaments

    A B C D E

ANSWERS, NOTES AND EXPLANATIONS

1. D, Only 4 is correct. The paired bulbourethral (Cowper's) glands are located within the substance of the sphincter urethrae. They are roundish glands about 0.5 to 1.5 cm in diameter. They each have a small duct which passes downward through the bulb of the penis to open into the penile urethra. These glands contribute a secretion to the semen.

2. A, 1, 2, and 3 are correct. The vagina is the female organ of copulation. The vagina lies in the midsagittal plane of the pelvis with the urinary bladder and urethra in front and the rectum behind. It forms the lower portion of the birth canal, whereas the uterus forms the upper portion. The cervix of the uterus projects down into the upper portion of the vagina. The vagina receives the erect penis during coitus and serves to conduct the semen to the external os of the uterus. The distal end of the vagina may or may not have a membrane (hymen) covering it. The barrel of the vagina also serves to carry the products of menstruation from the uterus to the opening of the vagina.

3. C, 2 and 4 are correct. Semen, containing seminal fluids and spermatozoa, is ejaculated from the penis during the final stages of the male sexual act. The seminal fluid makes up the larger portion of the semen and receives fluids from the vas deferens, the seminal vesicles, the prostate gland, and the bulbo-urethral glands. Secretions from these glands, with the exception of that from the prostate, are slightly acidic. The secretions from the prostate tends to neutralize the semen so that its pH is about 7.5. The fluids of the semen tend to act as nutrients for the spermatozoa, especially the secretions from the seminal vesicles. The average ejaculate of semen, at coitus, equals about 3.5 ml and each milliliter contains about 120 million spermatozoa. Even though it takes only one spermatozoon (sperm) to fertilize an ovum, if the number of spermatozoa per ml of ejaculate falls below 20 million, the individual is very likely to be infertile.

4. C, 2 and 4 are correct. The uterine (fallopian) tubes extend laterally from the upper portion of the uterus and distally come in close proximity to the ovaries. These hollow tubes are about 10 cm long and can be divided into four parts: a uterine part, an isthmus, an ampulla, and an infundibulum. The distal end or infundibulum has numerous finger-like processes called fimbriae that help direct the ovum into this segment of the tube. The muscle and mucous membrane layers of the uterine tube are responsible for passing the ovum towards the uterus. Fertilization of the ovum usually takes place in the distal one-third of the tube. So the uterine tube carries the ovum towards the uterus and sperm towards the ovary. The uterine tube also serves as a potential avenue for infectious bacteria to enter the abdominal cavity which may produce peritonitis.

5. A, 1, 2, and 3 are correct. The wall of the uterus consists of three components: an outer parietal peritoneal layer, a middle myometrial layer, and an inner endometrial layer. The structure of the endometrium beautifully reflects its functions of: preparing for implantation of the fertilized ovum, participation in implantation, and formation of the maternal portion of the placenta. The endometrium is a special mucous membrane that lines the uterus and consists of two distinct functional layers: 1) an outer thick layer, and 2) a deep thin basilar layer. The outer layer changes greatly during the menstrual cycle and at one point is almost completely lost and the deep layer remains to regenerate the superficial layer following menstruation. After menstruation a follicular (preovulatory) stage, lasting about 10 days, takes place where mitosis in the basilar layer regenerates the superficial layer of the endometrium. This phase is dominated by FSH but is assisted by small amounts of LH. This follicular phase gives way to the luteal (secretory, postovulatory) phase in which the lumina of the endometrial glands are filled with secretions. Ovulation takes place and then the superficial layer becomes ready for implantation of the fertilized ovum. This stage is stimulated by progesterone secretions from the corpus luteum. The premenstrual (late luteal) stage, which begins about three to four days prior to menstruation, shows degeneration of the connective tissue stroma of the superficial endometrium. The blood supply to the endometrium is diminished and the arteries become damaged so that they lose blood, along with the epithelial cells of the endometrium into the lumen of the uterus.

6. B, 1 and 3 are correct. The testes are contained within the scrotum, a sac suspended between the thighs from the abdominal wall. The testes perform two primary functions: produce spermatozoa (spermatogenesis) and secrete the male hormone, testosterone. The site of the male testes is extremely important, for the development of spermatozoa will occur only at scrotal temperatures, which are slightly lower than abdominal temperatures. So testes that have not descended into the scrotum will not produce effective numbers of viable spermatozoa and the individual will be sterile. The testes are controlled by two different hormones secreted by the anterior pituitary: ICSH (interstitial cell stimulating hormone) and FSH (follicule stimulating hormone). FSH stimulates

the seminiferous tubules to produce spermatozoa more rapidly and ICSH stimulates the interstitial cells of the testis to produce testosterone. Testosterone, among other functions, promotes 'maleness'.

7. B, 1 and 3 are correct. The sole function of the breast is to provide nourishment for the child. The breast tissue lies on top of the pectoralis major muscle and is separated from it by a fascial layer of connective tissue. The breast tissue itself consists of 15-20 lobes radiating about a centrally placed nipple. Each lobe has its own lactiferous duct which empties at the tip of the nipple. Surrounding the nipple is a pigmented area called the areola. The female sex hormones are involved in the development of the breast. Estrogen stimulates the growth of the duct system and progesterone stimulates development of the secretory tissue.

8. E, All are correct. The wall of the uterus consists of an inner mucous membrane layer (endometrium), a thick middle layer of smooth muscle (myometrium), and an outer thin layer of serosa (peritoneum). The outer layer of peritoneum blends with the broad ligament and the other ligaments that provide support to the uterus. The thick middle layer of smooth muscle undergoes tremendous hyperplasia during the progression of a pregnancy. At the termination of pregnancy or labor, this layer undergoes violent contractions which expel the fetus from the uterus into the vagina and through its orifice. The inner mucous membrane layer cycles roughly every 28 days, from being completely denuded of a superficial layer to having an extremely thick layer of edematous glandular tissue.

9. B, 1 and 3 are correct. During the long maturation process of sex cells, both primary oocytes and spermatocytes, there are two cellular divisions. The first division is a meiotic or reduction division. Here the normal chromosomal complement (2N) of the primary sex cells are halved to become haploid (1N). Later during the maturation process the haploid secondary sex cell will undergo an ordinary mitotic division producing two haploid gametes. In spermatogenesis, the primary spermatocytes following the two divisions produce four haploid gametes, whereas in oogenesis, the primary oocyte, following the two divisions, produces three polar bodies, with their haploid complement of chromosomes, and an ovum with its haploid complement of chromosomes.

10. A, 1, 2, and 3 are correct. At the time of ovulation, about day 14 of the 28-day cycle, a mature ovum is discharged from the follicular cells and is released from the surface of the ovary. At this time the cycling endometrium is passing from the proliferative stage to the secretory stage which means the endometrium is growing and has almost attained the stage where it can receive and nourish a zygote (fertilized ovum). At this stage the estrogen level, secreted by the rapidly growing ovarian follicle, is high and inhibits the follicule stimulating hormone (FSH) and increases the secretion of luteinizing hormone (LH). At ovulation the LH level is high and is thought to be responsible for stimulating the release of the ovum.

11. E, All are correct. Testosterone (androgen or masculinizing hormone) is produced by the interstitial (Leydig) cells of the testis. The interstitial cells are stimulated to secrete testosterone by increased blood levels of ICSH (interstitial cell stimulating hormone) from the anterior pituitary. High testosterone levels in the blood inhibit the secretion of ICSH so the control of these two hormones represents a negative feedback mechanism. Testosterone is implicated in the development and maintenance of male secondary sex characteristics, promotes growth of muscle and skeletal tissues through increased protein anabolism, has slight effects on kidney tubule reabsorption, and inhibits the secretion of gonadotropins (FSH and ICSH) by the anterior pituitary.

12. B, 1 and 3 are correct. It appears that both follicle-stimulating hormone (FSH) and luteinizing hormone (LH) are involved in promoting the secretion of estrogen, the female sex hormone. Estrogen has a similar role in developing the adult female body as does testosterone in the male. In the follicular phase of the menstrual cycle, the follicular cells of the growing ovarian follicle are stimulated by FSH and LH, two pituitary gonadotropins. Early in this phase, FSH is the predominating hormone, but it has been shown that neither FSH or LH acting alone can cause the production of estrogen. In the late stage of the follicular phase, FSH becomes slowly inhibited and LH secretion is greatly increased. This sudden increase in LH secretion causes ovulation and then stimulates the development of the corpus luteum in the early luteal phase. The corpus luteum produces some estrogen but its main secretory product is progesterone.

13. A, 1, 2, and 3 are correct. The sperm is the end product of spermatogenesis. The sperm is a haploid cell containing 22 autosomes and either an X sex chromosome or a Y sex chromosome. The type of sperm that fertilizes the ovum determines the sex of the zygote or early embryo. If an X-bearing sperm fertilizes the ovum, the embryo will be a female and if a Y-bearing sperm fertilizes the ovum, the embryo will be a male. The mature sperm contains little cytoplasm and has a head, middle piece, and a tail. The head contains the chromosomes which contribute to the genetic material of the zygote.

14. E, All are correct. Lactation is controlled by hormonal mechanisms. Estrogens and progesterone, secreted by the ovary, prepare the immature breast tissue so that it is structurally capable of secreting milk. These hormones also develop the ductal system and progesterone furthers the development of the alveoli, the secreting cells. When the baby is born, the loss of the placenta and its secretion of estrogens, causes the blood levels of estrogen to drop which in turn stimulates the anterior pituitary to secrete the lactogenic hormone. This hormone, along with oxytocin from the posterior pituitary, are also stimulated by actual suckling of the nipple of the breast. The lactogenic hormone stimulates the alveoli to secrete the milk and oxytocin is responsible for ejecting the milk into the ductal system.

15. C, 2 and 4 are correct. The paired ovaries are about 2-3 cm in length and each is supported in a position lateral to the uterus by an ovarian ligament and folds of the peritoneum. The ovary consists of a central core of connective tissue (medulla) which has a great many arteries and veins. The outer layer (cortex) of the ovary consists of a zone which contains the early primary (graafian) follicles and other follicles in different stages of maturation. The outermost layer of the cortex consists of a single layer of cuboidal cells (germinal epithelium) which covers the free surface of the ovary.

FIVE-CHOICE ASSOCIATION QUESTIONS

DIRECTIONS: Each group of questions below consists of a numbered list of descriptive words or phrases accompanied by a diagram with certain parts indicated by letters, or by a list of lettered headings. For each numbered word or phrase, SELECT THE LETTERED PART OR HEADING that matches it correctly. Then insert the letter in the space to the right of the appropriate number. Sometimes more than one numbered word may be correctly matched to the same lettered part or heading.

1. _____ The seminal vesicle
2. _____ The prostate gland
3. _____ The ejaculatory duct
4. _____ The bulbourethral gland
5. _____ The ductus deferens

A
B
C
D
E

6. _____ A condition when the prepuce cannot be drawn back
7. _____ An incision made in the perineum
8. _____ Usually will produce sterility
9. _____ Having both testicular and ovarian tissue present
10. _____ The end of the reproductive period in the female

A. Hermaphrodite
B. Phimosis
C. Cryptorchism
D. Episiotomy
E. None of the above

------------------------ ANSWERS, NOTES AND EXPLANATIONS ------------------------

1. B. The seminal vesicles are two pouches which produce a secretion that is added to the seminal fluid. They lie behind the bladder and the ureter and in front of the rectum, through which they can be palpated during a rectal examination. The medial or lower end of the seminal vesicle forms a duct which joins the ductus deferens to form the ejaculatory duct.

2. E. The prostate gland consists of glandular tissue surrounded by smooth muscle and fibrous connective tissue. The glands of the prostate secrete a large portion of the fluid which goes to form the seminal fluid. These secretions enter the urethra through many small openings. The prostate surrounds the first portion of the urethra, below the bladder, and sits on top of the sphincter urethrae. The size of the prostate varies and can be palpated during a rectal examination. The prostate may become enlarged and constrict the urethra. This gland may also become cancerous in older men.

3. C. The ejaculatory duct is formed by the union of the lower end of the ductus deferens and the duct of the seminal vesicle. This duct passes downward and enters the prostatic part of the urethra. The ejaculatory duct mixes the spermatozoa from the testis with the viscous fluid from the seminal vesicle and ejects them into the urethra.

4. D. The bulbourethral (Cowper's) glands are two small rounded glandular structures embedded in the substance of the sphincter urethrae. They secrete a mucus-like substance which contributes to the seminal fluid. The ducts from these glands pass downward and pass through the substance of the bulb of the penis to enter the urethra.

5. A. The ductus deferens (vas deferens) is a continuation of the duct of the epididymis and carries spermatozoa from the epididymis to the ejaculatory duct. At its distal end, it passes between the posterior surface of the bladder and the lower end of the ureter just before it enters the bladder. The ductus deferens, along with its artery and the pampiniform plexus of veins, are incorporated into the spermatic cord. The distal end of the ductus deferens widens to form an ampulla, just before it joins the duct of the seminal vesicle to form the ejaculatory duct.

6. B. Phimosis is the condition where the opening of the prepuce (foreskin) is too narrow to permit its retraction over the glans penis. This condition hampers maintenance of cleanliness of the glans and may be corrected by circumcision or removal of the prepuce.

7. D. Episiotomy is an incision made in the perineum prior to the passage of a fetus through the vagina. The purpose of this incision is to prevent tearing of the vaginal wall and perineal structures during childbirth.

8. C. Cryptorchism is the failure of the testes to descend from the abdomen into the scrotum. The testes will not form viable sperm in the abdominal cavity because the body temperature is too high. Therefore, the individual will be sterile. The scrotal temperature is slightly lower than the abdominal temperature and allows the production of viable sperm.

9. A. A hermaphrodite (intersexuality) is an individual that has both testicular and ovarian tissue in the gonads. This extremely rare condition differs from a pseudohermaphrodite where the individual has an ovary or a testis but the external genitalia is of the opposite sex. Pseudohermaphrodites are thought to be caused by an excessive or too little amount of androgens being produced when the sex organs are developing.

10. E. Menopause marks the complete cessation of menstruation and with it the end of the climacteric period which begins at the first signs of ovarian failure. This period usually begins in the mid-forties. The ovaries, at the end of this period, fail to produce both ova and female sex hormones. From this point in time the sex organs undergo gradual atrophy.

REVIEW EXAMINATION OF PART 4

INTRODUCTORY NOTE: This review examination consists of 100 multiple-choice questions based on the learning objectives listed for the following chapters of this Study Guide and Review Manual: 13. THE RESPIRATORY SYSTEM; 14. THE DIGESTIVE SYSTEM; 15. THE URINARY SYSTEM; 16. BODY FLUIDS AND ACID-BASE BALANCE; and 17. THE REPRODUCTIVE SYSTEM. Before beginning, tear out an answer sheet from the back of the book and read the directions on how to use it. The key to the correct responses is on page 340.

FIVE-CHOICE COMPLETION QUESTIONS

DIRECTIONS: Each of the following questions or incomplete statements is followed by five suggested answers or completions. SELECT THE SINGLE BEST ANSWER in each case and then blacken the appropriate space on the answer sheet.

1. THE PARANASAL AIR SINUSES ALL COMMUNICATE WITH THE:
   A. Nasopharynx
   B. Auditory tube
   C. Oral cavity
   D. Middle ear cavity
   E. None of the above

2. THE RENAL PYRAMID IS INDICATED BY _____.

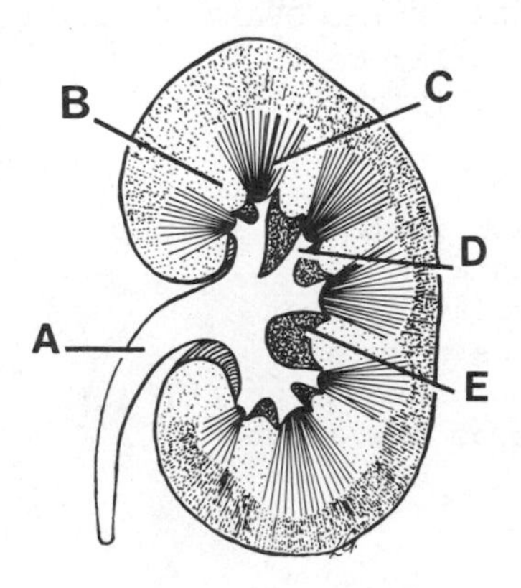

3. RENIN IS SECRETED BY THE:
   A. Podocytes
   B. Granular cells
   C. Glomerular capillaries
   D. Cells lining the loop of Henle
   E. None of the above

SELECT THE SINGLE BEST ANSWER

4. THE POUCH OF DOUGLAS IS INDICATED BY _____.

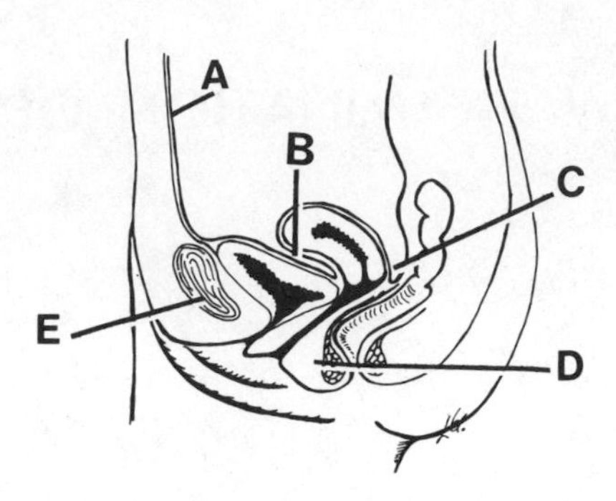

5. THE CAUDATE LOBE OF THE LIVER IS INDICATED BY ________.

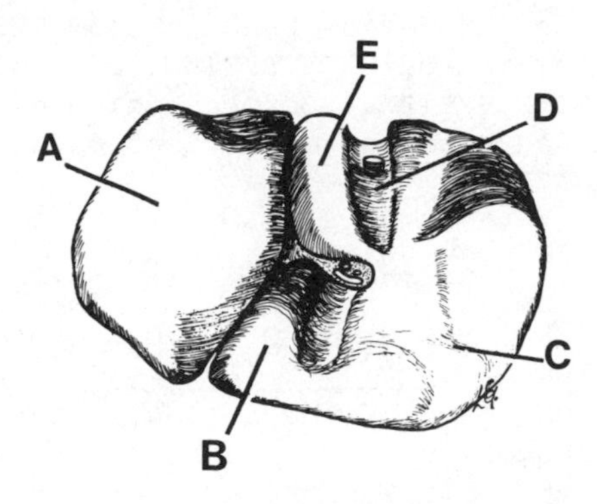

6. WHICH OF THE FOLLOWING STATEMENTS IS NOT CORRECT FOR THE PANCREAS?
   A. Secretes glucagon
   B. Contributes to the production of bile salts
   C. Empties its juices into the duodenum
   D. Has both exocrine and endocrine functions
   E. Its tail region may come in contact with the spleen

7. THE CELLS OF THE OVARY WHICH SECRETE ESTROGENS ARE THE:
   A. Follicular cells
   B. Interstitial cells
   C. Oocytes
   D. Germinal epithelial cells
   E. None of the above

8. THE ARROW INDICATES THE:
   A. Urethra
   B. Ejaculatory duct
   C. Ductus deferens
   D. Corpus cavernosum
   E. None of the above

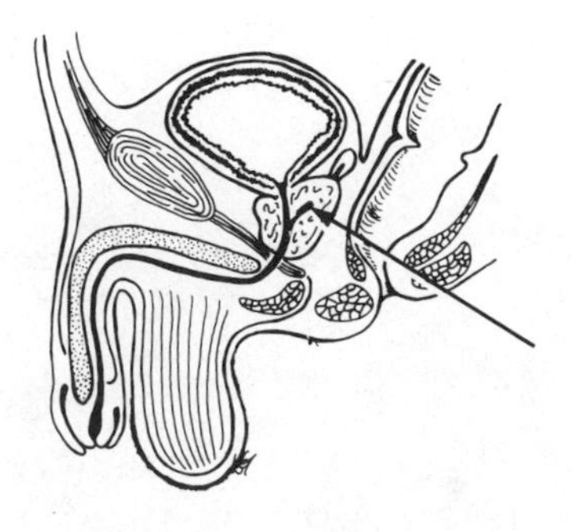

SELECT THE SINGLE BEST ANSWER

9. THE ARROW INDICATES THE _____.
   A. Middle lobe
   B. Oblique fissure
   C. Lower lobe
   D. Transverse fissure
   E. None of the above

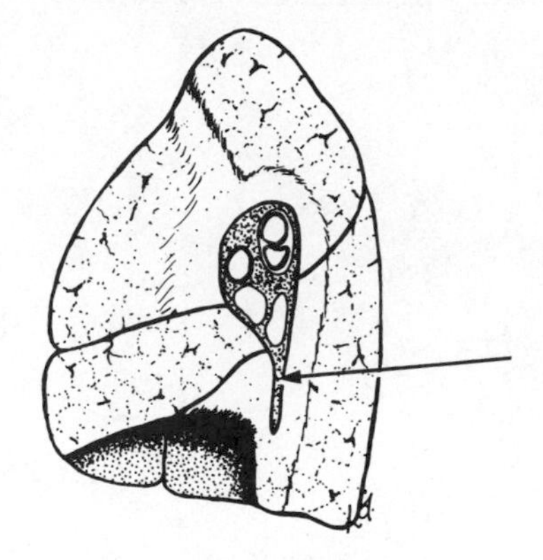

10. THE MAXIMUM NUMBER OF TEETH FOUND IN THE MAXILLA OF AN ADULT IS _____.
   A. 16
   B. 20
   C. 24
   D. 28
   E. 32

11. WHICH OF THE FOLLOWING LIGAMENTS DOES NOT SUPPORT THE UTERUS?
   A. Broad
   B. Round
   C. Anterior
   D. Posterior
   E. Triangular

12. FROM THE PAPILLA THE URINE WILL ENTER INTO THE:
   A. Ureter
   B. Renal pelvis
   C. Minor calyx
   D. Collecting duct
   E. None of the above

13. THE DUODENUM IS INDICATED BY _____.

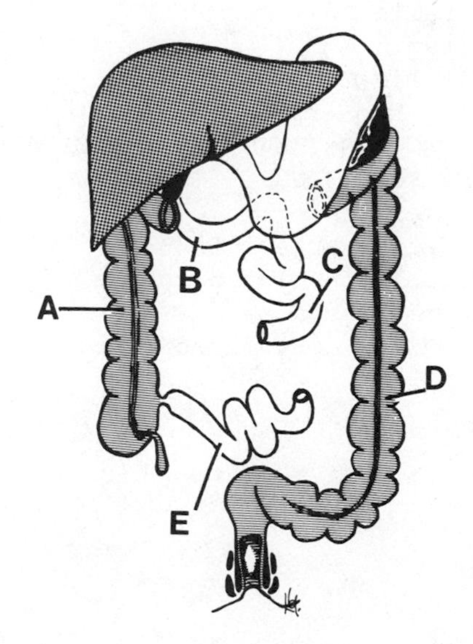

14. THE AIR PRESSURE BETWEEN THE TWO PLEURAL LAYERS IS CALLED THE _____ PRESSURE.
   A. Atmospheric
   B. Hydrostatic
   C. Intrathoracic
   D. Intrapulmonic
   E. None of the above

SELECT THE SINGLE BEST ANSWER

15. THE SEMINAL VESICLE IS INDICATED BY _____.

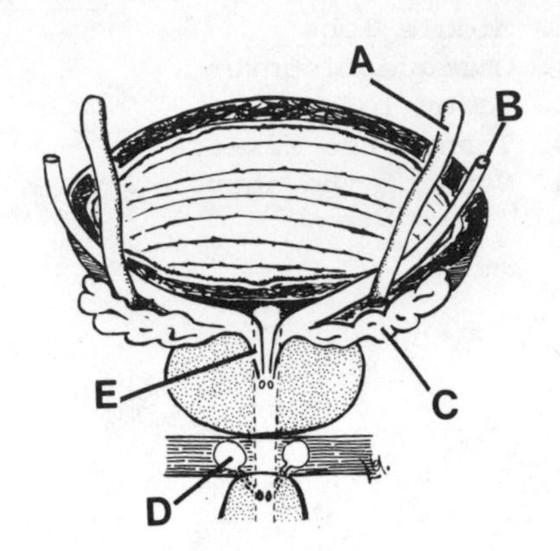

16. THE GLANDS FOUND ON EITHER SIDE OF THE VAGINAL ORIFICE ARE CALLED THE ______ GLANDS.

A. Ceruminous
B. Sebaceous
C. Meibomian
D. Greater vestibular
E. None of the above

17. THE URINARY BLADDER IS LINED BY _________ EPITHELIUM.

A. Transitional
B. Ciliated columnar
C. Low columnar
D. Stratified squamous
E. None of the above

18. THE BACTERIA OF THE LARGE INTESTINE PRODUCE QUANTITIES OF WHICH OF THE FOLLOWING VITAMINS?

A. Vitamin B_{12}
B. Vitamin A
C. Vitamin C
D. Vitamin D
E. Vitamin K

19. THE ARROW IN THE DIAGRAM INDICATES WHICH OF THE FOLLOWING?

A. Glomerulus
B. Efferent arteriole
C. Distal convoluted tubule
D. Bowman's space
E. Proximal convoluted tubule

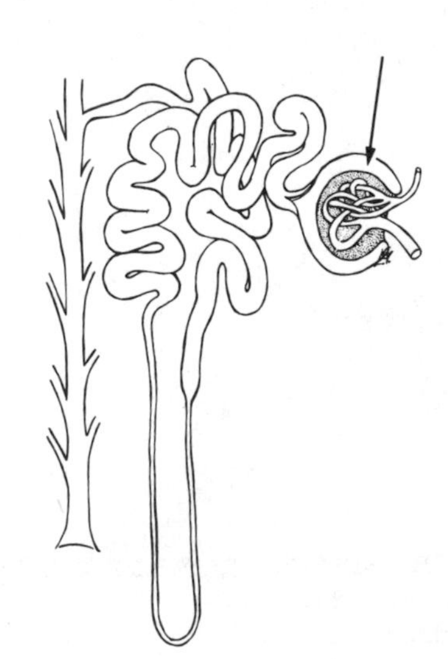

20. THE VOCAL CORDS ARE LOCATED IN THE ________.

A. Larynx
B. Pharynx
C. Trachea
D. Oropharynx
E. None of the above

SELECT THE SINGLE BEST ANSWER

21. WHICH OF THE FOLLOWING STATEMENTS IS CORRECT FOR THE BODY FLUIDS?
   A. The plasma volume equals 16 per cent of body weight
   B. The combined extracellular fluid is larger than the intracellular volume
   C. The intracellular fluid volume equals 40 liters
   D. Interstitial fluid forms the larger component of the extracellular volume
   E. None of the above

22. WHICH OF THE FOLLOWING IS NOT A COMPONENT OF THE SPERMATIC CORD?
   A. Lymphatic vessels
   B. Tail of the epididymis
   C. Testicular artery
   D. Ductus deferens
   E. Pampiniform plexus of veins

23. SECRETIN PRODUCED BY THE MUCOSA OF THE DUODENUM IS STIMULATED BY:
   A. Distention of the stomach
   B. Large quantities of bile
   C. A fatty chyme
   D. Acidity of the chyme
   E. None of the above

24. THE INFERIOR MESENTERIC VEIN IS INDICATED BY ________.

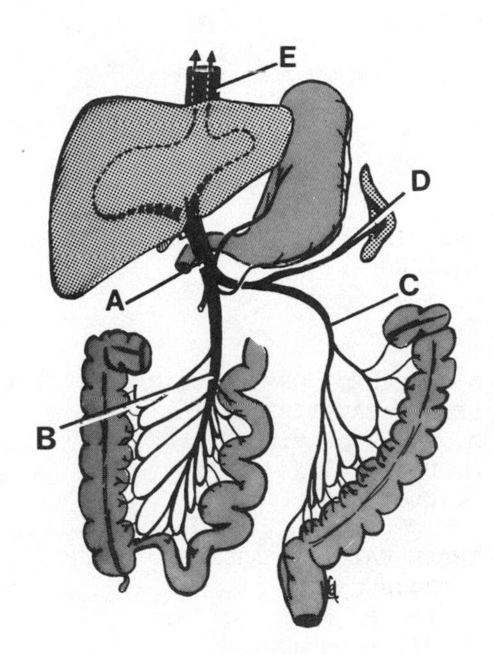

25. A CONDITION OF THE LUNG WHICH IS CHARACTERIZED BY A LOSS OF ELASTICITY AND A BRONCHIOLE OBSTRUCTION WITH A TRAPPING OF AIR DISTAL TO THE OBSTRUCTION IS CALLED:
   A. Asthma
   B. Emphysema
   C. Bronchitis
   D. Pneumonia
   E. None of the above

26. IN THE KIDNEY, ALDOSTERONE IS INVOLVED WITH:
   A. Sodium tubular reabsorption
   B. Renin synthesis
   C. Retention of urea
   D. Secretion of erythropoietin
   E. None of the above

SELECT THE SINGLE BEST ANSWER

27. HYPERKALEMIA IS AN INCREASE IN PLASMA ________.
A. Phosphate
B. Sodium
C. Magnesium
D. Potassium
E. None of the above

28. THE SPERMATOCYTES OF THE TESTIS ARE BOTH SUPPORTED AND SUPPLIED WITH NUTRIENTS BY THE ______ CELLS.
A. Graafian
B. Sertoli
C. Follicular
D. Interstitial cells
E. None of the above

29. THE ARROW INDICATES THE ________.
A. Epithelial lining
B. Serosa
C. Longitudinal layer
D. Muscularis mucosae
E. Submucosa

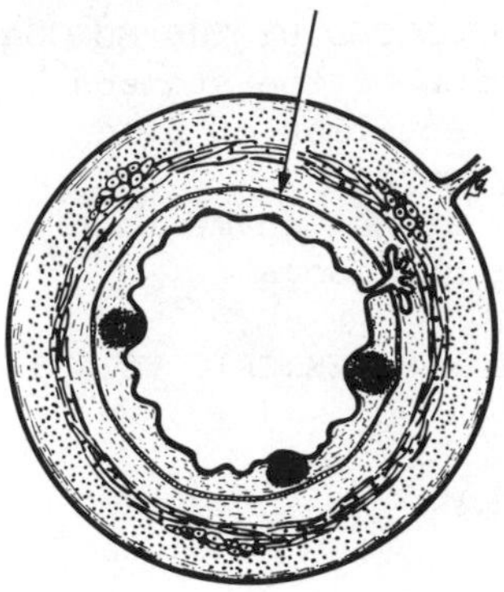

30. THE TIDAL VOLUME OF AIR EQUALS ________ ml.
A. 1000
B. 1500
C. 2000
D. 2500
E. None of the above

31. A COMPONENT OF THE URINARY SYSTEM NORMALLY FOUND IN BOTH THE ABDOMINAL AND PELVIC CAVITIES IS THE ________.
A. Renal pelvis
B. Ureter
C. Urethra
D. Bladder
E. None of the above

32. SMALL FAT TAGS ATTACHED TO THE LARGE INTESTINE ARE CALLED:
A. Haustra
B. Taenia coli
C. Veriform appendix
D. Transverse folds
E. None of the above

33. THE VENTRICULAR FOLD IS INDICATED BY ______.

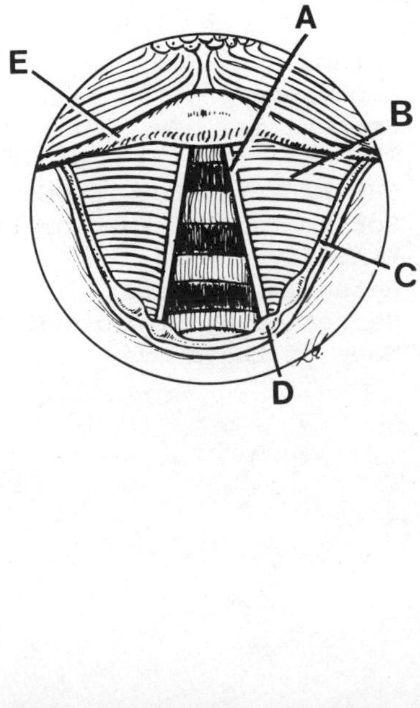

A
E
B
C
D

SELECT THE SINGLE BEST ANSWER

34. SPERMATOZOA ARE CONDUCTED FROM THE EPIDIDYMIS TO THE EJACULATORY DUCT BY THE ________.
A. Rete testis
B. Ductus deferens
C. Tunica vaginalis
D. Seminiferous tubules
E. None of the above

35. IN THE SMALL INTESTINE, MOST NEUTRAL FATS ARE ABSORBED INTO THE:
A. Lacunae
B. Lacteals
C. Capillaries
D. Venuoles
E. None of the above

36. WHICH OF THE FOLLOWING COMPONENTS OF THE NEPHRON IS LOCATED IN THE MEDULLA?
A. Glomerulus
B. Distal tubule
C. Bowman's capsule
D. Loop of Henle
E. Proximal tubule

37. A HORMONE SECRETED BY THE DUODENUM INHIBITING THE MOTILITY OF THE STOMACH IS THE ________.
A. Gastrin
B. Secretin
C. Pancreozymin
D. Cholecystokinin
E. None of the above

38. A CONDITION WHERE THE OPENING OF THE URETHRA IS ON THE VENTRAL SURFACE OF THE PENIS IS CALLED:
A. Epispadias
B. Hypospadias
C. Cryptorchism
D. Phimosis
E. None of the above

39. THE PAROTID DUCT EMPTIES SALIVA INTO THE ORAL CAVITY:
A. Beneath the tongue
B. Into the anterior portion of the vestibule
C. On to the dorsal surface of the tongue
D. Through the cheek at the level of the second upper molar tooth
E. None of the above

40. THE ENZYME SECRETED BY THE MUCOSA OF THE STOMACH IS THE ______.
A. Ptyalin
B. Pepsinogen
C. Steapsin
D. Trypsin
E. None of the above

41. WHICH OF THE FOLLOWING IS NOT A FUNCTION OF PHOSPHATE?
A. Involved in bone formation
B. Formation of nucleic acids
C. Neuronal transport
D. Functions as a coenzyme
E. Involved in acid-base balance

42. GLUCOSE PRODUCTION FROM PROTEINS IS CALLED ______.
A. Glycolysis
B. Gluconeogenesis
C. Glycogenesis
D. Glycogenolysis
E. None of the above

DIRECTIONS: In each of the following questions or incomplete statements, ONE OR MORE of the completions given is correct. On the answer sheet, blacken the space under A if 1, 2, and 3 are correct; B if 1 and 3 are correct; C if 2 and 4 are correct; D if only 4 is correct; and E if all are correct.

43. THE PROSTATE GLAND:
1. Is involved in expelling semen during intercourse
2. Lies below the neck of the urinary bladder
3. Provides a secretion that helps maintain an alkaline semen
4. Surrounds a segment of the urethra

44. THE RESPIRATORY MEMBRANE:
1. Is involved in warming the inspired air
2. Includes the surfactant layer
3. Contains a moderate amount of smooth muscle
4. Is located in part in the alveolus

45. WHICH OF THE FOLLOWING STATEMENTS IS(ARE) TRUE FOR THE LIVER?
1. Has four named lobes
2. Receives oxygenated blood from the hepatic arteries
3. Is not completely covered by peritoneum
4. Its bulk is found in the right hypochondrium

46. THE AMOUNT OF GLOMERULAR FILTRATE PRODUCED DOES NOT DEPEND ON WHICH OF THE FOLLOWING?
1. Volume of blood reaching the glomerulus
2. Glomerular capillary blood pressure
3. Permeability of the glomerular membrane
4. Adequate secretion of erythropoietin

47. SOURCES OF WATER LOSS INVOLVES WHICH OF THE FOLLOWING ORGANS?
1. Skin
2. Lungs
3. Kidneys
4. Intestines

48. LIPIDS ARE:
1. Important insulators against the cold
2. An important source of energy
3. Stored in large quantities in the body
4. Associated closely with the hepatic cells

49. THE SEMEN OF AN ADULT MALE WOULD CONSIST OF:
1. Prostatic secretions
2. Secretions of the seminal vesicles
3. Spermatozoa
4. Secretions from the greater vestibular glands

50. FUNCTIONS OF THE LIVER INCLUDE THE:
1. Storage of iron and vitamin B_{12}
2. Production of bile
3. Detoxification of substances that are toxic to the body
4. Catabolism of fatty foods

51. WHICH OF THE FOLLOWING FACTORS IS(ARE) INVOLVED WITH OXYGEN TRANSPORT?
1. Oxygen forms a loose chemical bond with a protein
2. High temperatures enhance the uptake of oxygen by the red blood cells
3. High P_{CO2} levels enhance the formation of reduced hemoglobin
4. Most oxygen is transported dissolved in solution

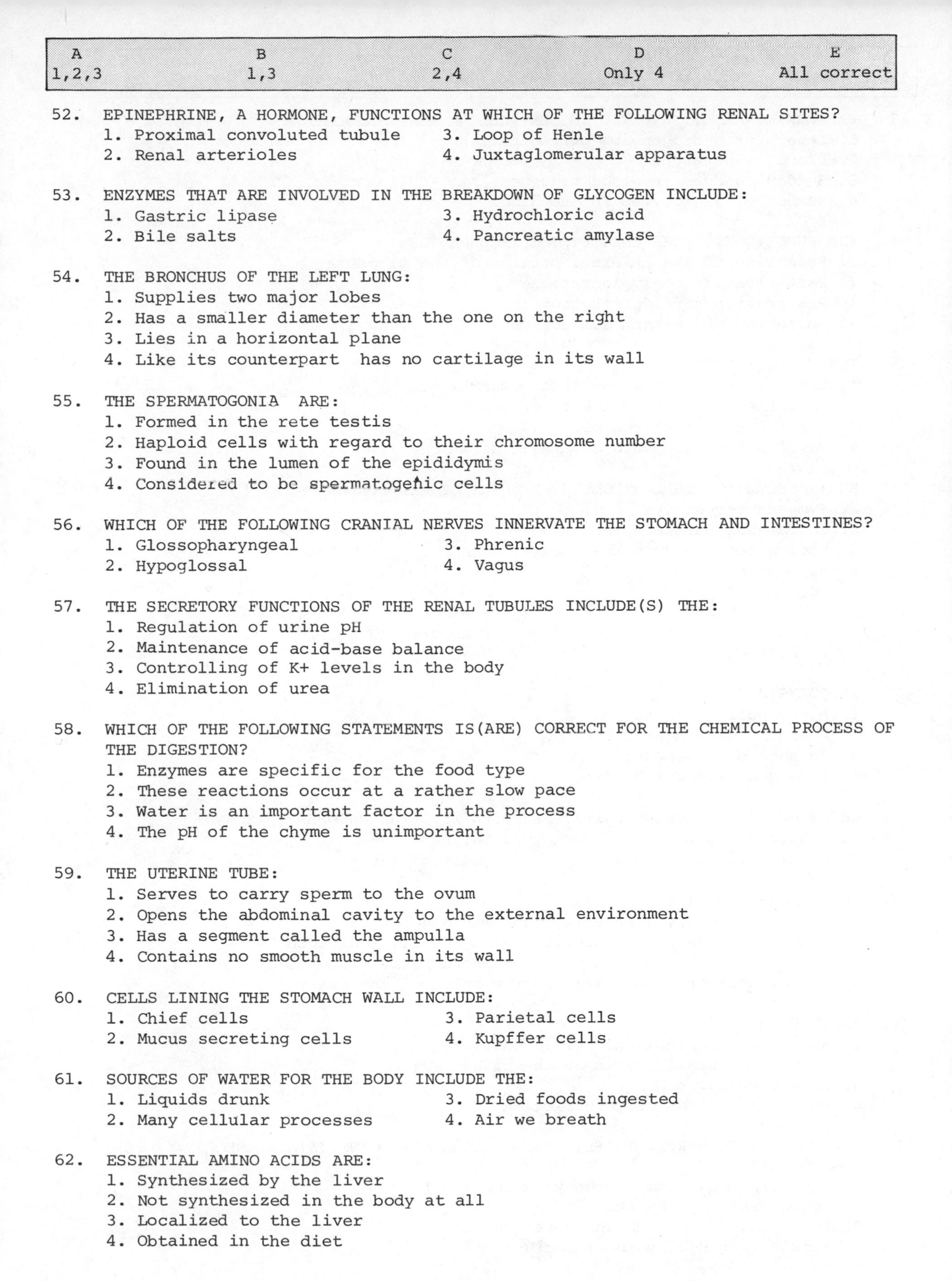

A	B	C	D	E
1,2,3	1,3	2,4	Only 4	All correct

52. EPINEPHRINE, A HORMONE, FUNCTIONS AT WHICH OF THE FOLLOWING RENAL SITES?
    1. Proximal convoluted tubule
    2. Renal arterioles
    3. Loop of Henle
    4. Juxtaglomerular apparatus

53. ENZYMES THAT ARE INVOLVED IN THE BREAKDOWN OF GLYCOGEN INCLUDE:
    1. Gastric lipase
    2. Bile salts
    3. Hydrochloric acid
    4. Pancreatic amylase

54. THE BRONCHUS OF THE LEFT LUNG:
    1. Supplies two major lobes
    2. Has a smaller diameter than the one on the right
    3. Lies in a horizontal plane
    4. Like its counterpart has no cartilage in its wall

55. THE SPERMATOGONIA ARE:
    1. Formed in the rete testis
    2. Haploid cells with regard to their chromosome number
    3. Found in the lumen of the epididymis
    4. Considered to be spermatogenic cells

56. WHICH OF THE FOLLOWING CRANIAL NERVES INNERVATE THE STOMACH AND INTESTINES?
    1. Glossopharyngeal
    2. Hypoglossal
    3. Phrenic
    4. Vagus

57. THE SECRETORY FUNCTIONS OF THE RENAL TUBULES INCLUDE(S) THE:
    1. Regulation of urine pH
    2. Maintenance of acid-base balance
    3. Controlling of K+ levels in the body
    4. Elimination of urea

58. WHICH OF THE FOLLOWING STATEMENTS IS(ARE) CORRECT FOR THE CHEMICAL PROCESS OF THE DIGESTION?
    1. Enzymes are specific for the food type
    2. These reactions occur at a rather slow pace
    3. Water is an important factor in the process
    4. The pH of the chyme is unimportant

59. THE UTERINE TUBE:
    1. Serves to carry sperm to the ovum
    2. Opens the abdominal cavity to the external environment
    3. Has a segment called the ampulla
    4. Contains no smooth muscle in its wall

60. CELLS LINING THE STOMACH WALL INCLUDE:
    1. Chief cells
    2. Mucus secreting cells
    3. Parietal cells
    4. Kupffer cells

61. SOURCES OF WATER FOR THE BODY INCLUDE THE:
    1. Liquids drunk
    2. Many cellular processes
    3. Dried foods ingested
    4. Air we breath

62. ESSENTIAL AMINO ACIDS ARE:
    1. Synthesized by the liver
    2. Not synthesized in the body at all
    3. Localized to the liver
    4. Obtained in the diet

A	B	C	D	E
1,2,3	1,3	2,4	Only 4	All correct

63. FACTORS NORMALLY INVOLVED IN CONTROLLING RESPIRATION ARE:
    1. Levels of hydrogen ion concentration
    2. Blood temperature
    3. Sudden changes in blood pressure
    4. Number of leukocytes in the blood

64. THE FUNCTIONS OF THE ENDOMETRIUM INCLUDE THE:
    1. Formation of the maternal portion of the placenta
    2. Participation in implantation
    3. Preparation for implantation
    4. Secretion of certain sex hormones

65. THE NASAL SEPTUM:
    1. Divides the nasopharynx into compartments
    2. Is formed in part by the vomer bone
    3. Has three lateral conchae
    4. Divides the nose into two cavities

66. RENIN SECRETION IS CONTROLLED BY WHICH OF THE FOLLOWING?
    1. Sympathetic nerves
    2. Intrarenal baroreceptors
    3. Sodium receptors in the macula densa
    4. Angiotensin

67. A DECREASE IN BLOOD pH STIMULATES THE RENAL TUBULES TO:
    1. Secrete NH_3
    2. Excrete H_2O
    3. Reabsorb $NaHCO_3$
    4. Reabsorb H+

68. GLYCOLYSIS IS:
    1. A catabolic process
    2. A process which produces energy
    3. An anaerobic process
    4. Involved with amino acids

69. WHICH OF THE FOLLOWING STRUCTURES OPEN INTO THE NASOPHARYNX?
    1. Oral cavity
    2. Nasal cavities
    3. Larynx
    4. Auditory tubes

70. THE PRIMARY OOCYTE:
    1. May not reach full maturity, but may degenerate
    2. Remains in the first meiotic division until adulthood
    3. Develops into a secondary oocyte and a polar body
    4. Is located within the medullary portion of the ovary

71. THE URETER:
    1. Has both abdominal and pelvic portions
    2. Leaves the renal pelvis at the hilus
    3. Is a muscular tube
    4. Has common sites of constriction

72. IN ADDITION TO BUFFER PAIRS, WHICH OF THE FOLLOWING IS(ARE) NECESSARY FOR pH HOMEOSTASIS?
    1. Expelling CO_2 from the body by the lungs
    2. Losing salt by the skin
    3. Excreting H+ through the renal tubules
    4. Eliminating H_2O by the feces

A	B	C	D	E
1,2,3	1,3	2,4	Only 4	All correct

73. SOURCES OF AMINO ACIDS FOR THE BLOOD INCLUDE:
    1. Catabolism of tissue proteins
    2. Those synthesized by the tissues of the body
    3. Absorption of ingested proteins
    4. Breakdown of complex lipids

74. THE FEMALE GONADS:
    1. Form the primary sex organs
    2. Are equivalent to the testis in the male
    3. Secrete various hormones
    4. Produce the mature ovum

75. THE MECHANICAL ACTIVITY OF THE MOUTH DEPENDS ON THE:
    1. Muscles of mastication
    2. Teeth
    3. Saliva
    4. Tongue

76. JUXTAGLOMERULAR (GRANULAR) CELLS ARE:
    1. Found in the tunica media of the efferent arteriole
    2. Controlled by the macula densa
    3. Thought to be modified epithelial cells
    4. Responsible for secreting renin, a proteolytic enzyme

77. THE RESPIRATORY CENTERS ARE LOCATED WITHIN THE:
    1. Midbrain
    2. Pons
    3. Cerebellum
    4. Medulla oblongata

78. THE CORPUS LUTEUM OF PREGNANCY SECRETES PROGESTERONE THAT IS RESPONSIBLE FOR WHICH OF THE FOLLOWING?
    1. Develops breast tissue for secretion
    2. Inhibits the myometrium
    3. Prevents ovulation
    4. Helps maintain the endometrium

79. THE ESOPHAGUS:
    1. Undergoes waves of contractions
    2. Produces no enzymes
    3. Secretes some mucus
    4. Contracts under voluntary control

80. WHICH OF THE FOLLOWING STATEMENTS ABOUT THE NEPHRON IS(ARE) CORRECT?
    1. Contains a vascular component
    2. Located in the cortex of the kidney
    3. Includes a tubular portion
    4. Total about one million for both kidneys

81. THE HYDROGEN ION (H+) CONCENTRATION (pH) OF A SOLUTION IS:
    1. Normally maintained within a narrow range
    2. Important only within intracellular solutions
    3. Effectively controlled by three mechanisms
    4. Normally about 7.54

A	B	C	D	E
1,2,3	1,3	2,4	Only 4	All correct

82. HIGHER BRAIN CENTERS EXERT CONTROL OF MICTURITION BY:
    1. Inhibiting the external urinary sphincter
    2. Overriding the micturition reflex once it has begun
    3. Initiating the reflex via the sacral micturition centers
    4. Partially inhibiting the micturition reflex at all times

83. THE VAGINA:
    1. Is about 8 cm in length
    2. Usually has longitudinal folds in its mucous membrane
    3. Has a recess about the cervix of the uterus
    4. Lies below the urinary bladder

84. THE MEDULLA OF THE KIDNEY CONTAINS THE:
    1. Distal convoluted tubule
    2. Renal glomeruli
    3. Proximal convoluted tubule
    4. Loop of Henle

85. VITAL CAPACITY:
    1. Depends greatly on the pulmonary compliance
    2. Averages about 5.1 liters in the female
    3. Is adversely affected by the paralysis of respiratory muscles
    4. Includes the expiratory reserve volume

86. WHICH OF THE FOLLOWING STATEMENTS IS(ARE) CORRECT FOR THE TCA (TRICARBOXYLIC ACID) CYCLE?
    1. Produces (anabolizes) fatty acids
    2. Provides heat as a biproduct
    3. Functions in anaerobic conditions
    4. Takes place within the mitochondria of cells

87. WHICH OF THE FOLLOWING IS(ARE) CORRECT FOR SODIUM IN THE FLUIDS OF THE BODY?
    1. Its balance is maintained primarily by sweating
    2. Its loss is accompanied by water loss
    3. Its depletion is due primarily to a decreased intake
    4. Its excess is associated with edema

88. WHICH OF THE FOLLOWING STATEMENTS IS(ARE) CORRECT FOR THE GALLBLADDER?
    1. Empties its bile directly into the bile duct
    2. Concentrates the bile salts by removing water
    3. Is suspended by a long mesentery
    4. Has a capacity of about 45 ml

89. WHICH OF THE FOLLOWING STATEMENTS IS(ARE) CORRECT FOR SPERMATOGENESIS?
    1. Begins in the testis normally at 11 years of age
    2. Is stimulated by luteinizing hormone
    3. Temperature is an unimportant variable
    4. Usually continues throughout the life of the individual

90. ABSORPTION TAKES PLACE THROUGH THE WALL OF THE SMALL INTESTINE BECAUSE OF WHICH OF THE FOLLOWING?
    1. The mucous membrane has an increased surface area
    2. Products of digestion are readily available
    3. Great length of this section of the gastrointestinal tract
    4. Venous and lymphatic drainage is adequate

F I V E - C H O I C E   A S S O C I A T I O N   Q U E S T I O N S

DIRECTIONS: Each group of questions below consists of a numbered list of descriptive words or phrases accompanied by a diagram with certain parts indicated by letters, or by a list of lettered headings. For each numbered word or phrase, SELECT THE LETTERED PART OR HEADING that matches it correctly. Then blacken the appropriate space on the answer sheet. Sometimes more than one numbered word or phrase may be correctly matched to the same lettered part or heading.

91. _____ The lesser omentum

92. _____ A retroperitoneal structure

93. _____ The lesser sac

94. _____ The greater omentum

95. _____ The greater sac

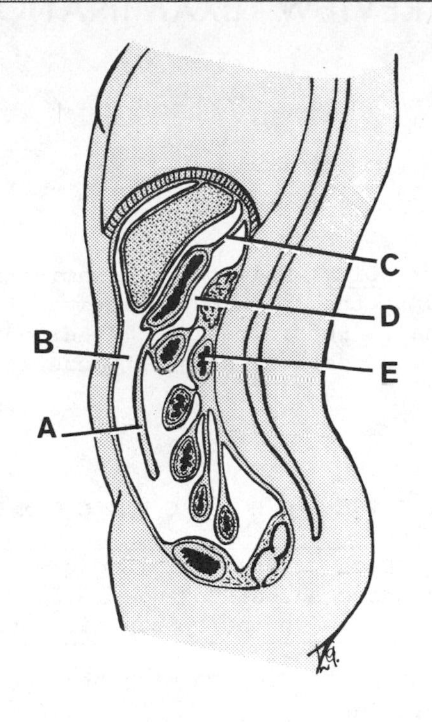

96. _____ Responsible for secondary male sexual characteristics

97. _____ Cells surrounding the oocyte secrete

98. _____ A hormone secreted by the seminal vesicle is called

99. _____ Ejection of milk from the ducts is stimulated by

100. _____ Stimulates the formation of the corpus luteum

A. Oxytocin

B. Estrogen

C. Testosterone

D. Follicular stimulating hormone

E. None of the above

FINAL REVIEW EXAMINATION

INTRODUCTORY NOTE: This review examination consists of 100 multiple-choice questions based on the learning objectives from all 17 chapters of this Study Guide and Review Manual. Before beginning tear out an answer sheet from the back of the book and read the directions on how to use it. The key to the correct responses is on page 340.

FIVE-CHOICE COMPLETIONS QUESTIONS

DIRECTIONS: Each of the following questions or incomplete statements is followed by five suggested answers or completions. SELECT THE SINGLE BEST ANSWER in each case and then blacken the appropriate space on the answer sheet.

1. THE ORGANELLE INVOLVED WITH PLACING A MEMBRANE AROUND SUBSTANCES IS CALLED:
   A. Nucleolus
   B. Lysosomes
   C. Mitochondria
   D. Golgi apparatus
   E. None of the above

2. THE STRUCTURE CARRYING NERVE IMPULSES AWAY FROM THE NERVE CELL BODY IS THE ________.
   A. Axon
   B. Dendrite
   C. Node of Ranvier
   D. Neurilemma
   E. None of the above

3. THE CILIARY BODY IS INDICATED BY _____.

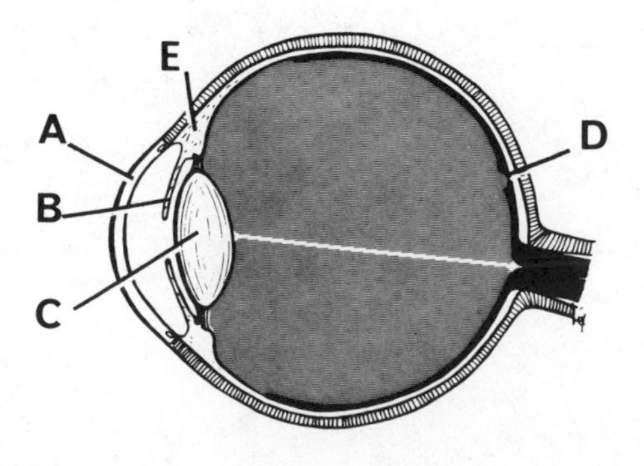

SELECT THE SINGLE BEST ANSWER

4. THE ARROW INDICATES THE ________.
   A. Recovery phase
   B. Depolarization of the atria
   C. Ventricular repolarization
   D. Depolarization of the ventricles
   E. None of the above

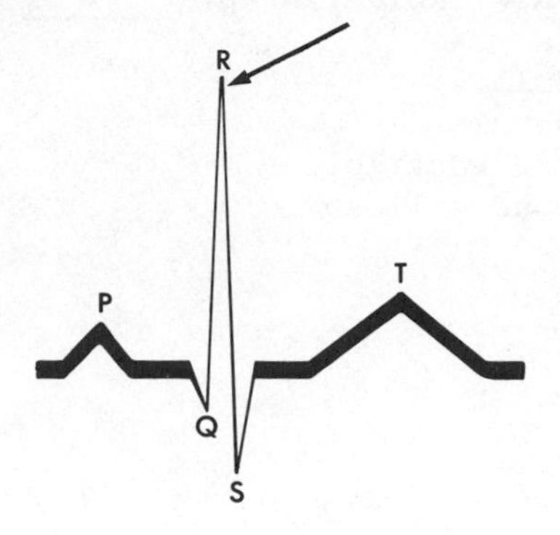

5. THE ARROW INDICATES THE ______.
   A. Distal phalanx
   B. Hamulus
   C. Metacarpal
   D. Pisiform
   E. None of the above

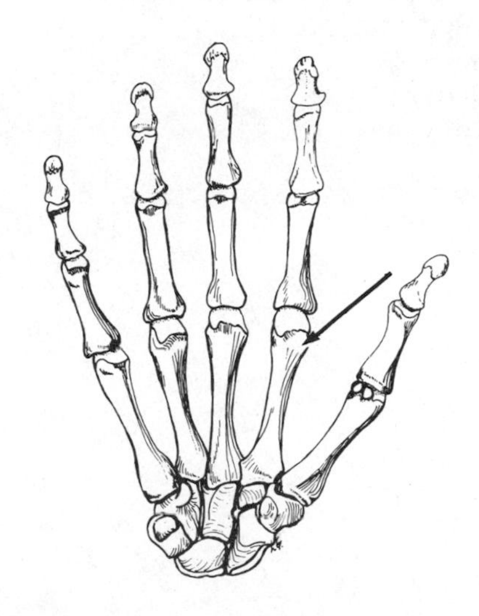

6. THE MOVEMENT OF A SUBSTANCE ACROSS A CELL MEMBRANE AGAINST A CONCENTRATION GRADIENT IS CALLED:
   A. Osmosis
   B. Diffusion
   C. Filtration
   D. Active transport
   E. Facilitated diffusion

7. THE SUPERIOR RECTUS MUSCLE MOVES THE EYEBALL SO THE PUPIL TURNS:
   A. Medially
   B. Laterally
   C. Downward
   D. Upward
   E. None of the above

8. THE ANTERIOR PITUITARY (ADENOHYPOPHYSIS) DOES NOT SECRETE WHICH OF THE FOLLOWING HORMONES?
   A. Growth
   B. Luteinizing
   C. Oxytocin
   D. Prolactin
   E. Thyroid-stimulating

9. REGULARLY ARRANGED DENSE CONNECTIVE TISSUE IS FOUND IN:
   A. Fascia
   B. Sheaths
   C. Tendons
   D. Joint capsules
   E. None of the above

SELECT THE SINGLE BEST ANSWER

10. THE ARROW INDICATES THE _______.
    A. Pons
    B. Medulla
    C. Midbrain
    D. Third ventricle
    E. Corpus callosum

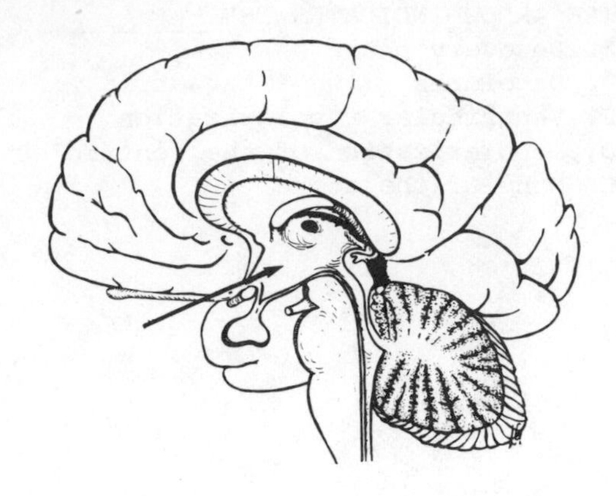

11. WHICH OF THE FOLLOWING ORGANS IS RESPONSIBLE FOR THE LARGEST AMOUNT OF WATER LOST FROM THE BODY?
    A. Lungs
    B. Kidneys
    C. Intestines
    D. Oral cavity
    E. None of the above

12. THE T-TUBULE IS INDICATED BY ______.

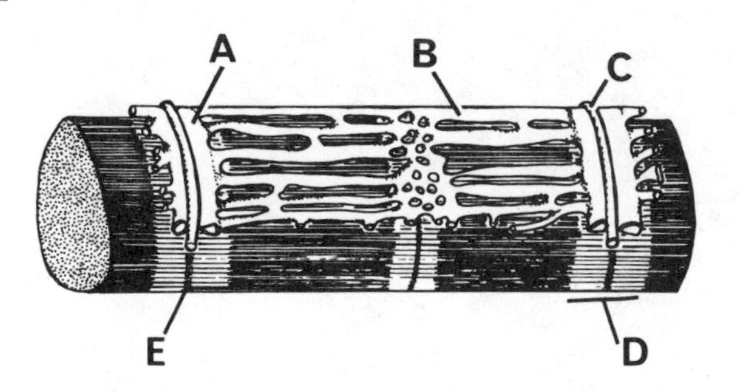

13. THE AVERAGE SYSTEMIC ARTERIAL DIASTOLIC PRESSURE IN THE YOUNG ADULT IS ______ mm Hg.
    A. 50-55
    B. 60-65
    C. 70-75
    D. 80-85
    E. 90-95

14. THE ARROW INDICATES THE:
    A. Pia mater
    B. Gray matter
    C. Dura mater
    D. White matter
    E. Arachnoid layer

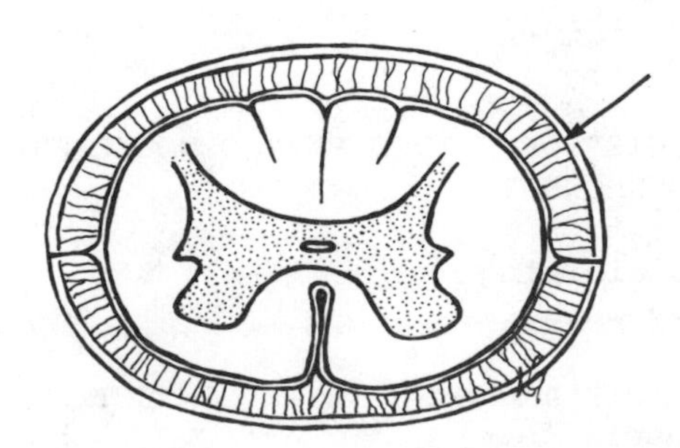

SELECT THE SINGLE BEST ANSWER

15. THE ARROW INDICATES _____.
    A. Ampulla
    B. Isthmus
    C. Infundibulum
    D. Round ligament
    E. Ovarian ligament

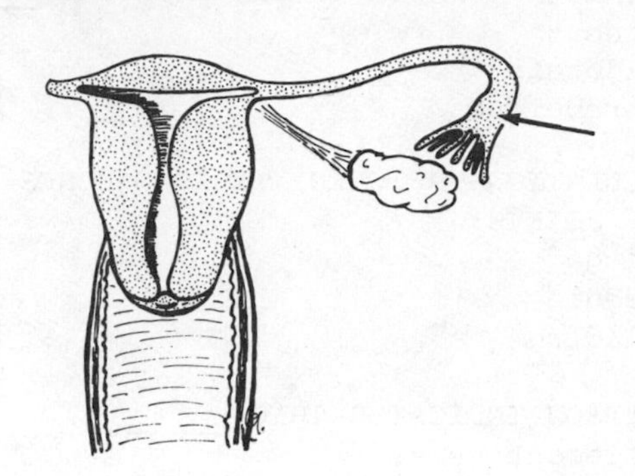

16. THE TRANSPLANTATION OF TISSUE BETWEEN TWO GENETICALLY IDENTICAL TWINS IS CALLED _____.
    A. Isograft
    B. Allograft
    C. Autograft
    D. Xenograft
    E. None of the above

17. THE ______ NERVE INNERVATES THE MUSCLES OF MASTICATION.
    A. Abducens
    B. Trigeminal
    C. Hypoglossal
    D. Facial
    E. Trochlear

18. THE MUSCLE RESPONSIBLE FOR A PARTICULAR DESIRED MOVEMENT IS CALLED A(AN) ____.
    A. Agonist
    B. Antagonist
    C. Synergist
    D. Fixator
    E. None of the above

19. IN A CROSS-SECTION OF A MATURE OVARY A WHITE SCAR-LIKE MASS WOULD REPRESENT A ______.
    A. Corpus luteum
    B. Primary follicle
    C. Mature follicle
    D. Corpus albicans
    E. None of the above

20. THIS DIAGRAM REPRESENTS _____.
    A. Hyaline cartilage
    B. Nerve cells
    C. Adipose tissue
    D. Skeletal muscle
    E. Areolar connective tissue

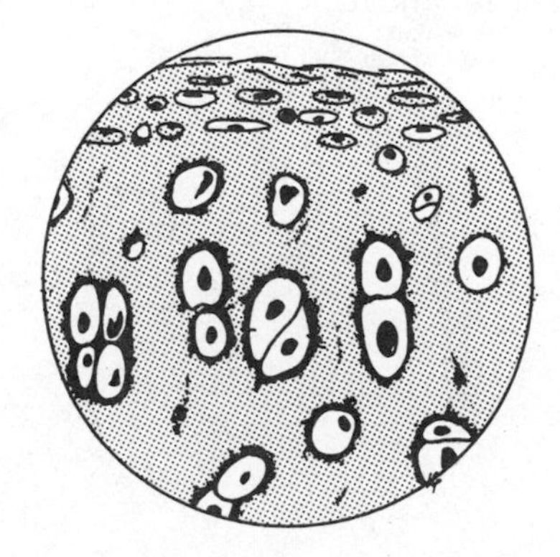

21. WHICH OF THE FOLLOWING IS NOT CONSIDERED A FUNCTION OF SKIN?
    A. Produces vitamin D
    B. Allows selective absorption
    C. Conducts electrical impulses
    D. Provides protection
    E. Mediates sensation

SELECT THE SINGLE BEST ANSWER

22. THE CARDIAC CENTER IS FOUND IN THE ______ OF THE BRAIN.
    A. Pons
    B. Midbrain
    C. Cerebellum
    D. Hypothalamus
    E. None of the above

23. ARTICULATIONS BETWEEN THE LONG BONES OF THE FINGERS ARE CONSIDERED TO BE _____ JOINTS.
    A. Hinge
    B. Saddle
    C. Gliding
    D. Ball and socket
    E. None of the above

24. THE ARROW INDICATES THE ________.
    A. Afferent vessel
    B. Germinal center
    C. Sinusoid
    D. Efferent vessel
    E. Trabecula

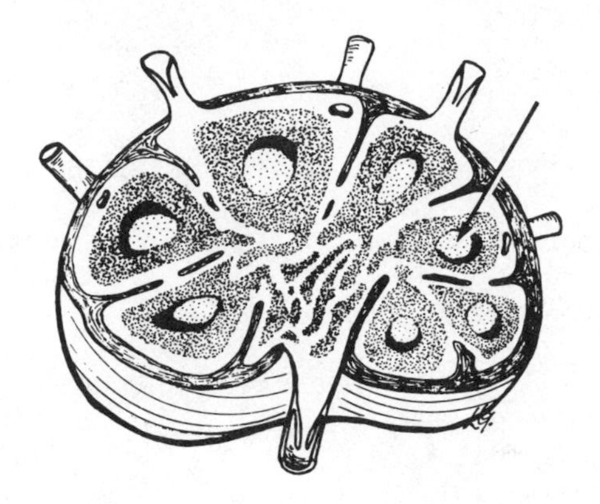

25. HORMONES FUNCTION BY WHICH OF THE FOLLOWING MECHANISMS?
    A. Induction of an action potential
    B. Stimulation of DNA synthesis
    C. Increasing the amount of ATP
    D. Activation of protein synthesis
    E. None of the above

26. THE ARROW INDICATES THE ______.
    A. Hair papilla
    B. Sebaceous gland
    C. Shaft of the hair
    D. Arrector muscle
    E. None of the above

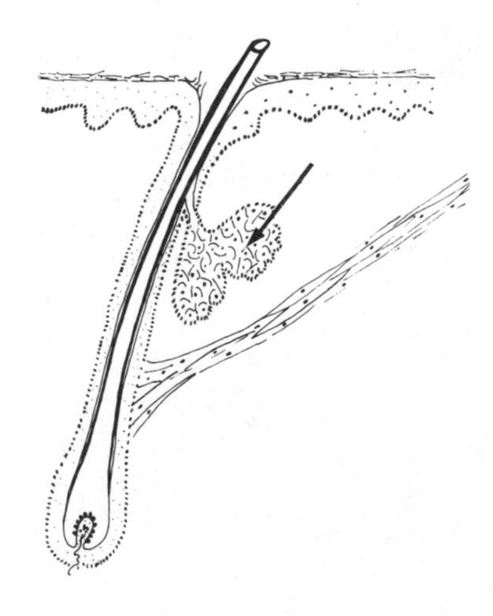

27. THE INTERSTITIAL CELLS ARE FOUND:
    A. In the rete testis
    B. Between the seminiferous tubules
    C. Lining the wall of the vas deferens
    D. Within the wall of the ductus epididymis
    E. None of the above

SELECT THE SINGLE BEST ANSWER

28. MOVEMENT OF A BONE ABOUT ITS LONG AXIS IS CALLED _____.
   A. Flexion
   B. Adduction
   C. Extension
   D. Circumduction
   E. None of the above

29. THE MAXILLA IS INDICATED BY ____.

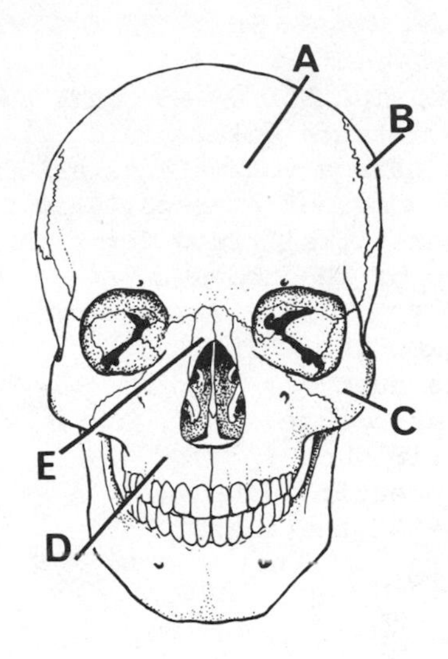

30. WHICH OF THE FOLLOWING IS A SECRETORY PRODUCT OF CELLS FOUND LINING THE STOMACH?
   A. Secretin
   B. Pancreozymin
   C. Cholecystokinin
   D. Enterogastrone
   E. None of the above

31. THE ARROW INDICATES THE:
   A. Trachea
   B. Parietal pleura
   C. Visceral pleura
   D. Primary bronchus
   E. Horizontal fissure

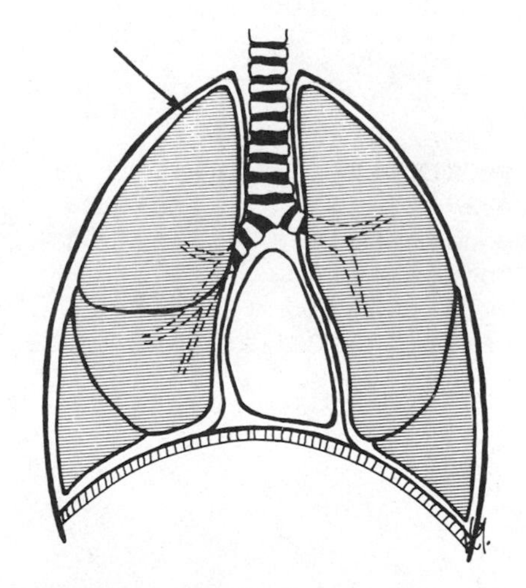

32. AN INFECTION IN THE SKIN OF THE BACK WOULD LIKELY CAUSE A SWELLING IN THE ______ LYMPH NODES.
   A. Axillary
   B. Inguinal
   C. Submental
   D. Superficial cubital
   E. Superficial cervical

33. WHICH OF THE FOLLOWING IS NOT A COMPONENT OF THE APPENDICULAR SKELETON?
   A. Clavicle
   B. Femur
   C. Fibula
   D. Skull
   E. Ulna

SELECT THE SINGLE BEST ANSWER

34. THE TOTAL VOLUME OF BLOOD REACHING THE KIDNEYS PER MINUTE EQUALS ABOUT ______ LITER(S).
A. 1
B. 2
C. 3
D. 4
E. 5

35. WHICH OF THE FOLLOWING STATEMENTS IS CORRECT FOR THE BODY FLUIDS?
A. The intracellular fluid volume equals 40 liters
B. The plasma volume equals 16 per cent of body weight
C. The combined extracellular fluid is larger than the intracellular volume
D. Interstitial fluid forms the smaller component of the extracellular volume
E. None of the above

36. THE ARROW INDICATES THE _____.
A. Bile duct
B. Hepatic duct
C. Cystic duct
D. Pancreatic duct
E. None of the above

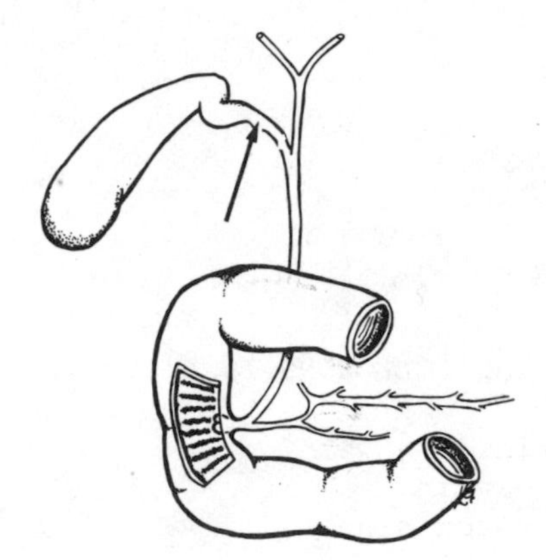

37. THE PROXIMAL END OF A MUSCLE IS CONSIDERED TO BE THE ________.
A. Belly
B. Aponeurosis
C. Insertion
D. Origin
E. None of the above

38. THE URINARY SPACE IS INDICATED BY ________.

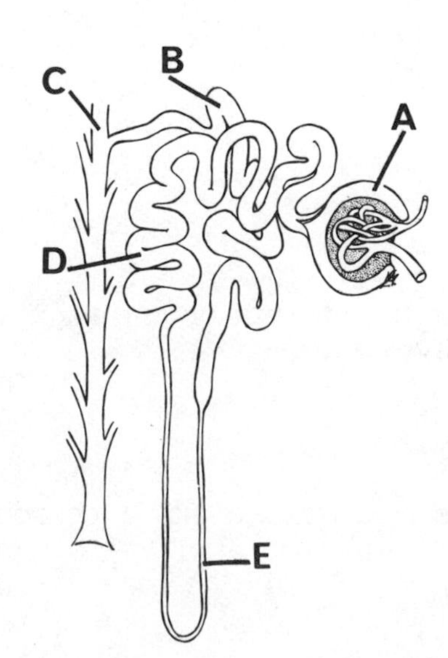

SELECT THE SINGLE BEST ANSWER

39. DECREASED BLOOD CALCIUM LEVELS ARE PRODUCED BY THE SECRETION OF ________.
    A. Calcitonin
    B. Oxytocin
    C. Thyroxine
    D. Parathyroid
    E. Epinephrine

40. WHICH OF THE FOLLOWING IS NOT A CAUSE OF HYPOXIA?
    A. Slow or decreased heart beat
    B. High altitude mountain climbing
    C. Production of erythropoietin by the kidney
    D. Decreased or low numbers of blood cells
    E. A substance combining chemically with hemoglobin

41. THE LAYER OF THE GUT WALL FORMED BY AN OUTER LONGITUDINAL AND AN INNER CIRCULAR LAYER OF SMOOTH MUSCLE IS CALLED _______.
    A. Seroa
    B. Subserosa
    C. Adventitia
    D. Muscularis mucosae
    E. None of the above

M U L T I - C O M P L E T I O N   Q U E S T I O N S

DIRECTIONS: In each of the following questions or incomplete statements, ONE OR MORE of the completions given is correct. On the answer sheet, blacken the space under A if 1, 2, and 3 are correct; B if 1 and 3 are correct; C if 2 and 4 are correct; D if only 4 is correct; and E if all are correct.

42. THE SOMATIC SENSORY AREA OF THE CEREBRUM:
    1. Interprets the position of the parts of the body
    2. Receives its input from the opposite side of the body
    3. Is located in the cortex of the postcentral gyrus
    4. Is connected by neurons to the hypothalamus

43. WHICH OF THE FOLLOWING SYSTEMS IS(ARE) INVOLVED IN CONTROLLING HOMEOSTASIS?
    1. Muscular
    2. Skeletal
    3. Lymphatic
    4. Endocrine

44. THE FEMALE BREAST:
    1. Is stimulated by oxytocin
    2. Functions only to produce milk
    3. Has a pigmented area called the areola
    4. Is divided by connective tissue into a number of lobes

45. WHICH OF THE FOLLOWING STATEMENTS IS(ARE) TRUE FOR PROTEIN METABOLISM?
    1. Takes place in the liver
    2. Produces a depletion of amino acids in the body
    3. Acts as a source of body energy
    4. Catabolism is more important than anabolism

46. WHICH OF THE FOLLOWING STATEMENTS CONCERNING THE CELL MEMBRANE IS(ARE) CORRECT?
    1. Contains proteins
    2. Measures about 75 angstroms across
    3. Is thought to be a trilaminar structure
    4. Functions as a non-selective permeable structure

47. WHICH OF THE FOLLOWING IS(ARE) NOT CLASSIFIED AS A CONNECTIVE TISSUE?
    1. Bone
    2. Lymph
    3. Adipose
    4. Endothelium

A	B	C	D	E
1,2,3	1,3	2,4	Only 4	All correct

48. WHICH OF THE FOLLOWING STATEMENTS IS(ARE) CORRECT FOR SMOOTH MUSCLE?
    1. Is considered to be voluntary
    2. Arranged in thin sheets
    3. Consists of striated fibers
    4. Found in the walls of hollow viscera

49. FUNCTIONS OF THE SKIN INCLUDE:
    1. Contains sensory cells
    2. Conducts electrical impulses
    3. Provides protection
    4. Synthesizes vitamin C

50. WHICH OF THE FOLLOWING STATEMENTS IS(ARE) TRUE FOR THE PORTAL VEIN OF THE ABDOMEN?
    1. Enters the liver sinusoids
    2. Formed in part by the superior mesenteric vein
    3. Carries the products of digestion
    4. Receives blood from the inferior vena cava

51. WHICH OF THE FOLLOWING STATEMENTS CONCERNING THE STERNUM IS(ARE) CORRECT?
    1. Has a spine
    2. Articulates with the clavicle
    3. Is considered a long bone
    4. Forms a synchrondrosis with some ribs

52. AT MENSTRUATION THE:
    1. Estrogen levels are relatively high
    2. Blood levels of LH is low
    3. Progesterone is being secreted at low levels
    4. Endometrium is being lost

53. THE DIGASTRIC MUSCLE:
    1. Consists of two bellies
    2. Has a tendon attached to the hyoid
    3. Contracts to raise the hyoid
    4. Attaches to the clavicle

54. A HINGE TYPE OF SYNOVIAL JOINT PERMITS:
    1. Gliding
    2. Flexion
    3. Rotation
    4. Extension

55. MECHANISMS OF RESPIRATION INVOLVED IN pH CONTROL INCLUDE:
    1. Neurons of the respiratory center detect increased CO_2 levels
    2. A blood pH of 7.47 causes hypoventilation
    3. Involve chemoreflexes of the carotid body
    4. Acidosis is produced by prolonged hyperventilation

56. SKELETAL MUSCLE IS:
    1. Found in hollow organs of the body
    2. Considered involuntary
    3. Continually contracting rhythmically
    4. Involved in maintaining posture

57. WHICH OF THE FOLLOWING ORGANS WOULD LIE WITHIN THE HYPOGASTRIC REGION OF THE ABDOMEN?
    1. Spleen
    2. Urinary bladder
    3. Gallbladder
    4. Small intestine

A	B	C	D	E
1,2,3	1,3	2,4	Only 4	All correct

58. THE AUDITORY TUBE:
    1. Enters the oropharynx
    2. Is continuous with the inner ear
    3. Controls the air pressure of the middle ear
    4. Usually is patent at all times

59. EPITHELIAL TISSUE:
    1. Has a low regenerative capacity
    2. Is anchored by a basement membrane
    3. Functions as a contracting tissue
    4. May have more than one cell layer

60. THE MUSCLES OF THE TONGUE:
    1. Consist of both skeletal and smooth muscle fibers
    2. Are involuntary
    3. Are innervated by the glossopharyngeal nerve
    4. Contain both extrinsic and intrinsic muscles

61. METABOLIC EFFECTS OF GROWTH HORMONE INCLUDE THE:
    1. Stimulation of protein synthesis
    2. Increased mobilization of stored fats
    3. Greater use of fats as a source of energy
    4. Decrease in the rate of carbohydrate utilization

62. TUBULAR REABSORPTION OF:
    1. Water is controlled by ADH
    2. Urea is carried out by active transport
    3. Glucose has a well defined threshold
    4. Water is minimal in most instances

63. THE ENDOCRINE SYSTEM:
    1. Consists of a number of ductless glands
    2. Uses the circulatory system to transport its secretions
    3. Has glands which are epithelial derivatives
    4. Produces chemical substances called enzymes

64. THE BASAL (NUCLEI) GANGLIA:
    1. Are located in the cerebral cortex
    2. Lie lateral to the thalamus
    3. Consist of white matter
    4. Includes the putamen

65. RESIDUAL AIR:
    1. Equals about 1600 ml
    2. Remains in the larynx and trachea
    3. May be expired with a great deal of effort
    4. Remains after the deepest expiration

66. CHRONIC EXCESSIVE AMOUNTS OF GLUCOCORTICOIDS PRODUCE:
    1. A mobilization of stored lipids
    2. Muscle tetany
    3. Cushing's syndrome
    4. Acceleration of rate of protein synthesis

A	B	C	D	E
1,2,3	1,3	2,4	Only 4	All correct

67. LONG BONE GROWTH IS:
    1. Located in the epiphyseal plates
    2. Because of chondrocyte stimulation
    3. Under control of somatotropin
    4. Over by 13 years of age in females

68. THE CISTERNA CHYLI:
    1. Is a dilated portion of lymphatic ductal system
    2. Drains directly into the right lymphatic duct
    3. Receives lymph from the lower limbs
    4. Is the main reservoir draining the head and neck

69. INSULIN:
    1. Stimulates the transport of glucose by facilitated diffusion
    2. Enhances the formation of glucose from glycogen
    3. Is secreted by the beta cells of the islet of Langerhans
    4. Promotes the release of free fatty acids from fat cells

70. BONES ARTICULATING WITH THE TIBIA TO FORM THE ANKLE JOINT INCLUDE THE:
    1. Cuboid
    2. Fibula
    3. Calcaneus
    4. Talus

71. THE MYOCARDIUM:
    1. Is the middle layer of the heart wall
    2. Forms part of the fibrous pericardium
    3. Consists of striated muscle fibers
    4. Usually is the thinnest layer of the heart wall

72. KIDNEY FUNCTION INCLUDES:
    1. Excretes waste products of digestion
    2. Production of glucose
    3. Helps maintain acid-base balance
    4. Involved in the production of vitamin K

73. SYNOVIAL FLUID:
    1. Consists mainly of water
    2. Is thought to act as a cushion
    3. Reduces friction between articulating surfaces
    4. Contains small numbers of blood cells

74. THE FOVEA CENTRALIS:
    1. May also be called the blind spot
    2. Consists only of cones
    3. Marks the site of entry for the optic nerve
    4. Is the most sensitive area of the retina

75. WHICH OF THE FOLLOWING STATEMENTS IS(ARE) CORRECT FOR THE RIGHT LUNG?
    1. Has three lobes
    2. Has a cardiac impression
    3. Has a large horizontal fissure
    4. Has a small diameter bronchus

76. TRUE STATEMENTS FOR FLUID BALANCE INCLUDE:
    1. Ideally intake equals output
    2. Control of the elimination of water is vital
    3. The distribution of fluids is important
    4. Blood volumes are maintained as a top priority

A	B	C	D	E
1,2,3	1,3	2,4	Only 4	All correct

77. ARTERIAL BLOOD PRESSURE IS MAINTAINED BY WHICH OF THE FOLLOWING?
    1. Viscosity of the blood
    2. Force of the ventricular contraction
    3. Elasticity of the walls of the large vessels
    4. Amount of peripheral resistance

78. A SPINAL NERVE:
    1. Contains both sensory and motor fibers
    2. Supplies efferent fibers to muscle
    3. Exits via an intervertebral foramen
    4. If severed produces a sensory loss

79. THE TRICARBOXYLIC ACID CYCLE:
    1. Is an aerobic process
    2. Produces large amounts of energy
    3. Takes place in mitochondria
    4. Is a step in the metabolism of glucose

80. THE UTRICLE AND SACCULE:
    1. Contain endolymph
    2. Both have a macula
    3. Are closely associated with righting reflexes
    4. Have continuous tectorial membranes

81. THE VITAL CAPACITY VOLUME IS COMPRISED OF WHICH OF THE FOLLOWING?
    1. Tidal volume
    2. Expiratory reserve volume
    3. Inspiratory reserve volume
    4. Maximal breathing capacity volume

82. THE ENDOCRINE PANCREAS SECRETES:
    1. Aldosterone
    2. Glucocorticoids
    3. Calcitonin
    4. Glucagon

83. IN URINATION THE:
    1. Detrusor muscle plays a small part in bladder emptying
    2. Tone of the trigonal smooth muscle is overcome
    3. Brain has no involvement in its control
    4. External sphincter of the bladder relaxes

84. WHICH OF THE FOLLOWING STATEMENTS IS(ARE) TRUE FOR THE PROSTATE GLAND?
    1. Secretes an alkaline fluid
    2. Is located below the neck of the bladder
    3. Expels the semen during ejaculation
    4. Contains the seminal vesicle

85. WHICH OF THE FOLLOWING STATEMENTS IS(ARE) TRUE FOR LYMPHATIC CAPILLARIES?
    1. Consist of tubules formed by endothelial cells
    2. Begin as blind tubes
    3. Have a slightly lower pressure than the capillaries
    4. Are more permeable than blood capillaries

DIRECTIONS: Each group of questions below consists of a numbered list of descriptive words or phrases accompanied by a diagram with certain parts indicated by letters, or by a list of lettered headings. For each numbered word or phrase, SELECT THE LETTERED PART OR HEADING that matches it correctly. Then blacken the appropriate space on the answer sheet. Sometimes more than one numbered word or phrase may be correctly matched to the same lettered part or heading.

86. _____ A vessel carrying oxygenated blood to the head and neck

87. _____ Receives blood from the liver

88. _____ The azygous vein

89. _____ Receives blood from the lungs

90. _____ A vessel carrying blood to the respiratory membrane

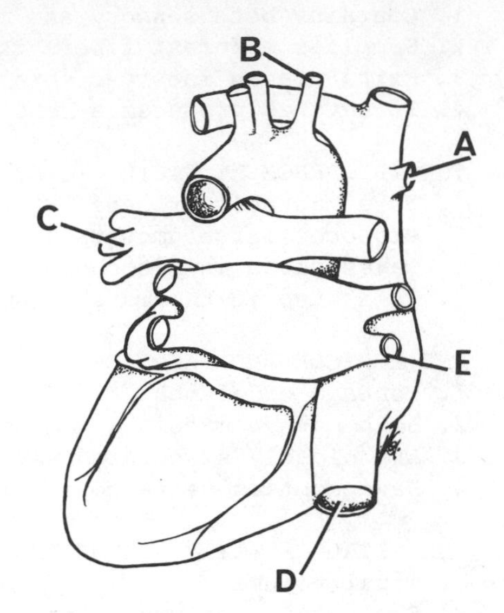

91. _____ A structure that stores bile

92. _____ A portion of the column supported by a mesentery

93. _____ The structure where most absorption takes place

94. _____ An organ that mixes the food and some gastric juices

95. _____ The descending colon

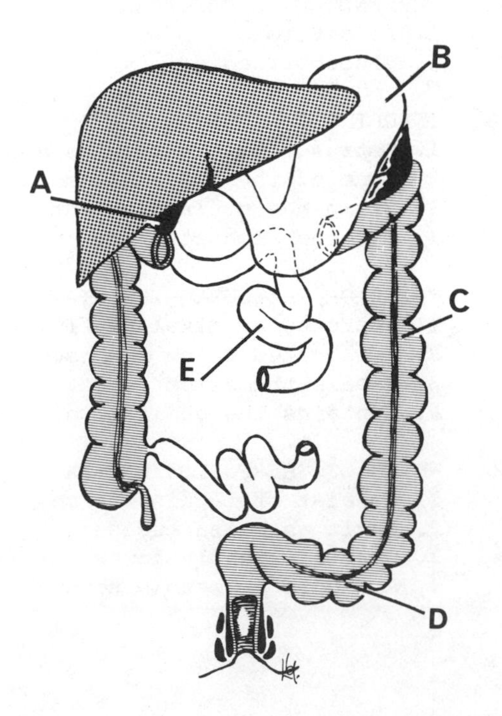

ASSOCIATION QUESTIONS

96. _____ The dorsal root

97. _____ A sympathetic chain ganglion

98. _____ Anterior primary ramus

99. _____ The gray rami communicans

100. _____ A nerve which supplies the dura

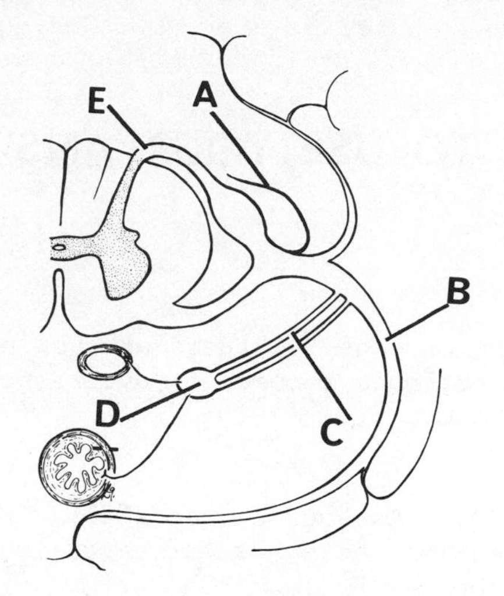

HOW TO USE YOUR ANSWER SHEETS

Beginning on page 341 there are copies of ANSWER SHEETS that you can tear out to use in keeping a record of your responses for the questions of the review examination.

DIRECTIONS: On your answer sheet, indicate for each question your choice of answer by blackening the space corresponding to the answer you have chosen. See the example below.

1. THE BASIC FUNCTIONAL COMPONENT OF THE HUMAN BODY IS AT THE _____ LEVEL.
   A. Organ
   B. Tissue
   C. Systems
   D. Cellular
   E. None of the above

1. A=== B=== C=== D▬▬ E===   35. A=== B=== C=== D=== E===   68. A=== B=== C=== D=== E===

After you have completed a review exam check your responses with the answers provided on pages 339 and 340.

INTERPRETATION OF SCORES

A student should obtain 85 per cent on any of the examinations in this Study Guide and Review Manual before he or she feels they have mastered the material. Any score less than this indicates that more study or clarification of concepts should be undertaken before rewriting any exam. In the following table, the number of correct responses equalling 85 per cent are given.

EXAM HAVING:	85 per cent equals:
50 questions	43 correct responses
80 questions	68 correct responses
100 questions	85 correct responses

KEYS TO REVIEW EXAMINATIONS

KEY - REVIEW EXAMINATION PART 1 (p.28).

1. B	2. A	3. D	4. D	5. B
6. E	7. B	8. B	9. E	10. B
11. B	12. E	13. E	14. B	15. D
16. E	17. A	18. A	19. A	20. D
21. D	22. C	23. C	24. E	25. B
26. D	27. B	28. C	29. E	30. D
31. E	32. B	33. C	34. A	35. D
36. E	37. C	38. E	39. E	40. C
41. A	42. B	43. D	44. D	45. A
46. E	47. C	48. B	49. D	50. B

KEY - REVIEW EXAMINATION PART 2 (p.111).

1. D	2. B	3. D	4. B	5. E
6. B	7. A	8. E	9. B	10. A
11. C	12. D	13. B	14. E	15. E
16. E	17. B	18. D	19. D	20. D
21. B	22. E	23. D	24. C	25. E
26. B	27. D	28. B	29. D	30. E
31. A	32. C	33. D	34. D	35. C
36. C	37. E	38. A	39. E	40. D
41. C	42. C	43. C	44. A	45. D
46. E	47. C	48. D	49. B	50. C
51. E	52. C	53. A	54. D	55. D
56. B	57. C	58. B	59. A	60. A
61. E	62. B	63. A	64. E	65. C
66. D	67. E	68. A	69. B	70. C
71. D	72. C	73. A	74. E	75. B
76. D	77. A	78. B	79. E	80. B

KEY - REVIEW EXAMINATION PART 3 (p.221).

1. A	2. C	3. A	4. A	5. D
6. E	7. B	8. D	9. A	10. C
11. A	12. D	13. B	14. A	15. D
16. D	17. C	18. B	19. E	20. D
21. D	22. A	23. C	24. C	25. C
26. B	27. C	28. D	29. A	30. D
31. D	32. D	33. B	34. C	35. A
36. B	37. B	38. B	39. B	40. C
41. C	42. C	43. E	44. A	45. C
46. B	47. A	48. E	49. E	50. D

51. E 52. E 53. E 54. D 55. D
56. C 57. C 58. B 59. B 60. C
61. C 62. B 63. E 64. E 65. B
66. D 67. B 68. E 69. C 70. D
71. E 72. D 73. A 74. C 75. D
76. B 77. A 78. A 79. C 80. A
81. B 82. E 83. E 84. E 85. E
86. E 87. B 88. A 89. C 90. D
91. B 92. A 93. E 94. C 95. D
96. B 97. C 98. A 99. D 100. E

KEY - REVIEW EXAMINATION PART 4 (p.311).

1. E 2. C 3. B 4. C 5. E
6. B 7. A 8. B 9. E 10. A
11. E 12. C 13. B 14. C 15. C
16. D 17. A 18. E 19. D 20. A
21. D 22. B 23. D 24. C 25. B
26. A 27. D 28. B 29. D 30. E
31. B 32. E 33. B 34. B 35. B
36. D 37. E 38. B 39. D 40. B
41. C 42. B 43. E 44. C 45. E
46. D 47. E 48. E 49. A 50. E
51. B 52. C 53. D 54. A 55. D
56. D 57. A 58. B 59. A 60. A
61. A 62. C 63. A 64. A 65. C
66. E 67. B 68. A 69. C 70. A
71. E 72. B 73. A 74. E 75. E
76. C 77. C 78. E 79. A 80. A
81. B 82. E 83. B 84. D 85. B
86. C 87. C 88. C 89. D 90. E
91. C 92. E 93. D 94. A 95. B
96. C 97. B 98. E 99. A 100. D

KEY - FINAL REVIEW EXAMINATION (p.324).

1. D 2. A 3. E 4. D 5. C
6. D 7. D 8. C 9. C 10. D
11. B 12. C 13. B 14. E 15. C
16. A 17. B 18. A 19. D 20. A
21. C 22. E 23. C 24. B 25. D
26. B 27. B 28. E 29. D 30. E
31. C 32. A 33. D 34. A 35. E
36. A 37. D 38. A 39. A 40. C
41. E 42. A 43. D 44. E 45. B
46. A 47. D 48. C 49. B 50. A
51. C 52. C 53. A 54. C 55. A
56. D 57. C 58. B 59. C 60. D
61. E 62. B 63. A 64. C 65. D
66. B 67. A 68. B 69. B 70. C
71. B 72. A 73. A 74. C 75. A
76. E 77. E 78. E 79. E 80. A
81. A 82. D 83. C 84. A 85. E
86. B 87. D 88. A 89. E 90. C
91. A 92. D 93. E 94. B 95. C
96. E 97. D 98. B 99. C 100. A

cut here

A N S W E R   S H E E T

Review Examination ______

1. A___ B___ C___ D___ E___
2. A___ B___ C___ D___ E___
3. A___ B___ C___ D___ E___
4. A___ B___ C___ D___ E___
5. A___ B___ C___ D___ E___
6. A___ B___ C___ D___ E___
7. A___ B___ C___ D___ E___
8. A___ B___ C___ D___ E___
9. A___ B___ C___ D___ E___
10. A___ B___ C___ D___ E___
11. A___ B___ C___ D___ E___
12. A___ B___ C___ D___ E___
13. A___ B___ C___ D___ E___
14. A___ B___ C___ D___ E___
15. A___ B___ C___ D___ E___
16. A___ B___ C___ D___ E___
17. A___ B___ C___ D___ E___
18. A___ B___ C___ D___ E___
19. A___ B___ C___ D___ E___
20. A___ B___ C___ D___ E___
21. A___ B___ C___ D___ E___
22. A___ B___ C___ D___ E___
23. A___ B___ C___ D___ E___
24. A___ B___ C___ D___ E___
25. A___ B___ C___ D___ E___
26. A___ B___ C___ D___ E___
27. A___ B___ C___ D___ E___
28. A___ B___ C___ D___ E___
29. A___ B___ C___ D___ E___
30. A___ B___ C___ D___ E___
31. A___ B___ C___ D___ E___
32. A___ B___ C___ D___ E___
33. A___ B___ C___ D___ E___
34. A___ B___ C___ D___ E___
35. A___ B___ C___ D___ E___
36. A___ B___ C___ D___ E___
37. A___ B___ C___ D___ E___
38. A___ B___ C___ D___ E___
39. A___ B___ C___ D___ E___
40. A___ B___ C___ D___ E___
41. A___ B___ C___ D___ E___
42. A___ B___ C___ D___ E___
43. A___ B___ C___ D___ E___
44. A___ B___ C___ D___ E___
45. A___ B___ C___ D___ E___
46. A___ B___ C___ D___ E___
47. A___ B___ C___ D___ E___
48. A___ B___ C___ D___ E___
49. A___ B___ C___ D___ E___
50. A___ B___ C___ D___ E___
51. A___ B___ C___ D___ E___
52. A___ B___ C___ D___ E___
53. A___ B___ C___ D___ E___
54. A___ B___ C___ D___ E___
55. A___ B___ C___ D___ E___
56. A___ B___ C___ D___ E___
57. A___ B___ C___ D___ E___
58. A___ B___ C___ D___ E___
59. A___ B___ C___ D___ E___
60. A___ B___ C___ D___ E___
61. A___ B___ C___ D___ E___
62. A___ B___ C___ D___ E___
63. A___ B___ C___ D___ E___
64. A___ B___ C___ D___ E___
65. A___ B___ C___ D___ E___
66. A___ B___ C___ D___ E___
67. A___ B___ C___ D___ E___
68. A___ B___ C___ D___ E___
69. A___ B___ C___ D___ E___
70. A___ B___ C___ D___ E___
71. A___ B___ C___ D___ E___
72. A___ B___ C___ D___ E___
73. A___ B___ C___ D___ E___
74. A___ B___ C___ D___ E___
75. A___ B___ C___ D___ E___
76. A___ B___ C___ D___ E___
77. A___ B___ C___ D___ E___
78. A___ B___ C___ D___ E___
79. A___ B___ C___ D___ E___
80. A___ B___ C___ D___ E___
81. A___ B___ C___ D___ E___
82. A___ B___ C___ D___ E___
83. A___ B___ C___ D___ E___
84. A___ B___ C___ D___ E___
85. A___ B___ C___ D___ E___
86. A___ B___ C___ D___ E___
87. A___ B___ C___ D___ E___
88. A___ B___ C___ D___ E___
89. A___ B___ C___ D___ E___
90. A___ B___ C___ D___ E___
91. A___ B___ C___ D___ E___
92. A___ B___ C___ D___ E___
93. A___ B___ C___ D___ E___
94. A___ B___ C___ D___ E___
95. A___ B___ C___ D___ E___
96. A___ B___ C___ D___ E___
97. A___ B___ C___ D___ E___
98. A___ B___ C___ D___ E___
99. A___ B___ C___ D___ E___
100. A___ B___ C___ D___ E___

ANSWER SHEET

Review Examination ______

1. A___ B___ C___ D___ E___
2. A___ B___ C___ D___ E___
3. A___ B___ C___ D___ E___
4. A___ B___ C___ D___ E___
5. A___ B___ C___ D___ E___
6. A___ B___ C___ D___ E___
7. A___ B___ C___ D___ E___
8. A___ B___ C___ D___ E___
9. A___ B___ C___ D___ E___
10. A___ B___ C___ D___ E___
11. A___ B___ C___ D___ E___
12. A___ B___ C___ D___ E___
13. A___ B___ C___ D___ E___
14. A___ B___ C___ D___ E___
15. A___ B___ C___ D___ E___
16. A___ B___ C___ D___ E___
17. A___ B___ C___ D___ E___
18. A___ B___ C___ D___ E___
19. A___ B___ C___ D___ E___
20. A___ B___ C___ D___ E___
21. A___ B___ C___ D___ E___
22. A___ B___ C___ D___ E___
23. A___ B___ C___ D___ E___
24. A___ B___ C___ D___ E___
25. A___ B___ C___ D___ E___
26. A___ B___ C___ D___ E___
27. A___ B___ C___ D___ E___
28. A___ B___ C___ D___ E___
29. A___ B___ C___ D___ E___
30. A___ B___ C___ D___ E___
31. A___ B___ C___ D___ E___
32. A___ B___ C___ D___ E___
33. A___ B___ C___ D___ E___
34. A___ B___ C___ D___ E___
35. A___ B___ C___ D___ E___
36. A___ B___ C___ D___ E___
37. A___ B___ C___ D___ E___
38. A___ B___ C___ D___ E___
39. A___ B___ C___ D___ E___
40. A___ B___ C___ D___ E___
41. A___ B___ C___ D___ E___
42. A___ B___ C___ D___ E___
43. A___ B___ C___ D___ E___
44. A___ B___ C___ D___ E___
45. A___ B___ C___ D___ E___
46. A___ B___ C___ D___ E___
47. A___ B___ C___ D___ E___
48. A___ B___ C___ D___ E___
49. A___ B___ C___ D___ E___
50. A___ B___ C___ D___ E___
51. A___ B___ C___ D___ E___
52. A___ B___ C___ D___ E___
53. A___ B___ C___ D___ E___
54. A___ B___ C___ D___ E___
55. A___ B___ C___ D___ E___
56. A___ B___ C___ D___ E___
57. A___ B___ C___ D___ E___
58. A___ B___ C___ D___ E___
59. A___ B___ C___ D___ E___
60. A___ B___ C___ D___ E___
61. A___ B___ C___ D___ E___
62. A___ B___ C___ D___ E___
63. A___ B___ C___ D___ E___
64. A___ B___ C___ D___ E___
65. A___ B___ C___ D___ E___
66. A___ B___ C___ D___ E___
67. A___ B___ C___ D___ E___
68. A___ B___ C___ D___ E___
69. A___ B___ C___ D___ E___
70. A___ B___ C___ D___ E___
71. A___ B___ C___ D___ E___
72. A___ B___ C___ D___ E___
73. A___ B___ C___ D___ E___
74. A___ B___ C___ D___ E___
75. A___ B___ C___ D___ E___
76. A___ B___ C___ D___ E___
77. A___ B___ C___ D___ E___
78. A___ B___ C___ D___ E___
79. A___ B___ C___ D___ E___
80. A___ B___ C___ D___ E___
81. A___ B___ C___ D___ E___
82. A___ B___ C___ D___ E___
83. A___ B___ C___ D___ E___
84. A___ B___ C___ D___ E___
85. A___ B___ C___ D___ E___
86. A___ B___ C___ D___ E___
87. A___ B___ C___ D___ E___
88. A___ B___ C___ D___ E___
89. A___ B___ C___ D___ E___
90. A___ B___ C___ D___ E___
91. A___ B___ C___ D___ E___
92. A___ B___ C___ D___ E___
93. A___ B___ C___ D___ E___
94. A___ B___ C___ D___ E___
95. A___ B___ C___ D___ E___
96. A___ B___ C___ D___ E___
97. A___ B___ C___ D___ E___
98. A___ B___ C___ D___ E___
99. A___ B___ C___ D___ E___
100. A___ B___ C___ D___ E___

cut here

cut here

ANSWER SHEET

Review Examination ______

1. A___B___C___D___E___
2. A___B___C___D___E___
3. A___B___C___D___E___
4. A___B___C___D___E___
5. A___B___C___D___E___
6. A___B___C___D___E___
7. A___B___C___D___E___
8. A___B___C___D___E___
9. A___B___C___D___E___
10. A___B___C___D___E___
11. A___B___C___D___E___
12. A___B___C___D___E___
13. A___B___C___D___E___
14. A___B___C___D___E___
15. A___B___C___D___E___
16. A___B___C___D___E___
17. A___B___C___D___E___
18. A___B___C___D___E___
19. A___B___C___D___E___
20. A___B___C___D___E___
21. A___B___C___D___E___
22. A___B___C___D___E___
23. A___B___C___D___E___
24. A___B___C___D___E___
25. A___B___C___D___E___
26. A___B___C___D___E___
27. A___B___C___D___E___
28. A___B___C___D___E___
29. A___B___C___D___E___
30. A___B___C___D___E___
31. A___B___C___D___E___
32. A___B___C___D___E___
33. A___B___C___D___E___
34. A___B___C___D___E___
35. A___B___C___D___E___
36. A___B___C___D___E___
37. A___B___C___D___E___
38. A___B___C___D___E___
39. A___B___C___D___E___
40. A___B___C___D___E___
41. A___B___C___D___E___
42. A___B___C___D___E___
43. A___B___C___D___E___
44. A___B___C___D___E___
45. A___B___C___D___E___
46. A___B___C___D___E___
47. A___B___C___D___E___
48. A___B___C___D___E___
49. A___B___C___D___E___
50. A___B___C___D___E___
51. A___B___C___D___E___
52. A___B___C___D___E___
53. A___B___C___D___E___
54. A___B___C___D___E___
55. A___B___C___D___E___
56. A___B___C___D___E___
57. A___B___C___D___E___
58. A___B___C___D___E___
59. A___B___C___D___E___
60. A___B___C___D___E___
61. A___B___C___D___E___
62. A___B___C___D___E___
63. A___B___C___D___E___
64. A___B___C___D___E___
65. A___B___C___D___E___
66. A___B___C___D___E___
67. A___B___C___D___E___
68. A___B___C___D___E___
69. A___B___C___D___E___
70. A___B___C___D___E___
71. A___B___C___D___E___
72. A___B___C___D___E___
73. A___B___C___D___E___
74. A___B___C___D___E___
75. A___B___C___D___E___
76. A___B___C___D___E___
77. A___B___C___D___E___
78. A___B___C___D___E___
79. A___B___C___D___E___
80. A___B___C___D___E___
81. A___B___C___D___E___
82. A___B___C___D___E___
83. A___B___C___D___E___
84. A___B___C___D___E___
85. A___B___C___D___E___
86. A___B___C___D___E___
87. A___B___C___D___E___
88. A___B___C___D___E___
89. A___B___C___D___E___
90. A___B___C___D___E___
91. A___B___C___D___E___
92. A___B___C___D___E___
93. A___B___C___D___E___
94. A___B___C___D___E___
95. A___B___C___D___E___
96. A___B___C___D___E___
97. A___B___C___D___E___
98. A___B___C___D___E___
99. A___B___C___D___E___
100. A___B___C___D___E___

A N S W E R   S H E E T

Review Examination ______

1. A___B___C___D___E___
2. A___B___C___D___E___
3. A___B___C___D___E___
4. A___B___C___D___E___
5. A___B___C___D___E___
6. A___B___C___D___E___
7. A___B___C___D___E___
8. A___B___C___D___E___
9. A___B___C___D___E___
10. A___B___C___D___E___
11. A___B___C___D___E___
12. A___B___C___D___E___
13. A___B___C___D___E___
14. A___B___C___D___E___
15. A___B___C___D___E___
16. A___B___C___D___E___
17. A___B___C___D___E___
18. A___B___C___D___E___
19. A___B___C___D___E___
20. A___B___C___D___E___
21. A___B___C___D___E___
22. A___B___C___D___E___
23. A___B___C___D___E___
24. A___B___C___D___E___
25. A___B___C___D___E___
26. A___B___C___D___E___
27. A___B___C___D___E___
28. A___B___C___D___E___
29. A___B___C___D___E___
30. A___B___C___D___E___
31. A___B___C___D___E___
32. A___B___C___D___E___
33. A___B___C___D___E___
34. A___B___C___D___E___
35. A___B___C___D___E___
36. A___B___C___D___E___
37. A___B___C___D___E___
38. A___B___C___D___E___
39. A___B___C___D___E___
40. A___B___C___D___E___
41. A___B___C___D___E___
42. A___B___C___D___E___
43. A___B___C___D___E___
44. A___B___C___D___E___
45. A___B___C___D___E___
46. A___B___C___D___E___
47. A___B___C___D___E___
48. A___B___C___D___E___
49. A___B___C___D___E___
50. A___B___C___D___E___
51. A___B___C___D___E___
52. A___B___C___D___E___
53. A___B___C___D___E___
54. A___B___C___D___E___
55. A___B___C___D___E___
56. A___B___C___D___E___
57. A___B___C___D___E___
58. A___B___C___D___E___
59. A___B___C___D___E___
60. A___B___C___D___E___
61. A___B___C___D___E___
62. A___B___C___D___E___
63. A___B___C___D___E___
64. A___B___C___D___E___
65. A___B___C___D___E___
66. A___B___C___D___E___
67. A___B___C___D___E___
68. A___B___C___D___E___
69. A___B___C___D___E___
70. A___B___C___D___E___
71. A___B___C___D___E___
72. A___B___C___D___E___
73. A___B___C___D___E___
74. A___B___C___D___E___
75. A___B___C___D___E___
76. A___B___C___D___E___
77. A___B___C___D___E___
78. A___B___C___D___E___
79. A___B___C___D___E___
80. A___B___C___D___E___
81. A___B___C___D___E___
82. A___B___C___D___E___
83. A___B___C___D___E___
84. A___B___C___D___E___
85. A___B___C___D___E___
86. A___B___C___D___E___
87. A___B___C___D___E___
88. A___B___C___D___E___
89. A___B___C___D___E___
90. A___B___C___D___E___
91. A___B___C___D___E___
92. A___B___C___D___E___
93. A___B___C___D___E___
94. A___B___C___D___E___
95. A___B___C___D___E___
96. A___B___C___D___E___
97. A___B___C___D___E___
98. A___B___C___D___E___
99. A___B___C___D___E___
100. A___B___C___D___E___

cut here

cut here

ANSWER SHEET

Review Examination ______

1. A___ B___ C___ D___ E___
2. A___ B___ C___ D___ E___
3. A___ B___ C___ D___ E___
4. A___ B___ C___ D___ E___
5. A___ B___ C___ D___ E___
6. A___ B___ C___ D___ E___
7. A___ B___ C___ D___ E___
8. A___ B___ C___ D___ E___
9. A___ B___ C___ D___ E___
10. A___ B___ C___ D___ E___
11. A___ B___ C___ D___ E___
12. A___ B___ C___ D___ E___
13. A___ B___ C___ D___ E___
14. A___ B___ C___ D___ E___
15. A___ B___ C___ D___ E___
16. A___ B___ C___ D___ E___
17. A___ B___ C___ D___ E___
18. A___ B___ C___ D___ E___
19. A___ B___ C___ D___ E___
20. A___ B___ C___ D___ E___
21. A___ B___ C___ D___ E___
22. A___ B___ C___ D___ E___
23. A___ B___ C___ D___ E___
24. A___ B___ C___ D___ E___
25. A___ B___ C___ D___ E___
26. A___ B___ C___ D___ E___
27. A___ B___ C___ D___ E___
28. A___ B___ C___ D___ E___
29. A___ B___ C___ D___ E___
30. A___ B___ C___ D___ E___
31. A___ B___ C___ D___ E___
32. A___ B___ C___ D___ E___
33. A___ B___ C___ D___ E___
34. A___ B___ C___ D___ E___
35. A___ B___ C___ D___ E___
36. A___ B___ C___ D___ E___
37. A___ B___ C___ D___ E___
38. A___ B___ C___ D___ E___
39. A___ B___ C___ D___ E___
40. A___ B___ C___ D___ E___
41. A___ B___ C___ D___ E___
42. A___ B___ C___ D___ E___
43. A___ B___ C___ D___ E___
44. A___ B___ C___ D___ E___
45. A___ B___ C___ D___ E___
46. A___ B___ C___ D___ E___
47. A___ B___ C___ D___ E___
48. A___ B___ C___ D___ E___
49. A___ B___ C___ D___ E___
50. A___ B___ C___ D___ E___
51. A___ B___ C___ D___ E___
52. A___ B___ C___ D___ E___
53. A___ B___ C___ D___ E___
54. A___ B___ C___ D___ E___
55. A___ B___ C___ D___ E___
56. A___ B___ C___ D___ E___
57. A___ B___ C___ D___ E___
58. A___ B___ C___ D___ E___
59. A___ B___ C___ D___ E___
60. A___ B___ C___ D___ E___
61. A___ B___ C___ D___ E___
62. A___ B___ C___ D___ E___
63. A___ B___ C___ D___ E___
64. A___ B___ C___ D___ E___
65. A___ B___ C___ D___ E___
66. A___ B___ C___ D___ E___
67. A___ B___ C___ D___ E___
68. A___ B___ C___ D___ E___
69. A___ B___ C___ D___ E___
70. A___ B___ C___ D___ E___
71. A___ B___ C___ D___ E___
72. A___ B___ C___ D___ E___
73. A___ B___ C___ D___ E___
74. A___ B___ C___ D___ E___
75. A___ B___ C___ D___ E___
76. A___ B___ C___ D___ E___
77. A___ B___ C___ D___ E___
78. A___ B___ C___ D___ E___
79. A___ B___ C___ D___ E___
80. A___ B___ C___ D___ E___
81. A___ B___ C___ D___ E___
82. A___ B___ C___ D___ E___
83. A___ B___ C___ D___ E___
84. A___ B___ C___ D___ E___
85. A___ B___ C___ D___ E___
86. A___ B___ C___ D___ E___
87. A___ B___ C___ D___ E___
88. A___ B___ C___ D___ E___
89. A___ B___ C___ D___ E___
90. A___ B___ C___ D___ E___
91. A___ B___ C___ D___ E___
92. A___ B___ C___ D___ E___
93. A___ B___ C___ D___ E___
94. A___ B___ C___ D___ E___
95. A___ B___ C___ D___ E___
96. A___ B___ C___ D___ E___
97. A___ B___ C___ D___ E___
98. A___ B___ C___ D___ E___
99. A___ B___ C___ D___ E___
100. A___ B___ C___ D___ E___

ANSWER SHEET

Review Examination ______

1. A___ B___ C___ D___ E___
2. A___ B___ C___ D___ E___
3. A___ B___ C___ D___ E___
4. A___ B___ C___ D___ E___
5. A___ B___ C___ D___ E___
6. A___ B___ C___ D___ E___
7. A___ B___ C___ D___ E___
8. A___ B___ C___ D___ E___
9. A___ B___ C___ D___ E___
10. A___ B___ C___ D___ E___
11. A___ B___ C___ D___ E___
12. A___ B___ C___ D___ E___
13. A___ B___ C___ D___ E___
14. A___ B___ C___ D___ E___
15. A___ B___ C___ D___ E___
16. A___ B___ C___ D___ E___
17. A___ B___ C___ D___ E___
18. A___ B___ C___ D___ E___
19. A___ B___ C___ D___ E___
20. A___ B___ C___ D___ E___
21. A___ B___ C___ D___ E___
22. A___ B___ C___ D___ E___
23. A___ B___ C___ D___ E___
24. A___ B___ C___ D___ E___
25. A___ B___ C___ D___ E___
26. A___ B___ C___ D___ E___
27. A___ B___ C___ D___ E___
28. A___ B___ C___ D___ E___
29. A___ B___ C___ D___ E___
30. A___ B___ C___ D___ E___
31. A___ B___ C___ D___ E___
32. A___ B___ C___ D___ E___
33. A___ B___ C___ D___ E___
34. A___ B___ C___ D___ E___
35. A___ B___ C___ D___ E___
36. A___ B___ C___ D___ E___
37. A___ B___ C___ D___ E___
38. A___ B___ C___ D___ E___
39. A___ B___ C___ D___ E___
40. A___ B___ C___ D___ E___
41. A___ B___ C___ D___ E___
42. A___ B___ C___ D___ E___
43. A___ B___ C___ D___ E___
44. A___ B___ C___ D___ E___
45. A___ B___ C___ D___ E___
46. A___ B___ C___ D___ E___
47. A___ B___ C___ D___ E___
48. A___ B___ C___ D___ E___
49. A___ B___ C___ D___ E___
50. A___ B___ C___ D___ E___
51. A___ B___ C___ D___ E___
52. A___ B___ C___ D___ E___
53. A___ B___ C___ D___ E___
54. A___ B___ C___ D___ E___
55. A___ B___ C___ D___ E___
56. A___ B___ C___ D___ E___
57. A___ B___ C___ D___ E___
58. A___ B___ C___ D___ E___
59. A___ B___ C___ D___ E___
60. A___ B___ C___ D___ E___
61. A___ B___ C___ D___ E___
62. A___ B___ C___ D___ E___
63. A___ B___ C___ D___ E___
64. A___ B___ C___ D___ E___
65. A___ B___ C___ D___ E___
66. A___ B___ C___ D___ E___
67. A___ B___ C___ D___ E___
68. A___ B___ C___ D___ E___
69. A___ B___ C___ D___ E___
70. A___ B___ C___ D___ E___
71. A___ B___ C___ D___ E___
72. A___ B___ C___ D___ E___
73. A___ B___ C___ D___ E___
74. A___ B___ C___ D___ E___
75. A___ B___ C___ D___ E___
76. A___ B___ C___ D___ E___
77. A___ B___ C___ D___ E___
78. A___ B___ C___ D___ E___
79. A___ B___ C___ D___ E___
80. A___ B___ C___ D___ E___
81. A___ B___ C___ D___ E___
82. A___ B___ C___ D___ E___
83. A___ B___ C___ D___ E___
84. A___ B___ C___ D___ E___
85. A___ B___ C___ D___ E___
86. A___ B___ C___ D___ E___
87. A___ B___ C___ D___ E___
88. A___ B___ C___ D___ E___
89. A___ B___ C___ D___ E___
90. A___ B___ C___ D___ E___
91. A___ B___ C___ D___ E___
92. A___ B___ C___ D___ E___
93. A___ B___ C___ D___ E___
94. A___ B___ C___ D___ E___
95. A___ B___ C___ D___ E___
96. A___ B___ C___ D___ E___
97. A___ B___ C___ D___ E___
98. A___ B___ C___ D___ E___
99. A___ B___ C___ D___ E___
100. A___ B___ C___ D___ E___

cut here

cut here

ANSWER SHEET

Review Examination ______

1. A___ B___ C___ D___ E___
2. A___ B___ C___ D___ E___
3. A___ B___ C___ D___ E___
4. A___ B___ C___ D___ E___
5. A___ B___ C___ D___ E___
6. A___ B___ C___ D___ E___
7. A___ B___ C___ D___ E___
8. A___ B___ C___ D___ E___
9. A___ B___ C___ D___ E___
10. A___ B___ C___ D___ E___
11. A___ B___ C___ D___ E___
12. A___ B___ C___ D___ E___
13. A___ B___ C___ D___ E___
14. A___ B___ C___ D___ E___
15. A___ B___ C___ D___ E___
16. A___ B___ C___ D___ E___
17. A___ B___ C___ D___ E___
18. A___ B___ C___ D___ E___
19. A___ B___ C___ D___ E___
20. A___ B___ C___ D___ E___
21. A___ B___ C___ D___ E___
22. A___ B___ C___ D___ E___
23. A___ B___ C___ D___ E___
24. A___ B___ C___ D___ E___
25. A___ B___ C___ D___ E___
26. A___ B___ C___ D___ E___
27. A___ B___ C___ D___ E___
28. A___ B___ C___ D___ E___
29. A___ B___ C___ D___ E___
30. A___ B___ C___ D___ E___
31. A___ B___ C___ D___ E___
32. A___ B___ C___ D___ E___
33. A___ B___ C___ D___ E___
34. A___ B___ C___ D___ E___
35. A___ B___ C___ D___ E___
36. A___ B___ C___ D___ E___
37. A___ B___ C___ D___ E___
38. A___ B___ C___ D___ E___
39. A___ B___ C___ D___ E___
40. A___ B___ C___ D___ E___
41. A___ B___ C___ D___ E___
42. A___ B___ C___ D___ E___
43. A___ B___ C___ D___ E___
44. A___ B___ C___ D___ E___
45. A___ B___ C___ D___ E___
46. A___ B___ C___ D___ E___
47. A___ B___ C___ D___ E___
48. A___ B___ C___ D___ E___
49. A___ B___ C___ D___ E___
50. A___ B___ C___ D___ E___
51. A___ B___ C___ D___ E___
52. A___ B___ C___ D___ E___
53. A___ B___ C___ D___ E___
54. A___ B___ C___ D___ E___
55. A___ B___ C___ D___ E___
56. A___ B___ C___ D___ E___
57. A___ B___ C___ D___ E___
58. A___ B___ C___ D___ E___
59. A___ B___ C___ D___ E___
60. A___ B___ C___ D___ E___
61. A___ B___ C___ D___ E___
62. A___ B___ C___ D___ E___
63. A___ B___ C___ D___ E___
64. A___ B___ C___ D___ E___
65. A___ B___ C___ D___ E___
66. A___ B___ C___ D___ E___
67. A___ B___ C___ D___ E___
68. A___ B___ C___ D___ E___
69. A___ B___ C___ D___ E___
70. A___ B___ C___ D___ E___
71. A___ B___ C___ D___ E___
72. A___ B___ C___ D___ E___
73. A___ B___ C___ D___ E___
74. A___ B___ C___ D___ E___
75. A___ B___ C___ D___ E___
76. A___ B___ C___ D___ E___
77. A___ B___ C___ D___ E___
78. A___ B___ C___ D___ E___
79. A___ B___ C___ D___ E___
80. A___ B___ C___ D___ E___
81. A___ B___ C___ D___ E___
82. A___ B___ C___ D___ E___
83. A___ B___ C___ D___ E___
84. A___ B___ C___ D___ E___
85. A___ B___ C___ D___ E___
86. A___ B___ C___ D___ E___
87. A___ B___ C___ D___ E___
88. A___ B___ C___ D___ E___
89. A___ B___ C___ D___ E___
90. A___ B___ C___ D___ E___
91. A___ B___ C___ D___ E___
92. A___ B___ C___ D___ E___
93. A___ B___ C___ D___ E___
94. A___ B___ C___ D___ E___
95. A___ B___ C___ D___ E___
96. A___ B___ C___ D___ E___
97. A___ B___ C___ D___ E___
98. A___ B___ C___ D___ E___
99. A___ B___ C___ D___ E___
100. A___ B___ C___ D___ E___

A N S W E R   S H E E T

Review Examination ______

1. A___B___C___D___E___
2. A___B___C___D___E___
3. A___B___C___D___E___
4. A___B___C___D___E___
5. A___B___C___D___E___
6. A___B___C___D___E___
7. A___B___C___D___E___
8. A___B___C___D___E___
9. A___B___C___D___E___
10. A___B___C___D___E___
11. A___B___C___D___E___
12. A___B___C___D___E___
13. A___B___C___D___E___
14. A___B___C___D___E___
15. A___B___C___D___E___
16. A___B___C___D___E___
17. A___B___C___D___E___
18. A___B___C___D___E___
19. A___B___C___D___E___
20. A___B___C___D___E___
21. A___B___C___D___E___
22. A___B___C___D___E___
23. A___B___C___D___E___
24. A___B___C___D___E___
25. A___B___C___D___E___
26. A___B___C___D___E___
27. A___B___C___D___E___
28. A___B___C___D___E___
29. A___B___C___D___E___
30. A___B___C___D___E___
31. A___B___C___D___E___
32. A___B___C___D___E___
33. A___B___C___D___E___
34. A___B___C___D___E___
35. A___B___C___D___E___
36. A___B___C___D___E___
37. A___B___C___D___E___
38. A___B___C___D___E___
39. A___B___C___D___E___
40. A___B___C___D___E___
41. A___B___C___D___E___
42. A___B___C___D___E___
43. A___B___C___D___E___
44. A___B___C___D___E___
45. A___B___C___D___E___
46. A___B___C___D___E___
47. A___B___C___D___E___
48. A___B___C___D___E___
49. A___B___C___D___E___
50. A___B___C___D___E___
51. A___B___C___D___E___
52. A___B___C___D___E___
53. A___B___C___D___E___
54. A___B___C___D___E___
55. A___B___C___D___E___
56. A___B___C___D___E___
57. A___B___C___D___E___
58. A___B___C___D___E___
59. A___B___C___D___E___
60. A___B___C___D___E___
61. A___B___C___D___E___
62. A___B___C___D___E___
63. A___B___C___D___E___
64. A___B___C___D___E___
65. A___B___C___D___E___
66. A___B___C___D___E___
67. A___B___C___D___E___
68. A___B___C___D___E___
69. A___B___C___D___E___
70. A___B___C___D___E___
71. A___B___C___D___E___
72. A___B___C___D___E___
73. A___B___C___D___E___
74. A___B___C___D___E___
75. A___B___C___D___E___
76. A___B___C___D___E___
77. A___B___C___D___E___
78. A___B___C___D___E___
79. A___B___C___D___E___
80. A___B___C___D___E___
81. A___B___C___D___E___
82. A___B___C___D___E___
83. A___B___C___D___E___
84. A___B___C___D___E___
85. A___B___C___D___E___
86. A___B___C___D___E___
87. A___B___C___D___E___
88. A___B___C___D___E___
89. A___B___C___D___E___
90. A___B___C___D___E___
91. A___B___C___D___E___
92. A___B___C___D___E___
93. A___B___C___D___E___
94. A___B___C___D___E___
95. A___B___C___D___E___
96. A___B___C___D___E___
97. A___B___C___D___E___
98. A___B___C___D___E___
99. A___B___C___D___E___
100. A___B___C___D___E___

cut here

A N S W E R   S H E E T

Review Examination ______

1. A___ B___ C___ D___ E___
2. A___ B___ C___ D___ E___
3. A___ B___ C___ D___ E___
4. A___ B___ C___ D___ E___
5. A___ B___ C___ D___ E___
6. A___ B___ C___ D___ E___
7. A___ B___ C___ D___ E___
8. A___ B___ C___ D___ E___
9. A___ B___ C___ D___ E___
10. A___ B___ C___ D___ E___
11. A___ B___ C___ D___ E___
12. A___ B___ C___ D___ E___
13. A___ B___ C___ D___ E___
14. A___ B___ C___ D___ E___
15. A___ B___ C___ D___ E___
16. A___ B___ C___ D___ E___
17. A___ B___ C___ D___ E___
18. A___ B___ C___ D___ E___
19. A___ B___ C___ D___ E___
20. A___ B___ C___ D___ E___
21. A___ B___ C___ D___ E___
22. A___ B___ C___ D___ E___
23. A___ B___ C___ D___ E___
24. A___ B___ C___ D___ E___
25. A___ B___ C___ D___ E___
26. A___ B___ C___ D___ E___
27. A___ B___ C___ D___ E___
28. A___ B___ C___ D___ E___
29. A___ B___ C___ D___ E___
30. A___ B___ C___ D___ E___
31. A___ B___ C___ D___ E___
32. A___ B___ C___ D___ E___
33. A___ B___ C___ D___ E___
34. A___ B___ C___ D___ E___
35. A___ B___ C___ D___ E___
36. A___ B___ C___ D___ E___
37. A___ B___ C___ D___ E___
38. A___ B___ C___ D___ E___
39. A___ B___ C___ D___ E___
40. A___ B___ C___ D___ E___
41. A___ B___ C___ D___ E___
42. A___ B___ C___ D___ E___
43. A___ B___ C___ D___ E___
44. A___ B___ C___ D___ E___
45. A___ B___ C___ D___ E___
46. A___ B___ C___ D___ E___
47. A___ B___ C___ D___ E___
48. A___ B___ C___ D___ E___
49. A___ B___ C___ D___ E___
50. A___ B___ C___ D___ E___
51. A___ B___ C___ D___ E___
52. A___ B___ C___ D___ E___
53. A___ B___ C___ D___ E___
54. A___ B___ C___ D___ E___
55. A___ B___ C___ D___ E___
56. A___ B___ C___ D___ E___
57. A___ B___ C___ D___ E___
58. A___ B___ C___ D___ E___
59. A___ B___ C___ D___ E___
60. A___ B___ C___ D___ E___
61. A___ B___ C___ D___ E___
62. A___ B___ C___ D___ E___
63. A___ B___ C___ D___ E___
64. A___ B___ C___ D___ E___
65. A___ B___ C___ D___ E___
66. A___ B___ C___ D___ E___
67. A___ B___ C___ D___ E___
68. A___ B___ C___ D___ E___
69. A___ B___ C___ D___ E___
70. A___ B___ C___ D___ E___
71. A___ B___ C___ D___ E___
72. A___ B___ C___ D___ E___
73. A___ B___ C___ D___ E___
74. A___ B___ C___ D___ E___
75. A___ B___ C___ D___ E___
76. A___ B___ C___ D___ E___
77. A___ B___ C___ D___ E___
78. A___ B___ C___ D___ E___
79. A___ B___ C___ D___ E___
80. A___ B___ C___ D___ E___
81. A___ B___ C___ D___ E___
82. A___ B___ C___ D___ E___
83. A___ B___ C___ D___ E___
84. A___ B___ C___ D___ E___
85. A___ B___ C___ D___ E___
86. A___ B___ C___ D___ E___
87. A___ B___ C___ D___ E___
88. A___ B___ C___ D___ E___
89. A___ B___ C___ D___ E___
90. A___ B___ C___ D___ E___
91. A___ B___ C___ D___ E___
92. A___ B___ C___ D___ E___
93. A___ B___ C___ D___ E___
94. A___ B___ C___ D___ E___
95. A___ B___ C___ D___ E___
96. A___ B___ C___ D___ E___
97. A___ B___ C___ D___ E___
98. A___ B___ C___ D___ E___
99. A___ B___ C___ D___ E___
100. A___ B___ C___ D___ E___

cut here

ANSWER SHEET

Review Examination ______

1. A___B___C___D___E___
2. A___B___C___D___E___
3. A___B___C___D___E___
4. A___B___C___D___E___
5. A___B___C___D___E___
6. A___B___C___D___E___
7. A___B___C___D___E___
8. A___B___C___D___E___
9. A___B___C___D___E___
10. A___B___C___D___E___
11. A___B___C___D___E___
12. A___B___C___D___E___
13. A___B___C___D___E___
14. A___B___C___D___E___
15. A___B___C___D___E___
16. A___B___C___D___E___
17. A___B___C___D___E___
18. A___B___C___D___E___
19. A___B___C___D___E___
20. A___B___C___D___E___
21. A___B___C___D___E___
22. A___B___C___D___E___
23. A___B___C___D___E___
24. A___B___C___D___E___
25. A___B___C___D___E___
26. A___B___C___D___E___
27. A___B___C___D___E___
28. A___B___C___D___E___
29. A___B___C___D___E___
30. A___B___C___D___E___
31. A___B___C___D___E___
32. A___B___C___D___E___
33. A___B___C___D___E___
34. A___B___C___D___E___
35. A___B___C___D___E___
36. A___B___C___D___E___
37. A___B___C___D___E___
38. A___B___C___D___E___
39. A___B___C___D___E___
40. A___B___C___D___E___
41. A___B___C___D___E___
42. A___B___C___D___E___
43. A___B___C___D___E___
44. A___B___C___D___E___
45. A___B___C___D___E___
46. A___B___C___D___E___
47. A___B___C___D___E___
48. A___B___C___D___E___
49. A___B___C___D___E___
50. A___B___C___D___E___
51. A___B___C___D___E___
52. A___B___C___D___E___
53. A___B___C___D___E___
54. A___B___C___D___E___
55. A___B___C___D___E___
56. A___B___C___D___E___
57. A___B___C___D___E___
58. A___B___C___D___E___
59. A___B___C___D___E___
60. A___B___C___D___E___
61. A___B___C___D___E___
62. A___B___C___D___E___
63. A___B___C___D___E___
64. A___B___C___D___E___
65. A___B___C___D___E___
66. A___B___C___D___E___
67. A___B___C___D___E___
68. A___B___C___D___E___
69. A___B___C___D___E___
70. A___B___C___D___E___
71. A___B___C___D___E___
72. A___B___C___D___E___
73. A___B___C___D___E___
74. A___B___C___D___E___
75. A___B___C___D___E___
76. A___B___C___D___E___
77. A___B___C___D___E___
78. A___B___C___D___E___
79. A___B___C___D___E___
80. A___B___C___D___E___
81. A___B___C___D___E___
82. A___B___C___D___E___
83. A___B___C___D___E___
84. A___B___C___D___E___
85. A___B___C___D___E___
86. A___B___C___D___E___
87. A___B___C___D___E___
88. A___B___C___D___E___
89. A___B___C___D___E___
90. A___B___C___D___E___
91. A___B___C___D___E___
92. A___B___C___D___E___
93. A___B___C___D___E___
94. A___B___C___D___E___
95. A___B___C___D___E___
96. A___B___C___D___E___
97. A___B___C___D___E___
98. A___B___C___D___E___
99. A___B___C___D___E___
100. A___B___C___D___E___

cut here

NOTES

NOTES

NOTES

NOTES

NOTES

NOTES

NOTES

NOTES